Kallikrein-related peptidases
Novel cancer-related biomarkers
Magdolen, Sommerhoff, Fritz and Schmitt (Eds.)

Kallikrein-related peptidases

—

Volume 1

Characterization, regulation, and interactions within the protease web

Viktor Magdolen, Christian P. Sommerhoff, Hans Fritz, and Manfred Schmitt (Eds.)

ISBN 978-3-11-026036-6
e-ISBN 978-3-11-026037-3

De Gruyter, Berlin 2012

Volume 2

Novel cancer-related biomarkers

Viktor Magdolen, Christian P. Sommerhoff, Hans Fritz, and Manfred Schmitt (Eds.)

ISBN 978-3-11-030358-2
e-ISBN 978-3-11-030366-7

De Gruyter, Berlin 2012

Kallikrein-related peptidases

Volume 2

Novel cancer-related biomarkers

Editors

Viktor Magdolen, Christian P. Sommerhoff,
Hans Fritz, and Manfred Schmitt

DE GRUYTER

Editors

Prof. Dr. Viktor Magdolen
Frauenklinik der TU München
Klinikum rechts der Isar
Ismaninger Str. 22
81675 München
viktor.magdolen@lrz.tum.de

Prof. Dr. Christian P. Sommerhoff
Institut für Laboratoriumsmedizin
Klinikum der LMU München
Nussbaumstraße 20
80336 München
sommerhoff@med.uni-muenchen.de

Prof. Dr. Hans Fritz
LMU München / Klinikum Innenstadt
Abteilung für Klinische Chemie und Klinische
Biochemie in der Chirurgischen Klinik und
Poliklinik
Nussbaumstraße 20
80336 München
fritz@med.uni-muenchen.de

Prof. Dr. Manfred Schmitt
Frauenklinik der TU München
Klinikum rechts der Isar
Ismaninger Straße 22
81675 München
manfred.schmitt@lrz.tum.de

Cover image: Undifferentiated ovarian cancer tumor tissue specimen stained for KLK6 with mono-specific polyclonal rabbit antibody 623A (immunized and affinity-purified with a synthetic peptide representing amino acids 109-119 of KLK6) and the DAKO EnVision peroxidase polymer system (brown color). Nuclei were counterstained with hematoxylin (blue color). Reference: Seiz et al. (2012). Stromal cell-associated expression of kallikrein-related peptidase 6 (KLK6) indicates poor prognosis of ovarian cancer patients. Biol. Chem. 393, 391-401.

This book contains 16 figures and 11 tables.

ISBN 978-3-11-048153-2
e-ISBN 978-3-11-030366-7

Library of Congress Cataloging-in-Publication Data
A CIP catalog record for this book has been applied for at the Library of Congress.

Bibliographic information published by the Deutsche Nationalbibliothek
The Deutsche Nationalbibliothek lists this publication in the Deutsche Nationalbibliografie; detailed bibliographic data are available in the internet at http://dnb.dnb.de.

© 2012 Walter de Gruyter GmbH, Berlin/Boston
The publisher, together with the authors and editors, has taken great pains to ensure that all information presented in this work (programs, applications, amounts, dosages, etc.) reflects the standard of knowledge at the time of publication. Despite careful manuscript preparation and proof correction, errors can nevertheless occur. Authors, editors and publisher disclaim all responsibility and for any errors or omissions or liability for the results obtained from use of the information, or parts thereof, contained in this work.

Illustrations: Andreas Hoffmann, Berlin
Typesetting: Beltz Bad Langensalza GmbH, Bad Langensalza
Printing and binding: Hubert & Co. GmbH & Co. KG, Göttingen

Printed on acid-free paper
Printed in Germany
www.degruyter.com

Preface

It was challenging to bring some semblance of order to the increasing body of information on kallikrein-related peptidases (KLKs) in normal physiology and under patho-physiological conditions. We therefore decided to cover contributions of authors, each of them an expert in the field, in two volumes of the book on *Kallikrein-related peptidases*. Volume 1, containing fifteen review chapters, highlights the genomic and proteomic organization of KLKs, including 3D-structures, substrate and inhibitor specificity of KLKs, interaction and regulation of KLKs within the proteolytic web, and the importance of KLKs for tooth development, the physiological function of seminal KLKs, and for non-malignant diseases such as those of the skin and the central nervous system.

Different from Volume 1, the second volume presents a selection of review articles focusing on the clinical utility of KLKs in various types of solid malignant tumors. The ten chapters contained in Volume 2 summarize the clinical importance of diverse KLKs in several major types of cancer, e.g. that of the lung, the gastrointestinal and urogenital tract, the breast, and cancers of the head & neck. Each chapter is organized such that it provides an overview and interpretation of the clinical impact of certain KLKs in these types of malignancies.

All chapters of Volume 1 and 2 are comprehensive, with an extensive list of cited literature references and each is designed to stand on itself, so that the reader does not need to refer back to previous reports for background information. We wish you enjoyable reading!

Manfred Schmitt, Christian P. Sommerhoff, Hans Fritz, and Viktor Magdolen

List of contributing authors

Jane Bayani
Department of Pathology and Laboratory
Medicine
Mount Sinai Hospital
Toronto, Canada
e-mail: jane.bayani@utoronto.ca
Chapter 9

Judith A. Clements
Institute of Health and Biomedical Innovation
Queensland University of Technology
Kelvin Grove, Australia
e-mail: j.clements@qut.edu.au
Chapter 5

Yves Courty
INSERM U1100
Faculty of Medicine
University F. Rabelais
Tours, France
e-mail: courty@univ-tours.fr
Chapter 1

Marina Devetzi
Department of Cellular Physiology
G. Papanicolaou Research Center of Oncology
Saint Savvas Cancer Hospital
Athens, Greece
and
Department of Biological Applications
and Technology
Faculty of Science and Technology
University of Ioannina
Ioannina, Greece
e-mail: marinadevetzi@yahoo.gr
Chapter 2

Eleftherios P. Diamandis
Department of Pathology & Laboratory Medicine
Mount Sinai Hospital
Toronto, Canada
e-mail: ediamandis@mtsinai.on.ca
Chapter 6, 7, 9

Ying Dong
Institute of Health and Biomedical Innovation
Queensland University of Technology
Kelvin Grove, Australia
e-mail: y.dong@qut.edu.au
Chapter 5

Julia Dorn
Frauenklinik der Technischen Universität
München
Klinikum rechts der Isar
Munich, Germany
e-mail: julia.dorn@lrz.tum.de
Chapter 6, 7

Ruth A. Fuhrman-Luck
Institute of Health and Biomedical Innovation
Queensland University of Technology
Kelvin Grove, Australia
e-mail: r.fuhrmanluck@qut.edu.au
Chapter 5

Nathalie Heuzé-Vourc'h
INSERM U1100/EA6305
Faculty of Medicine
University F. Rabelais
Tours, France
e-mail: nathalie.vourch@med.univ-tours.fr
Chapter 1

Dietmar W. Hutmacher
Faculty of Built Environment and Engineering
Institute of Health and Biomedical Innovation
and
Australian Prostate Cancer Research Centre
Queensland
Queensland University of Technology
e-mail: dietmar.hutmacher@qut.edu.au
Chapter 5

Rong Jiang
Department of Human Genetics
Emory University
Atlanta, GA, USA
e-mail: rong.jiang@emory.edu
Chapter 3

Jeffrey Johnson
Department of Chemistry & Biochemistry
Harper Cancer Research Institute
University of Notre Dame
Notre Dame, IN, USA
e-mail: jjohns39@nd.edu
Chapter 3

Marion Kiechle
Frauenklinik der Technischen Universität
München
Munich, Germany
e-mail: marion.kiechle@lrz.tum.de
Chapter 6

Hannu Koistinen
Department of Clinical Chemistry
University of Helsinki
Helsinki, Finland
e-mail: hannu.k.koistinen@helsinki.fi
Chapter 4

Vathany Kulasingam
Department of Clinical Biochemistry
University Health Network
Toronto, Canada
e-mail: Dr.Vathany.Kulasingam@uhn.on.ca
Chapter 7

Daniela Loessner
Institute of Health and Biomedical Innovation
Queensland University of Technology
Kelvin Grove, Australia
e-mail: daniela.lossner@qut.edu.au
Chapter 5

Liu-Ying Luo
Research and Diagnostic Systems
Minneapolis, MN, USA
e-mail: Yvonne.Luo@rndsystems.com
Chapter 6

Viktor Magdolen
Frauenklinik der Technischen Universität
München
Klinikum rechts der Isar
Munich, Germany
e-mail: viktor.magdolen@lrz.tum.de
Chapter 5

Konstantinos Mavridis
Department of Biochemistry and Molecular
Biology
University of Athens
Athens, Greece
e-mail: mavridiskos@gmail.com
Chapter 10

Valentina Milou
Ontario Institute for Cancer Research
Toronto, Canada
e-mail: valentina.milou@oicr.on.ca
Chapter 7

Barbara Schmalfeldt
Frauenklinik der Technischen Universität
München
Klinikum rechts der Isar
Munich, Germany
e-mail: barbara.schmalfeldt@lrz.tum.de
Chapter 7

Manfred Schmitt
Frauenklinik der Technischen Universität
München
Klinikum rechts der Isar
Munich, Germany
e-mail: manfred.schmitt@lrz.tum.de
Chapter 2, 6, 7, 10

Andreas Scorilas
Department of Biochemistry and Molecular
Biology
University of Athens
Athens, Greece
e-mail: ascorilas@biol.uoa.gr
Chapter 10

List of contributing authors IX

Zonggao Shi
Harper Cancer Research Institute
University of Notre Dame
Notre Dame, IN, USA
e-mail: zshi2@nd.edu
Chapter 3

Shirly Sieh
Faculty of Built Environment and Engineering
Australian Prostate Cancer Research Centre
Queensland
Queensland University of Technology
Kelvin Grove, Australia
e-mail: s.sieh@qut.edu.au
Chapter 5

M. Sharon Stack
Department of Chemistry and Biochemistry
Harper Cancer Research Institute
University of Notre Dame
Notre Dame, IN, USA
e-mail: Sharon.Stack.11@nd.edu
Chapter 3

Ulf-Håkan Stenman
Department of Clinical Chemistry
University of Helsinki
Helsinki, Finland
email: ulf-hakan.stenman@helsinki.fi
Chapter 4

Anna Taubenberger
Faculty of Built Environment and Engineering
Institute of Health and Biomedical Innovation
Kelvin Grove, Australia
e-mail: anna.taubenberger@qut.edu.au
Chapter 5

Maroulio Talieri
Department of Cellular Physiology
G. Papanicolaou Research Center of Oncology
Saint Savvas Cancer Hospital
Athens, Greece
e-mail: talieri@agsavvas-hosp.gr
Chapter 2, 3

Nicole M.A. White
St. Michael's Hospital
Toronto, Canada
e-mail: whiteni@smh.ca
Chapter 8

George M. Yousef
Department of Laboratory Medicine
St. Michael's Hospital
Toronto, Canada
e-mail: YousefG@smh.ca
Chapter 8

Table of Contents

Manfred Schmitt, Christian P. Sommerhoff, Hans Fritz,
and Viktor Magdolen

Introduction to Volume 2: Kallikrein-related Peptidases. Novel Cancer-related Biomarkers

Volume 2 of this book, entitled *"Kallikrein-related Peptidases. Novel cancer-related biomarkers"*, encompasses 10 chapters, authored by prominent researchers in the field of KLK research, focusing on the current knowledge about the role of KLKs in malignant disorders. These chapters summarize and discuss the potential use of various cancer-associated KLKs as prognostic or predictive cancer biomarkers, as monitoring and imaging tools, and as novel targets, so as to allow tailored, patient-oriented therapy. Such a comprehensive overview on the clinical relevance of KLKs in cancer has not been published so far, although the rate of publications demonstrates a rapid increase in knowledge about these important biomarkers (**Fig. 0.1**).

KLKs are primarily known for their biomarker value in prostate and ovarian cancer, with regard to predicting the course of the disease (prognosis) and/or the patient's response to therapy (Avgeris et al., 2012; Lawrence et al., 2010; Lilja et al., 2008; Yousef and Diamandis, 2009). Yet, more recent data points to an important role of certain KLKs for improving individualized cancer therapy also in several other malignancies, including those of the gastrointestinal and urogenital tract, the breast, lung, brain, and head & neck, (Schmitt and Magdolen, 2009). Pioneering work in this direction has been conducted by several authors contributing to volume 2 of this book, repre-

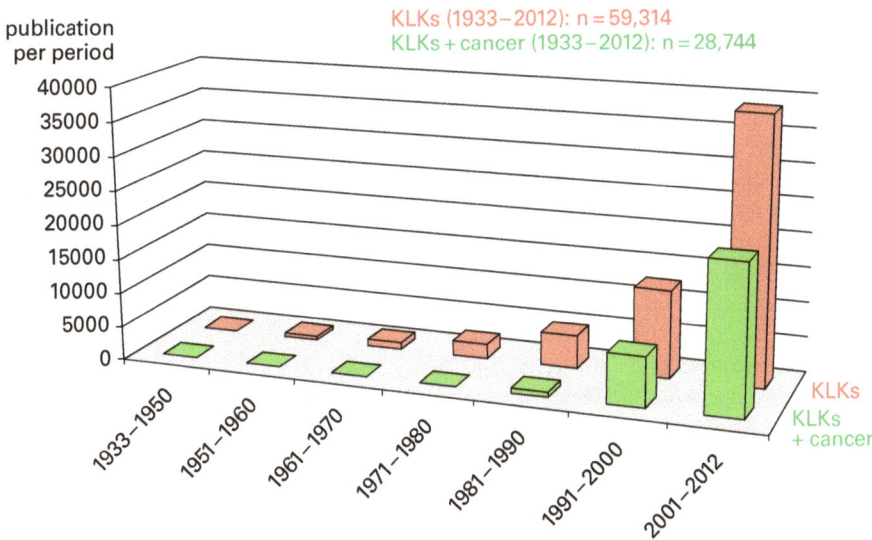

Fig. 0.1 Articles published between 1933 and 2012 relevant to KLKs and cancer.

senting clinical research centers in Canada, Australia, Greece, Germany, and France. Numerous of their studies have demonstrated that KLKs represent a rich source of serum and tissue biomarkers that allow molecular classification, early diagnosis, and prognosis of human malignancies. However, KLKs may also be indicators of the efficacy of anticancer drugs (Avgeris *et al.*, 2010; Borgoño and Diamandis, 2004; Yousef and Diamandis, 2009).

Such a KLK-biomarker-driven approach has immense potential for predicting the clinical outcome for a cancer patient and for providing customized treatment options to those patients who are likely to respond, while directing those who would not benefit towards alternative therapeutic options. Unquestionably, KLKs have already proven their cancer biomarker capacity and thus represent valuable utensils in the toolbox of personalized cancer medicine (Mavridis and Scorilas, 2010). We thus anticipate that the articles contained in volume 2 of this book will give the reader quick access to the methodologies, tools, reagents, and statistics for assessing early and late expression of the KLKs in tumor tissues and bodily fluids of different origin.

However, in order to interpret the clinical impact of different KLKs on the broad range of malignancies, the exact mechanisms of action of KLKs in distinct tumor entities and their microenvironments will need to be further elucidated. Eventually this will tell us, for a particular cancer disease, which of the KLKs should be employed for which patient so as to serve as a tailored cancer biomarker to predict the clinical outcome of the cancer patient or response to cancer therapy.

Bibliography

Avgeris, M., Mavridis, K., and Scorilas, A. (2010). Kallikrein-related peptidase genes as promising biomarkers for prognosis and monitoring of human malignancies. Biol. Chem. 391, 505–511.

Avgeris, M., Mavridis, K., and Scorilas A. (2012). Kallikrein-related peptidases in prostate, breast, and ovarian cancers: from pathobiology to clinical relevance. Biol. Chem. 393, 301–317.

Borgoño, C.A., and Diamandis, E.P. (2004). The emerging roles of human tissue kallikreins in cancer. Nat. Rev. Cancer 4, 876–890.

Lawrence, M.G., Lai, J., and Clements, J.A. (2010). Kallikreins on steroids: structure, function, and hormonal regulation of prostate-specific antigen and the extended kallikrein locus. Endocr. Rev. 31, 407–446.

Lilja, H., Ulmert, D., and Vickers, A.J. (2008). Prostate-specific antigen and prostate cancer: prediction, detection and monitoring. Nat. Rev. Cancer 8, 268–278.

Mavridis, K., and Scorilas, A. (2010). Prognostic value and biological role of the kallikrein-related peptidases in human malignancies. Future Oncol. 6, 269–285.

Schmitt, M., and Magdolen, V. (2009). Using kallikrein-related peptidases (KLK) as novel cancer biomarkers. Thromb. Haemost. 101, 222–224.

Yousef, G.M., and Diamandis, E.P. (2009). The human kallikrein gene family: new biomarkers for ovarian cancer. Cancer Treat. Res. 149, 165–187.

Nathalie Heuzé-Vourc'h, and Yves Courty

1 Pathophysiology of Kallikrein-related Peptidases in Lung Cancer

1.1 Introduction

The lungs have to deal with the complex physical challenges posed by massive changes in pressure and volume and also with environmental challenges arising from the vast array of pathogens and particles to which they are exposed. The lung tissues and cells must therefore ensure both respiratory function and protect the body. Proteases play a key role in this homeostasis by regulating processes such as regeneration and repair following injury. Chronic inflammatory lung diseases and lung cancer are often associated with above-normal levels of proteases. One group of proteases, the kallikrein-related peptidases (KLK), contains 15 secreted serine proteinases with trypsin-like and/or chymotrypsin-like activity. The *KLK* genes are located in tandem on chromosome 19q13.4 and are expressed throughout the human body, particularly in the respiratory tract. They have recently been identified as important actors in cancer. The main goal of this chapter is to provide a foundation for the scientific investigation and the dialogue required to better understand the functions of KLKs in the pathophysiology of the lung.

1.2 Expression pattern of KLKs in the normal lung

KLK1, a member of the KLK family, was first identified by Christiansen *et al.* (1987) in bronchoalveolar lavage fluid (BALF) obtained from asthmatic lung patients. Later studies confirmed the presence of KLK1 in the airways and located this protease in the serous cell granules of the submucosal glands of the trachea and bronchi (Poblete *et al.*, 1993; Proud and Vio, 1993). Several more recent studies have used RT-PCR or ELISA to examine the expression of *KLK* genes in normal lung tissue analyses (Shaw and Diamandis, 2007). We and others have demonstrated that the *KLK* genes are moderately or poorly expressed in normal lungs, except for *KLK2, 4,* and *15* (**Tab. 1.1**). Tissue distribution of several KLKs (KLK5–7, and 10–14) was also examined by immunohistochemistry (Petraki *et al.*, 2001; 2002; 2003; 2006; Planque *et al.*, 2008a). The eight KLKs were found in the epithelium of the trachea, the bronchial tree, and in the submucosal glands, but not in the alveolar epithelium. Thus, the KLKs appear to be restricted to the airways in the normal human lung and these structures (surface epithelium of the bronchial tree and bronchial submucosal glands) produce several KLKs simultaneously. Because these structures discharge their secretions into the lumen of the airways, KLKs are probably components of the lung epithelial lining fluid. This is

Tab. 1.1 Expression of human KLKs in the lung.

KLK	Localization / cancer type	Data	References
1	Normal lung	KLK1 in submucosal glands of trachea and bronchi by IHC.	Poblete et al., 1993 Proud and Vio, 1993 Chee et al., 2008
	NSCLC & SCLC	Cytoplasmic immunostaining in 53–80% of cancer cells and nuclear detection in 20–30% of cancer cells.	
	Plasma	No difference between NSCLC patients and healthy individuals.	Planque et al., 2008b
2	Normal lung	Weak expression in trachea (RT-PCR), no expression in lung tissue (RT-PCR, ELISA).	Olsson et al., 2005 Shaw and Diamandis, 2007
	NSCLC	No expression.	Personal communication
3	Normal lung NSCLC	Weak expression (ELISA, RT-PCR). mRNA expression in about 60% of malignant specimens. No association with any clinico-pathologic variables.	Shaw and Diamandis, 2007 Zarghami et al., 1997
4	Normal lung	No expression detected by PCR, IHC, ELISA.	Seiz et al., 2010 Shaw and Diamandis, 2007
	Plasma	No difference between NSCLC patients and healthy individuals.	Planque et al., 2008b
5	Normal lung	Expressed (qRT-PCR, WB, IHC and ELISA).	Petraki et al., 2006 Planque et al., 2005 Shaw and Diamandis, 2007
	NSCLC & SCLC	mRNA expression in 93% of NSCLCs and over-expression associated with SqCC subtype. Cytoplasmic immunostaining in 50–90% of cancer cells (SqCC, carcinoid tumor and SCLC). Strong nuclear labeling in carcinoid tumor cells.	Planque et al., 2005 Singh et al., 2008
	Plasma	Serum level reduced in NSCLC patients compared to healthy individuals.	Planque et al., 2008b
6	Normal lung	Expression in glandular epithelium of normal airways (IHC, qRT-PCR).	Heuze-Vourc'h et al., 2009 Petraki et al., 2001
	NSCLC & SCLC	mRNA upregulated in NSCLC and high KLK6 mRNA associated with lower survival. Cytoplasmic immunostaining in 70–90% of cancer cells in the SqCC, carcinoid tumor and SCLC subtypes.	Heuze-Vourc'h et al., 2009 Singh et al., 2008
	Plasma	No difference between NSCLC patients and healthy individuals.	Planque et al., 2008b
7	Normal lung	Expression determined by qRT-PCR, WB, ELISA	Planque et al., 2005 Shaw and Diamandis, 2007

Tab. 1.1 (continued)

KLK	Localization / cancer type	Data	References
	NCSLC & SCLC	mRNA expression in 100% of malignant specimens. Decreased expression in ADC. Cytoplasmic immunostaining in 40, 70, and 90% of cancer cells in SqCC, carcinoid tumor, and SCLC subtypes, respectively. Strong nuclear labeling in carcinoid tumor cells.	Planque *et al.*, 2005 Singh *et al.*, 2008
	Plasma	Serum level reduced in NSCLC patients compared to healthy individuals.	Planque *et al.*, 2008b
8	Normal lung	Expression determined by qRT-PCR. Early-stage NSCLC patients with high *KLK8* expression have a lower recurrence rate.	Planque *et al.*, 2010 Sher *et al.*, 2006
	NSCLC & SCLC	Global activity increased in NSCLC but without association with the overall survival. *KLK8-T4* splice variant is an independent marker of unfavorable outcome. Cytoplasmic immunostaining in 60–90% of cancer cells (SqCC, carcinoid tumor and SCLC).	Sher *et al.*, 2006 Planque *et al.*, 2010 Singh *et al.*, 2008
	Plasma	Serum level reduced in NSCLC patients compared to healthy individuals.	Planque *et al.*, 2008b
9	Normal lung	*KLK9* mRNA weakly expressed.	Shaw and Diamandis, 2007
10	Normal lung	Detection in lung tissue (ELISA, qRT-PCR). IHC expression in glandular epithelium of airways.	Luo *et al.*, 2001 Planque *et al.*, 2006 Petraki *et al.*, 2002
	NSCLC tissue	*KLK10* mRNA expression in 100% of malignant specimens. mRNA downregulated in about 43% of NSCLC specimens. High expression associated with the SqCC subtype. *KLK10* mRNA downregulated in 57% of NSCLC due to CpG island hypermethylation.	Planque *et al.*, 2006 Zhang *et al.*, 2009
	Plasma	Serum level reduced in NSCLC patients compared to healthy individuals.	Planque *et al.*, 2008b
11	Normal lung	Detection in glandular epithelium of normal airways (IHC, ELISA, qRT-PCR).	Petraki *et al.*, 2006 Planque *et al.*, 2006 Shaw and Diamandis, 2007
	NSCLC	mRNA decreased in tumor tissues compared to adjacent non-malignant tissues. Expression down-regulated in about 40% of patients with NSCLC.	Planque *et al.*, 2006 Sasaki *et al.*, 2006
	Plasma	Serum level increased in NSCLC patients compared with healthy individuals.	Planque *et al.*, 2008b

Tab. 1.1 (continued)

KLK	Localization / cancer type	Data	References
12	Normal lung	Determined by qRT-PCR.	Guillon-Munos *et al.*, 2011
	NSCLC	Upregulation.	Guillon-Munos *et al.*, 2011
	Plasma	Serum level reduced in NSCLC patients compared to healthy individuals.	Planque *et al.*, 2008b
13	Normal lung	Expression in glandular epithelium of normal airways (IHC, qRT-PCR).	Petraki *et al.*, 2003 Planque *et al.*, 2008a
	NSCLC	Increased expression in about 1/3 of patients, and decreased in about 1/3 of patients. *KLK13* mRNA overexpression correlates with a positive nodal status and lower overall survival. Cytoplasmic immunostaining present in 68% of NSCLC specimens and associated with the ADC histotype. Upregulation in NSCLC cell lines with invasive potential due to demethylation of its upstream region.	Planque *et al.*, 2008a Chou *et al.*, 2011
	Plasma	Serum level increased in NSCLC patients compared to healthy individuals.	Planque *et al.*, 2008b
14	Normal lung	Expressed in glandular epithelium of normal airways (IHC, qRT-PCR).	Petraki *et al.*, 2006 Planque *et al.*, 2008a
	NSCLC	Expression increased in about 1/3 of patients and decreased in about 1/3 of patients.	Planque *et al.*, 2008a
	Serum	Serum level increased in NSCLC patients compared to healthy individuals.	Planque *et al.*, 2008b
15	Normal lung	Not expressed.	Shaw and Diamandis, 2007

Abbreviations: ADC, adenocarcinoma; IHC, immunohistochemistry; NSCLC, non-small-cell lung carcinoma; qRT-PCR, quantitative reverse transcription PCR; SCLC, small-cell lung carcinoma; SqCC, squamous cell carcinoma; WB, Western blot

supported by the finding of KLK1 and KLK11 in BALF and sputum (Christiansen *et al.*, 1987; Nicholas *et al.*, 2006).

1.3 KLKs in lung cancer

It is clear that many KLK genes are differentially expressed in a wide variety of carcinomas and there is evidence that KLKs are implicated in cancer progression (reviewed in Avgeris *et al.*, 2010; Borgoño and Diamandis, 2004; Borgoño *et al.*, 2004; Paliouras *et al.*, 2007). Although not as extensively studied, KLKs have been found to undergo distinctive differential expression in lung cancer. The mRNA and protein patterns of KLKs in lung cancer, analyzed using several approaches, are summarized in **Tab. 1.1.** To date, eleven *KLK* genes *(KLK1, 3, 5–8,* and *10–14)* have been found to be expressed in lung cancers. Almost all these studies have been on non-small-cell lung carcinomas (NSCLCs), and little is known about KLKs in small-cell lung carcinomas (SCLCs). However, immunochemical studies showed KLK1, and 5–8 in specimens from patients with SCLCs (Chee *et al.*, 2008; Singh *et al.*, 2008). KLK immunostaining has generally been found in the cytoplasm of both NSCLC and SCLC tumor cells. This location is consistent with the secretory nature of KLKs. The nuclei of some malignant cells may also stain positively for KLK1, 5, 7, and 8 in some lung cancer subtypes (Chee *et al.*, 2008; Singh *et al.*, 2008), but the biological significance of this labeling is not clear. Stromal cells located at the periphery of tumor nodules may sometimes contain staining for KLKs (Planque *et al.*, 2008a). At the mRNA level, *KLK6, 8,* and *12* are clearly up-regulated in NSCLC tissues compared to adjacent non-malignant tissues. High KLK6 mRNA was associated with lower overall survival, whereas patients presenting early-stage tumors and high *KLK8* expression had survived disease-free for a longer time than those with low *KLK8* expression (Heuze-Vourc'h *et al.*, 2009; Sher *et al.*, 2006). The expression of other *KLK* genes may be predictable in subgroups of patients within each lung cancer subtype (**Tab. 1.1**). NSCLCs are known to be heterogeneous cancers, and genetic and molecular analyses have revealed distinct differences within subtypes. We therefore need to examine the patterns of KLK gene expression with reference to key "driver" mutations that have been discovered in specific subgroups of NSCLC patients (reviewed in Sanders and Albitar, 2010).

Clinical and preclinical data indicates that KLKs play two roles in carcinogenesis. They can be either cancer promoters or cancer inhibitors, depending on the tissue type and the tumor microenvironment (reviewed in Borgoño and Diamandis, 2004; Emami and Diamandis, 2007; Sotiropoulou *et al.*, 2009). KLK6 and KLK13 promote cancer progression in the lung, whereas KLK8 and KLK10 may be inhibitors (Chou *et al.*, 2011; Heuze-Vourc'h *et al.*, 2009; Sher *et al.*, 2006; Zhang *et al.*, 2009). Recent findings indicate that KLKs may contribute to NSCLC progression by regulating the proliferation of cancer cells. The cell cycle of KLK6-producing NSCLC cells is accelerated between the G1 and S phases. This is accompanied by a marked increase in cyclin E and c-Myc, and a decrease in p21 (Heuze-Vourc'h *et al.*, 2009). In contrast, KLK10 suppresses NSCLC cell proliferation *in vitro* and *in vivo* (Zhang *et al.*, 2009). This suppressive action of KLK10 is supported by clinical data showing that *KLK10* is frequently down-regulated in NSCLC (Planque *et al.*, 2006; Zhang *et al.*, 2009). KLK8 also has a suppressive action in the early stages of NSCLC (Sher *et al.*, 2006). Overproduction of KLK8 decreases the invasiveness of NSCLC cells *in vitro*, and tumor growth and cell invasion *in vivo*. These effects were attributed in part to the degradation of fibronectin by KLK8. Other findings indicate that KLKs take part in the metastatic cascade by facilitating tumor cell detachment, cell motility, and by enabling invasion across the extracellular matrix (ECM) barriers. For example, KLK13 can induce N-cadherin synthesis by NSCLC cells. N-cadherin is a Ca^{2+}-dependent adhesion molecule that is essential to cell-cell interaction and has been reported to play a pivotal role in the extravasation of cancerous cells. In addition, KLK13 facilitates the invasive and metastatic behavior of lung tumor cells both *in vitro* and *in vivo*, by proteolyzing laminin (Chou *et al.*, 2011).

Several KLKs, either alone or in combination, are promising biomarkers for some malignancies (reviewed in Avgeris *et al.*, 2010; Paliouras *et al.*, 2007). However, the limited numbers of patients examined so far has made it impossible to decide whether KLK proteases are potential markers for lung cancer. However, several studies have pointed to some candidates. For example, the increased transcription of *KLK11* in the C2 subgroup of neuroendocrine tumor tissues has been associated with poor patient outcome (Bhattacharjee *et al.*, 2001). Similarly, a high concentration of *KLK6* mRNA and the *KLK8* type 4 splice variant were identified as independent indicators of poor patient outcome in NSCLC (Heuze-Vourc'h *et al.*, 2009; Planque *et al.*, 2010). Finally, KLKs 7, 8, and 12–14 have been proposed as potential serum protein markers of lung cancer. The combination of several KLKs (KLK4, 8, 10–14) may increase the accuracy of detecting NSCLC (Planque *et al.*, 2008b).

1.4 Regulation of KLKs in the lung

Few studies have examined the mechanisms underlying the deregulation of KLKs in lung cancer. This paragraph examines some regulatory mechanisms found in other physiological/pathological contexts and their potential relevance to the lungs.

1.4.1 Control of gene transcription

The *KLK* locus at 19q13.3/4 is highly unstable and the copy number of the *KLK* genes can vary greatly. Although it appears that the copy number did not directly regulate expression of the *KLK* genes in ovarian cancer, genomic changes at 19q13 may be a contributing factor (Bayani *et al.*, 2011). Alterations to 19q13 were also observed in lung cancer and were associated with poor survival of the patients (Kim *et al.*, 2005). It is therefore conceivable that aberrations in regions of the chromosome bearing the KLK genes contribute to their deregulation in lung cancer. The expression of *KLK* genes could also be regulated through epigenetic factors, such as DNA methylation. There is evidence that the *KLK10* gene is highly methylated in NSCLC tissue samples, but not in non-cancer lung tissues. Moreover, the methylation of *KLK10* exon 3 was found to be specifically associated with a low level of *KLK10* transcripts (Zhang *et al.*, 2009). *KLK13* was found to be regulated by DNA-methylation in a number of lung cancer cell lines and hypomethylation during NSCLC progression could account for increased *KLK13* expression identified in late-stage NSCLC patients (Chou *et al.*, 2011; Planque *et al.*, 2008a).

Most of the processes that drive the profound alterations in the transcription of the *KLK* genes in many malignant diseases remain elusive. However, in several malignancies such as cancer of the breast, ovary, and prostate, the expression of many KLKs is clearly dependent on steroid hormones, with the sex hormones (estrogens, androgens, and progestins) being much more important than glucocorticoids or mineralocorticoids (reviewed in Lawrence *et al.*, 2010). Sex steroid hormones may also play a role in NSCLC. Diverse observations in the population suggest that factors other than tobacco, namely female sex hormones, are involved in regulating NSCLC progression (Yano *et al.*, 2011). There is a growing body of pre-clinical and clinical data to support this concept. Aromatase, a key enzyme in estrogen synthesis, is present in NSCLC (Miki *et al.*, 2010; Siegfried *et al.*, 2009) and the outcome for women with high concentrations of aromatase in their tumors is poor (Verma *et al.*, 2011). About 73% of NSCLCs from both male and female patients have higher concentrations of estradiol in the tumor than in the paired non-neoplastic lung tissue (Niikawa *et al.*, 2008). Thus, lung cancer cells produce their own estrogens (reviewed in Verma *et al.*, 2011). Estrogen receptors (ERα, ERβ) have been found in both extranuclear and nuclear sites in most NSCLCs (Marquez-Garban *et al.*, 2011; Siegfried *et al.*, 2009), and estrogens promote the transcription of estrogen-responsive genes in NSCLC cells and stimulate cell proliferation (Hershberger *et al.*, 2005; Stabile *et al.*, 2002).

Like estrogens, androgens and progestins regulate the expression of many KLKs in diverse tissues. The progesterone receptor (PR) has been frequently detected in specimens of NSCLC, mostly in the nucleus. Its presence is associated with positive clinical-pathological features (Ishibashi *et al.*, 2005). The proliferation of PR-positive NSCLC cells is inhibited by progesterone, in a dose-dependent manner, both *in vitro* and *in vivo* (Ishibashi *et al.*, 2005). The androgen receptor (AR) is present in normal adult lungs, mainly in the bronchial epithelium and pneumocyte type II cells. Several types of NSCLCs are also AR-positive, especially squamous cell carcinomas (Mikkonen *et al.*, 2009). In summary, while the lung was not previously considered to be a target organ for sex steroids, recent findings clearly indicate that these hormones can influence the biology of human lung cancers. Therefore, their involvement in the regulation of KLK genes should be analyzed.

Tumorigenesis and the progression of lung cancer depend on a variety of cross-reacting, growth-stimulating pathways. Some of the best-studied signal transduction pathways in lung cancer are the Ras/Raf/MAPK pathway, the phosphoinositol 3-kinase (PI3K) pathway, and the phospholipase C/protein kinase C pathway (reviewed in Pallis *et al.*, 2010). Some transcription factors targeted by these pathways probably contribute to the control of *KLK* gene transcription in lung cancers. Moreover, there are links between the RAS/MEK/ERK and PI3K/AKT signalling pathways and the activation of *KLK3, 10,* and *11* in prostate and breast cancers (Bakin *et al.*, 2003; Paliouras and Diamandis, 2008a and b). Similarly, the activation of the RAS signalling pathway has been associated with the upregulation of *KLK6* in colon cancer (Henkhaus *et al.*, 2008). Finally, several studies have shown that a great variety of transcription factors bind to the promoter regions of *KLK* genes and regulate their expression (Dong *et al.*, 2008; Lawrence *et al.*, 2010; Paliouras and Diamandis, 2008b).

MicroRNAs (miRNAs) have recently been put forward as new regulators of *KLKs*. Using bioinformatics, Chow *et al.* (2008) identified 96 strong *KLK*/miRNA interactions. *KLK10* was the most frequently targeted *KLK*, followed by *KLK5* and *KLK13* (Chow *et al.*, 2008). The transfection of breast and ovarian cancer cells with particular miRNAs significantly decreased the concentrations of KLK6 and KLK10 proteins in them (Chow *et al.*, 2008; White *et al.*, 2010). Interestingly, some miRNAs that have been predicted to regulate *KLKs* are dysregulated in lung cancer (Chow *et al.*, 2008).

1.4.2 Post-translational control of KLK function

The protease activity of KLKs is thought to be mostly regulated by controlling zymogen activation, by endogenous inhibitors, and by self-degradation (Borgoño and Diamandis, 2004). Most KLK zymogens are converted to active enzymes by limited proteolysis (Borgoño *et al.*, 2004; Emami and Diamandis, 2007). Several *in vitro* studies have reported activation of pro-KLKs by other serine proteinases and by MMPs (Brattsand *et al.*, 2005; Emami and Diamandis, 2008; Lovgren *et al.*, 1997; Yoon *et al.*, 2007; 2008). Several lines of evidence indicate that KLKs exert their normal and pathological functions through proteolytic cascades involving the serial activation of KLK zymogens by other mature KLKs or other endopeptidases (Brattsand *et al.*, 2005; Eissa *et al.*, 2011; Eissa and Diamandis, 2008; Emami and Diamandis, 2008; Pampalakis and Sotiropoulou, 2007; Shaw and Diamandis, 2008). The presence of several KLKs in normal and cancerous lung tissue prompted us to speculate that there are similar proteolytic cascades in lung tissue and/or in the lung epithelial lining fluid.

Activated, mature KLKs can be inactivated by endogenous inhibitors (reviewed in Goettig *et al.*, 2010), in particular serpins (serine peptidase inhibitors), Kazal-type inhibitors and the macroglobulins. The serpins, especially α1-antitrypsin (α1-AT), are the major peptidase inhibitors in the lung (Askew and Silverman, 2008) and individuals with hereditary α1-AT deficiency are at risk of developing lung diseases such as emphysema and chronic obstructive pulmonary disease (COPD) (reviewed in Gooptu *et al.*, 2009). This serpin inhibits several KLKs, especially KLK4 and KLK7 (Goettig *et al.*, 2010). Several other serpins, such as PAI-1 (plasminogen activator inhibitor type-1), ATIII (antithrombin III), PCI (protein C inhibitor), which may modulate the activities of KLKs, are functional in normal and diseased lungs (reviewed in Askew and Silverman, 2008). Kazal-type inhibitors are important regulators of KLKs in the skin (Eissa and Diamandis, 2008; Meyer-Hoffert *et al.*, 2010), but there is no published evidence that they are present in lungs.

Bikunin, one of the two Kunitz-type inhibitors of KLKs, is an endogenous inhibitor in the lungs. Bikunin binds to KLK1 in the picomolar range and is associated with this KLK in BALF from asthma patients, after an allergen challenge (Forteza *et al.*, 2007). Bikunin has also been found to be associated with KLK6 and KLK10 in ovarian cancer ascitic fluid (Oikonomopoulou *et al.*, 2010a). In NSCLC, bikunin was detected in infiltrating cells surrounding the tumor islets and in highly differentiated cancerous cells (Bourguignon *et al.*, 1999). Spontaneous lung metastases are more frequent in bikunin-deficient mice than in Bik+/+ mice, suggesting that a bikunin deficiency increases the sensitivity of mice to lung metastases, due to a lack of circulating inter-α inhibitors (Yagyu *et al.*, 2006).

Clearly, further studies are required to identify the multiple mechanisms involved in the control of KLK activities in the lungs. An important parameter that must be determined is the spatial distribution of the various actors. Several findings indicate that compartmentalization plays a major role in the control of KLK activity in the lungs. For example, the glycosaminoglycan hyaluronan (HA) immobilizes inactive bronchial KLK1 at the epithelial surface and within the lumen of submucosal glands, creating a pool of readily available but inactive KLK1 at the surface of the airways. The breakdown of HA by reactive oxygen species (ROS) results in a dramatic increase in KLK1 proteolytic activity and the subsequent activation of KLK1-dependent cell signalling pathways (Casalino-Matsuda *et al.*, 2006). This phenomenon is believed to play a major role in the increased airway KLK1 activity in lung diseases, associated with reactive oxygen species (ROS) production.

1.5 Potential KLK targets in the lung

In vitro and *in silico* studies have identified many molecules as potential substrates of KLKs (reviewed in Debela *et al.*, 2008; Emami and Diamandis, 2007; Lawrence *et al.*, 2010). This section focuses on those potential KLK substrates that appear to be relevant to lung physiology and lung tumorigenesis.

1.5.1 Substrates involved in host defense

The lung epithelium protects the airways by providing a barrier against injury from external insult. This barrier is augmented by a layer of mucus in the large airways, where it is essential for airway integrity and pulmonary defense. Inhaled particles are trapped in the mucus and are removed from the airways by mucociliary clearance. The major structural components of normal mucus are mucins, high-molecular-weight glycoproteins with extensive O-glycosidic-linked oligosaccharides (up to 500 kDa) that are either secreted or membrane-bound (Thornton *et al.*, 2008). The protein backbone is encoded by a specific group of mucin genes (MUC). The mRNAs corresponding to at least 12 human mucin genes have been found in the lungs of healthy individuals, but MUC2, MUC4, MUC5AC, and MUC5B are usually considered to be airway mucins and these airway mucins are overproduced by patients with chronic lung diseases or lung cancer (Evans et al., 2009; Thai et al., 2008).

Shaw and Diamandis (2008) recently showed that KLK5 and KLK12, two KLKs also found in the lung, cleave MUC4 and MUC5B. In contrast KLK6, 11, and 13 do not degrade these mucins (Shaw et al., 2008). These findings suggest that KLKs are also implicated in the disruption of the mucus layer that protects the epithelium in lung cancer, rendering the airways more susceptible to infections. Mucins also capture biologically active molecules, some of which are overproduced in lung adenocarcinomas and are associated with tumorigenesis (dos Santos Silva et al., 2000; Hollingsworth and Swanson, 2004; Radiloff et al., 2011). Moreover, membrane-associated mucins like MUC4 can activate some of the ERBB tyrosine kinase receptors, either directly or by exposing the EGF-like domain (Evans and Koo, 2009; Hollingsworth and Swanson, 2004). Thus, the cleavage of mucins by KLKs may modulate the availability of sequestered factors or regulate signaling pathways, thereby contributing to lung pathophysiology.

The antimicrobial peptides produced by human leukocytes and epithelial cells are important to innate immunity. They can kill a broad spectrum of microorganisms by disrupting their membranes. The main antimicrobial peptides in the respiratory tract are the defensins and the cathelicidins (Herr et al., 2007; Tecle et al., 2010). The 18 kDa human cationic antimicrobial protein (hCAP-18) is the precursor of the 37-residue cathelicidin peptide LL-37 that is released by proteolytic cleavage (Herr et al., 2007). KLK5 cleaves the hCAP-18 precursor to produce cathelicidin in the skin, while KLK7 cleaves hCAP18 to produce multiple products. Along with KLK7, KLK5 further breaks down the LL-37 peptide into various shorter antimicrobial peptides (Yamasaki et al., 2006). In the normal lung, hCAP-18 is produced by inflammatory cells, the epithelial cells of conducting airways, and submucosal glands. As KLK5 and KLK7 are produced at the same location, they are probably involved in hCAP-18 processing in the airways. LL-37/hCAP-18 is also produced by lung cancer cells, and several studies have shown that LL-37/hCAP-18 regulates the tumor-forming activity of malignant lung cells (von Haussen et al., 2008; Wu et al., 2010). Thus, KLKs are likely to contribute to lung cancer progression by processing hCAP-18 to LL-37, which then acts as an autocrine growth factor, released from lung tumor epithelial cells.

LL-37 is not the only antimicrobial peptide that has been found to activate lung epithelial cancer cells. Human α-defensin 1 stimulates human lung cancer cells to proliferate in vitro, although this effect is still a matter of debate (Aarbiou et al., 2002; Xu et al., 2008). The defensins are divided into two major subgroups, α-defensins and β-defensins (Tecle et al., 2010). Although the human β-defensins 1 and 2 are constitutive components of the respiratory tract epithelial cells, leukocytes, and dendritic cells, they are not yet considered to be substrates of KLKs. Indeed, β-defensin 1 is not cleaved by KLK5, 6, or 11–13 in vitro (Shaw et al., 2008). Human α-defensins 1–4 are produced by neutrophils and their precursors are packaged in primary (azurophil) granules. The processing of these defensin precursors to mature forms involves at least three proteolytic cleavages. Transient intermediates are released through constitutive secretion, indicating that the proteolytic processing also occurs post-secretion. The processing of α-defensins by KLKs, in order to release a mature antimicrobial C-terminal peptide has been studied by Shaw *et al.* (2008), who found that KLK5 processes α-defensin 1, while KLK6, and 11–13 do not cleave this defensin *in vitro*.

1.5.2 Cytokines and growth factors

The synthesis and release of pro-inflammatory cytokines is an essential step in the activation of an effective innate lung defense and the subsequent modulation of the adaptive immune responses. Interleukin-1β (IL-1β) is a potent pro-inflammatory cytokine with pleiotropic action, primarily in the inflammatory and immune responses. IL-1β also has homeostatic functions in the airways, including the repair of epithelial cells and the stimulation of mucin secretion (Crosby and Waters, 2010; Fujisawa *et al.*, 2009). IL-1β is produced as an inactive precursor, pro-IL-1β, activated either intracellularly by caspase-1/inflammasome (Netea *et al.*, 2010) or extracellularly (van de Veerdonk *et al.*, 2011). KLKs can activate pro-IL-1β *in vitro,* as shown for KLK7 and KLK13 (Nylander-Lundqvist and Egelrud, 1997; Yao *et al.*, 2006). As these KLKs are present in normal and malignant lung tissues, the extracellular activation of pro- IL-1β by KLKs may make a contribution to the IL-1β available in the lung. IL-1β immunoreactivity has been detected in human lung tumors, where the concentrations of IL-1β were significantly higher than in normal tissues (Colasante *et al.*, 1997; Landvik *et al.*, 2009). The secretion of IL-1β is often enhanced during lung tumor progression in experimental tumor models and in cancer patients, and is usually directly correlated with a bad prognosis (Apte and Voronov, 2008). As this is also true for KLK13 (Chou *et al.*, 2011; Planque *et al.*, 2008a), it seems likely that the KLK13-mediated activation of IL-1β contributes to lung tumor progression and invasiveness. IL-1β stimulates tumor progression and the metastasis of NSCLCs via several mechanisms. It promotes tumor invasiveness by stimulating the synthesis of proteases (Cho *et al.*, 2011). It may also contribute to angiogenesis via the IL-1β-dependent production of IL-8 and TGF-β1 in NSCLC cells (Colasante *et al.*, 1997).

Transforming growth factor β (TGF-β) also is a pleiotropic cytokine in the airways, which is critical for limiting some inflammatory reactions, and plays a key role in mediating tissue remodeling and repair (reviewed in Santibanez *et al.*, 2011). Three members of the TGF-β family (TGF-β1, TGF-β2, and TGF-β3) have been identified in humans. Although both TGF-β1 and TGF-β3 have been implicated in lung homeostasis, most studies of the lungs have focused on TGF-β1 only (Camoretti-Mercado and Solway, 2005). TGF-β is synthesized as a precursor and secreted as a homodimer of mature TGF-β, non-covalently associated with the latency-associated protein (LAP), or with latent-TGF-β-binding protein (LTBP), which is important for targeting TGF-β to the ECM. Activation of TGF-β requires the release of TGF-β from the LAP and the

LTBP, by proteolysis or a conformational change. KLK1, 2, 5, and 14 all activate TGF-β by cleaving LAP and/or LTBP (Emami and Diamandis, 2010). Activated TGF-β acts by binding to the TGF-β type I (TGF-βRI) and type II (TGF-βRII) receptors present in the lungs (Camoretti-Mercado and Solway, 2005; Santibanez et al., 2011). TGF-β may also interact with matricellular proteins of the CCN family (see below), which serve as chaperon proteins to improve the presentation of TGF-β1 to TGF-β receptors (Abreu et al., 2002). We recently showed that KLKs modify the interaction of TGF-β with CCNs through limited proteolysis of these binding proteins. This is likely to modulate both the bioavailability and the activity of TGF-β (Guillon-Munos et al., 2011). Several tumors, including those arising in the lungs, contain high concentrations of TGF-β, and this concentration is correlated with tumor progression and clinical prognosis. The plasma concentration of TGF-β is elevated in all forms of lung cancer (Toonkel et al., 2010). TGF-β can be a tumor-suppressor in early stages of tumor development, by virtue of its potent growth inhibitory effect on lung epithelial cells (Colasante et al., 1997; Santibanez et al., 2011). However, lung cancer cells can escape from this suppressive action (Toonkel et al., 2010).

Paradoxically, TGF-β becomes a pro-oncogenic factor that stimulates tumor cell growth and invasiveness at later stages of tumorigenesis, stimulating the production of autocrine mitogenic growth factors and the epithelial-mesenchymal transition (EMT), the generation of myofibroblasts (also called cancer-associated fibroblasts) that facilitate tumor cell proliferation and invasion as well as promoting neoangiogenesis, and helping cancer cells evade immune surveillance (Santibanez et al., 2011). In summary, the KLK-mediated regulation of TGF-β signaling, which has two time-dependent actions on cancer progression, may either favor or limit the progression of lung tumors.

The components of the insulin-like growth factor (IGF) axis were among the first substrates described for KLKs (Cohen et al., 1992; Rehault et al., 2001). IGFs (IGF-I and IGF-II) are important mitogenic peptides involved in regulating the proliferation, differentiation, apoptosis, and transformation of normal and malignant cells. IGFs can only act after they have been released from IGF-binding proteins (IGFBPs). These IGFBPs form a family of six proteins that antagonize the binding of IGFs to the IGF-I receptor (IGF-IR). Several KLKs (KLK1-5, 11, and 14) can cleave IGFBPs, hence decreasing their affinity for IGFs and increasing the bioavailability of these growth factors (Borgoño et al., 2007; Maeda et al., 2009; Michael et al., 2006; Rehault et al., 2001; Sano et al., 2007).

Since the components of the IGF-system are ubiquitous, we need to examine the implication of KLKs in the IGF-mediated progression of lung cancer. A high serum IGF-I concentration is a risk factor for lung cancer, and lung airway epithelial cells produce IGFs in an autocrine manner, leading to deregulation of IGF-IR activation (Kim et al., 2011a and b). It has also been shown that over-production of IGF-I in the lungs of mice promotes the development and progression of lung tumors. This process is accelerated by tobacco carcinogens. IGFBP-3 has been reported to suppress tumor

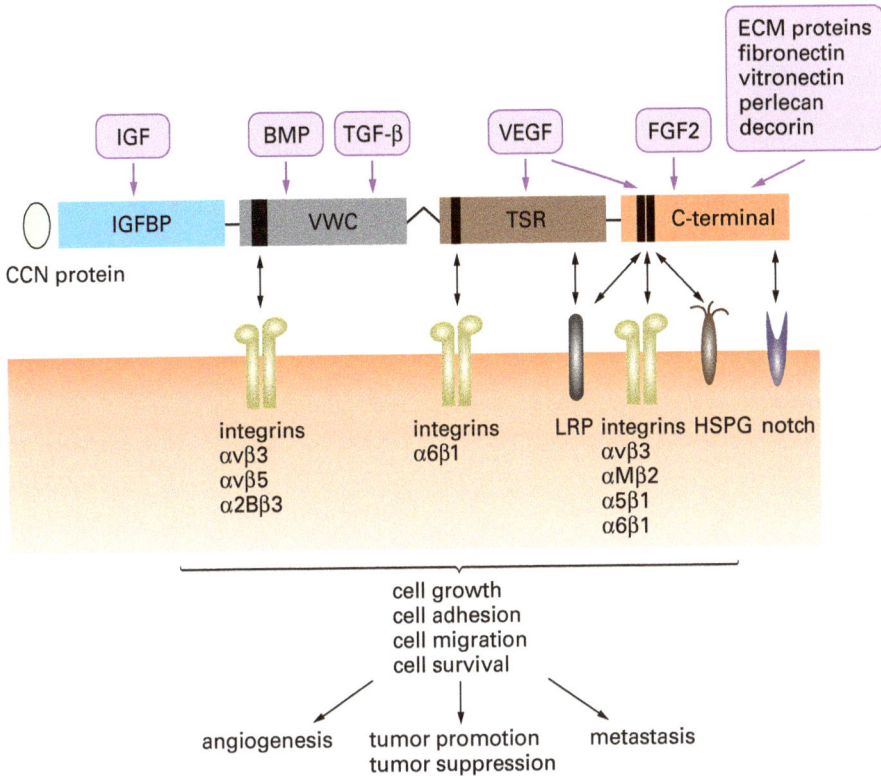

Fig. 1.1 Schematic of CCN protein structure and interaction of CCN proteins with other molecules. The six CCN proteins are composed of an amino-terminal secretory peptide and four conserved modular domains: the insulin-like growth factor binding protein (IGFBP) domain, the von Willebrand factor C (VWC) domain, the thrombospondin type 1 repeat (TSR) domain, and the carboxy-terminal domain (C-terminal). CCN5 uniquely lacks the C-terminal domain. The modular property of CCN proteins gives them the ability to bind and interact with a broad range of factors. It is known that these modules can bind to cell-surface molecules such as heparan sulfate proteoglycans (HSPG), integrins, lipoprotein receptor-related proteins (LRP), and Notch. CCN proteins also physically interact with several extracellular matrix (ECM) proteins and growth factors (including insulin-like growth factor (IGF), vascular endothelial growth factor (VEGF), fibroblast growth factor 2 (FGF2), transforming growth factor-β (TGF-β), and bone morphogenetic proteins (BMP)). The ability of CCN proteins to physically interact with various regulatory partners places them at the interface of key signaling pathways. The CCN proteins trigger signal transduction events that culminate in the regulation of cell proliferation, adhesion, migration, differentiation, and survival. The role of CCN proteins is highly complex, particularly in cancer. Depending on the carcinogenesis of specific sites and types, expression levels may vary. CCN proteins can influence tumor progression in several ways, such as cell survival, angiogenesis, and metastasis. Interestingly, many studies reported anti-cancer effects or a tumor suppressive role of CCN proteins (reviewed in Zuo *et al.*, 2010).

growth and angiogenesis by both IGF-dependent and IGF-independent mechanisms (Kim *et al.*, 2011a and b). This IGFBP is targeted by KLK5, 11, and 14 present in lung cancer.

1.5.3 Pericellular and membrane-associated substrates

Numerous *in vitro* studies have shown that many components (collagens, fibronectin, etc) of the ECM are broken down by KLKs (reviewed in Debela *et al.*, 2008; Emami and Diamandis, 2007; Lawrence *et al.*, 2010; Sotiropoulou and Pampalakis, 2010). We can infer from this that KLKs are involved in tissue remodeling, either directly, by cleaving ECM components, or indirectly, via the activation of peptidases such as pro-uPA, pro-MMP-2 or pro-MMP-9. This is probably also the case in the lung, as this tissue contains both enzymes and substrates. Several recent findings support this. Sher *et al.* (2006) showed that KLK8 modifies the extracellular microenvironment in NSCLC by cleaving the ECM component fibronectin. This suppresses integrin signaling and retards cancer cell motility by inhibiting actin polymerization. KLK13 can also degrade laminin, which will facilitate the migration and invasion of NSCLC cells (Chou *et al.*, 2011).

We recently identified a new class of substrates for KLKs, the CCN matricellular proteins (**Fig. 1.1**; Guillon-Munos *et al.*, 2011). There are six of them, CCN1–6. The CCNs are all secreted ECM-associated proteins and are involved in internal and external cellular signaling to regulate cell adhesion, migration, mitogenesis, differentiation, and survival (reviewed in Chen and Lau, 2009; Holbourn *et al.*, 2008; Jun and Lau, 2011; Zuo *et al.*, 2010). They also regulate angiogenesis and could be a new class of inflammation modulator (Kular *et al.*, 2010). Most of the above effects are believed to be mediated by binding with integrins, which are receptors that mediate cell-cell and cell-ECM interactions. The members of the CCN family also modulate the functions

Fig. 1.2 **KLK12 mobilizes growth factors from growth factor-CCN complexes. (a)** Left panel shows analysis of KLK 1, 5, 12, and 14 hydrolysis of recombinant CCN1 on silver-stained 4–12% SDS-PAGE gels. CCN1 was incubated with the KLKs (enzyme:substrate ratio of 1:100, w/w) for 2 hr. Right panel shows mass spectrometry analysis (MALDI-TOF) of the CCN1 fragments generated with KLK12 and KLK14. Molecular mass of the main fragments is given in Daltons. **(b)** Edman degradation of the amino-terminal amino acids of the fragments generated with KLK12 revealed five sequences consistent with cleavage either after a lysine (K) or an arginine (R). These predicted cleavage sites are located within the hinge region between the VWC and TSR domains and within the IGFBP and C-terminal domains (arrows). **(c)** CCN1 was complexed with growth factors on microtiter plates for 12 hr at 4 °C and prior to a 3 hr incubation with activated KLK12 (KLK2:CCN ratio of 1:100, w/w), at 37 °C. Specific ELISAs were used to quantify the residual growth factors bound to CCN1 after KLK12 cleavage. (+) designates a statistically significant difference in residual binding (p < 0.5). Fragmentation of CCN1 by KLK12 triggered the release of BMP-2, VEGF, and TGF-β from the growth factor-CCN complexes.

(a)

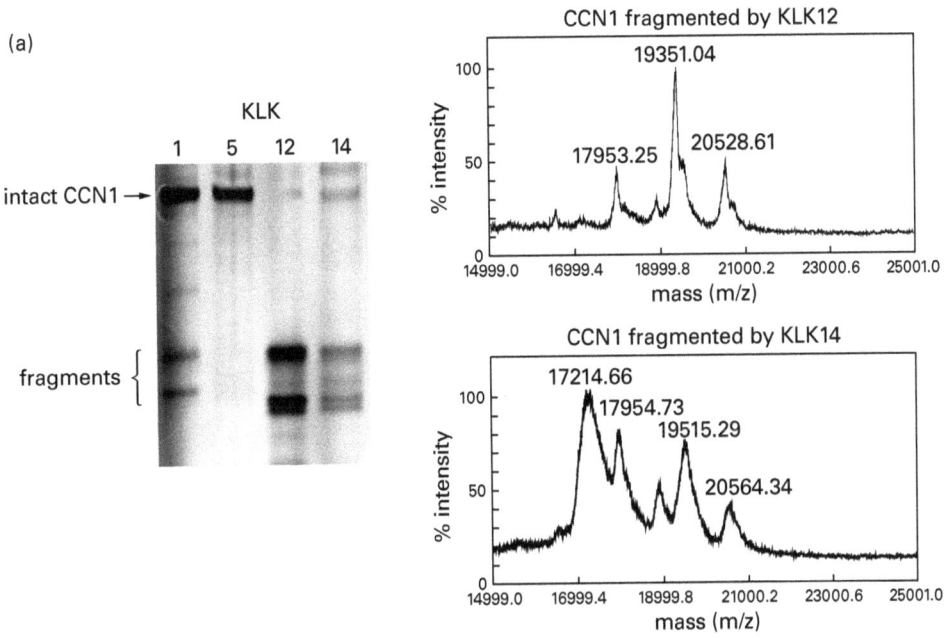

(b)

1: K88-R89
2: K208-K209
3: R218-I219 (main cleavage site)
4: K291-K292
5: K345-N346

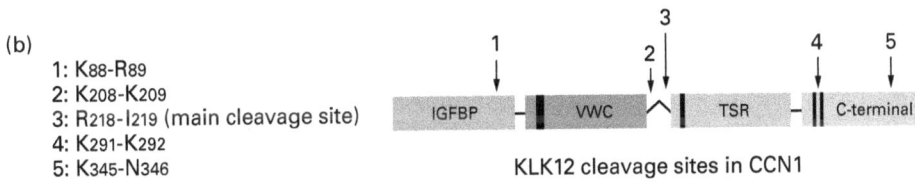

KLK12 cleavage sites in CCN1

(c)

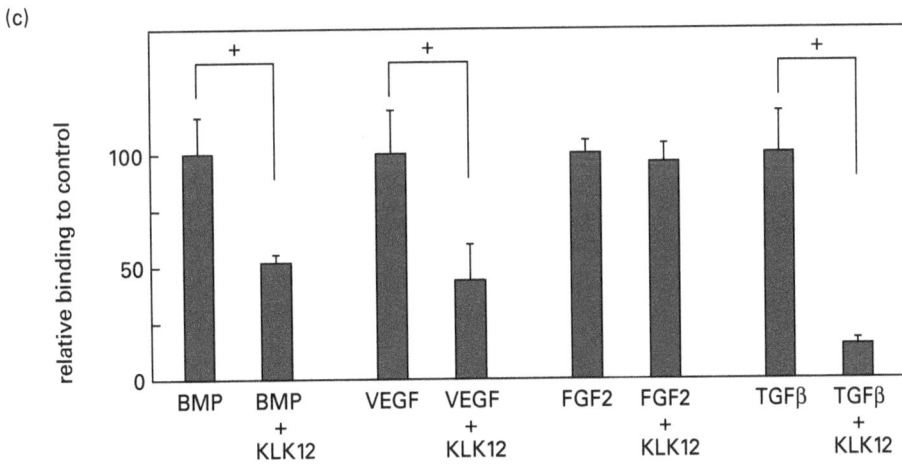

of several important growth factors and related pathways, such as the IGFs, VEGF, TGF-β, BMPs (bone morphogenetic proteins), FGFs (fibroblast growth factors), and Wnt signalling. The CCN proteins are sensitive to proteolysis by KLK1, 5, 12, and 14, but not by KLK6, 8, 11, and 13. The fragmentation of CCNs alters the way these multifunctional proteins interact with growth factors (**Fig. 1.2**; Guillon-Munos *et al.*, 2011). Our findings suggest that KLKs regulate the bioavailability and activity of several important growth factors by processing their CCN binding partners. The synthesis of CCN proteins is deregulated in chronic lung diseases and in lung cancer (Chang *et al.*, 2004; Chen *et al.*, 2007; Guillon-Munos *et al.*, 2011). The fragmentation of CCNs by KLKs probably plays an important role in several lung diseases, since the lungs contain KLKs, CCNs, and growth factors.

Proteinase-activated receptors (PARs) are a special class of G-protein-coupled receptors (GPCR) that are mainly activated by serine proteases and matrix metalloproteinases. To date, four PARs (PAR1–4) have been identified. All subtypes are present in different cells in the human respiratory tract (Peters and Henry, 2009). Activation of PARs by KLKs has been reported (Oikonomopoulou *et al.*, 2006; Stefansson *et al.*, 2008) (see Chapter 15, Volume 1). KLK4-6, and 14 all activate PAR1, PAR2, and PAR4 by partial cleavage of their extracellular domain, and the KLK-mediated activation of the PARs results in Ca^{2+} mobilization. KLK14 preferentially cleaves PAR2, and there is evidence for a negative regulatory feedback system through which KLK14 deactivates PAR1. The impact of KLK-mediated activation of PARs on cancer has been discussed (Hollenberg *et al.*, 2008; Oikonomopoulou *et al.*, 2010b), but little work has been done to determine the role of PARs in lung cancer. A high concentration of PAR1 is accompanied by an aggressive phenotype and poor outcome in patients with NSCLC (Cisowski *et al.*, 2011).

Also, a significant correlation is found between PAR1 expression and VEGF in human lung cancer tissues (Ghio *et al.*, 2006). Moreover, activation of PAR1 was associated with enhanced motility of lung cancer cells and angiogenesis in human lung tumors grown in nude mice, by stimulating VEGF production (Cisowski *et al.*, 2011). Lastly, the stimulation of PAR2 in A549 cancer cells led to the release of pro-inflammatory mediators, including PGE2, IL-6, and IL-8 (Moriyuki *et al.*, 2009). IL-8 is known to promote angiogenic responses in endothelial cells, increase the proliferation and survival of endothelial and cancer cells, and to potentiate the migration of cancer cells (Waugh and Wilson, 2008).

1.6 Conclusion

KLKs are an emerging family of serine proteases in the lung. In this organ, as in many tissues, the synthesis of KLKs is probably tightly regulated and understanding its regulation, as well as identifying endogenous substrates, could shed some light on their biological function in such normal developmental processes as lung branching morphogenesis, and in many pathological conditions, including cancer, asthma, and chronic obstructive lung disease.

Acknowledgements

The English text was edited by Dr. Owen Parkes. This work was supported by a grant (KalliCap) from the Region Centre (France).

Bibliography

Aarbiou, J., Ertmann, M., van Wetering, S., van Noort, P., Rook, D., Rabe, K.F., Litvinov, S. V., van Krieken, J.H., de Boer, W.I., and Hiemstra, P.S. (2002). Human neutrophil defensins induce lung epithelial cell proliferation *in vitro*. J. Leukoc. Biol. 72, 167–174.

Abreu, J.G., Ketpura, N.I., Reversade, B., and De Robertis, E.M. (2002). Connective-tissue growth factor (CTGF) modulates cell signalling by BMP and TGF-beta. Nat. Cell Biol. 4, 599–604.

Apte, R.N., and Voronov, E. (2008). Is interleukin-1 a good or bad 'guy' in tumor immunobiology and immunotherapy? Immunol. Rev. 222, 222–241.

Askew, D.J., and Silverman, G.A. (2008). Intracellular and extracellular serpins modulate lung disease. J. Perinatol. 28, S127–135.

Avgeris, M., Mavridis, K., and Scorilas, A. (2010). Kallikrein-related peptidase genes as promising biomarkers for prognosis and monitoring of human malignancies. Biol. Chem. 391, 505–511.

Bakin, R.E., Gioeli, D., Sikes, R.A., Bissonette, E.A., and Weber, M.J. (2003). Constitutive activation of the Ras/mitogen-activated protein kinase signaling pathway promotes androgen hypersensitivity in LNCaP prostate cancer cells. Cancer Res. 63, 1981–1989.

Bayani, J., Marrano, P., Graham, C., Zheng, Y., Li, L., Katsaros, D., Lassus, H., Butzow, R., Squire, J.A., and Diamandis, E.P. (2011). Genomic instability and copy-number heterogeneity of chromosome 19q, including the kallikrein locus, in ovarian carcinomas. Mol. Oncol. 5, 48–60.

Bhattacharjee, A., Richards, W.G., Staunton, J., Li, C., Monti, S., Vasa, P., Ladd, C., Beheshti, J., Bueno, R., Gillette, M., Loda, M., Weber, G., Mark, E.J., Lander, E.S., Wong, W., Johnson, B.E., Golub, T.R., Sugarbaker, D.J., and Meyerson, M. (2001). Classification of human lung carcinomas by mRNA expression profiling reveals distinct adenocarcinoma subclasses. Proc. Natl. Acad. Sci. USA 98, 13790–13795.

Borgoño, C.A., and Diamandis, E.P. (2004). The emerging roles of human tissue kallikreins in cancer. Nat. Rev. Cancer 4, 876–890.

Borgoño, C.A., Michael, I.P., and Diamandis, E.P. (2004). Human tissue kallikreins: physiologic roles and applications in cancer. Mol. Cancer Res. 2, 257–280.

Borgoño, C.A., Michael, I.P., Shaw, J.L., Luo, L.Y., Ghosh, M.C., Soosaipillai, A., Grass, L., Katsaros, D., and Diamandis, E.P. (2007). Expression and functional characterization of the cancer-related serine protease, human tissue kallikrein 14. J. Biol. Chem. 282, 2405–2422.

Bourguignon, J., Borghi, H., Sesboue, R., Diarra-Mehrpour, M., Bernaudin, J. F., Metayer, J., Martin, J.P., and Thiberville, L. (1999). Immunohistochemical distribution of inter-alpha-trypsin inhibitor chains in normal and malignant human lung tissue. J. Histochem. Cytochem. 47, 1625–1632.

Brattsand, M., Stefansson, K., Lundh, C., Haasum, Y., and Egelrud, T. (2005). A proteolytic cascade of kallikreins in the stratum corneum. J. Invest. Dermatol. 124, 198–203.

Camoretti-Mercado, B., and Solway, J. (2005). Transforming growth factor-beta1 and disorders of the lung. Cell. Biochem. Biophys. 43, 131–148.

Casalino-Matsuda, S.M., Monzon, M.E., and Forteza, R.M. (2006). Epidermal growth factor receptor activation by epidermal growth factor mediates oxidant-induced goblet cell metaplasia in human airway epithelium. Am. J. Respir. Cell Mol. Biol. 34, 581–591.

Chang, C.C., Shih, J.Y., Jeng, Y.M., Su, J.L., Lin, B.Z., Chen, S.T., Chau, Y.P., Yang, P.C., and Kuo, M.L. (2004). Connective tissue growth factor and its role in lung adenocarcinoma invasion and metastasis. J. Natl. Cancer Inst. 96, 364–375.

Chee, J., Naran, A., Misso, N.L., Thompson, P.J., and Bhoola, K.D. (2008). Expression of tissue and plasma kallikreins and kinin B1 and B2 receptors in lung cancer. Biol. Chem. 389, 1225–1233.

Chen, C.C., and Lau, L.F. (2009). Functions and mechanisms of action of CCN matricellular proteins. Int. J. Biochem. Cell Biol. 41, 771–783.

Chen, P.P., Li, W.J., Wang, Y., Zhao, S., Li, D.Y., Feng, L.Y., Shi, X.L., Koeffler, H.P., Tong, X.J., and Xie, D. (2007). Expression of Cyr61, CTGF, and WISP-1 correlates with clinical features of lung cancer. PLoS One 2, e534.

Cho, W.C., Kwan, C.K., Yau, S., So, P.P., Poon, P.C., and Au, J.S. (2011). The role of inflammation in the pathogenesis of lung cancer. Expert Opin. Ther. Targets 15, 1127–1137.

Chou, R. H., Lin, S.C., Wen, H.C., Wu, C.W., and Chang, W.S. (2011). Epigenetic activation of human kallikrein 13 enhances malignancy of lung adenocarcinoma by promoting N-cadherin expression and laminin degradation. Biochem. Biophys. Res. Commun. 409, 442–447.

Chow, T.F., Crow, M., Earle, T., El-Said, H., Diamandis, E.P., and Yousef, G.M. (2008). Kallikreins as microRNA targets: an in silico and experimental-based analysis. Biol. Chem. 389, 731–738.

Christiansen, S.C., Proud, D., and Cochrane, C.G. (1987). Detection of tissue kallikrein in the bronchoalveolar lavage fluid of asthmatic subjects. J. Clin. Invest. 79, 188–197.

Cisowski, J., O'Callaghan, K., Kuliopulos, A., Yang, J., Nguyen, N., Deng, Q., Yang, E., Fogel, M., Tressel, S., Foley, C., Agarwal, A., Hunt, S.W. 3rd, McMurry, T., Brinckerhoff, L., and Covic, L. (2011). Targeting protease-activated receptor-1 with cell-penetrating pepducins in lung cancer. Am. J. Pathol. 179, 513–523.

Cohen, P., Graves, H., Peehl, D., Kamarei, M., Giudice, L., and Rosenfeld, R. (1992). Prostate-specific antigen (PSA) is an insulin-like growth factor binding protein-3 protease found in seminal plasma. J. Clin. Endocrinol. Metab. 75, 1046–1053.

Colasante, A., Mascetra, N., Brunetti, M., Lattanzio, G., Diodoro, M., Caltagirone, S., Musiani, P., and Aiello, F.B. (1997). Transforming growth factor beta 1, interleukin-8 and interleukin-1, in non-small-cell lung tumors. Am. J. Respir. Crit. Care Med. 156, 968–973.

Crosby, L.M., and Waters, C.M. (2010). Epithelial repair mechanisms in the lung. Am. J. Physiol. Lung Cell. Mol. Physiol. 298, L715–731.

Debela, M., Beaufort, N., Magdolen, V., Schechter, N.M., Craik, C.S., Schmitt, M., Bode, W., and Goettig, P. (2008). Structures and specificity of the human kallikrein-related peptidases KLK 4, 5, 6, and 7. Biol. Chem. 389, 623–632.

Dong, Y., Matigian, N., Harvey, T.J., Samaratunga, H., Hooper, J.D., and Clements, J.A. (2008). Tissue-specific promoter utilisation of the kallikrein-related peptidase genes, KLK5 and KLK7, and cellular localisation of the encoded proteins suggest roles in exocrine pancreatic function. Biol. Chem. 389, 99–109.

dos Santos Silva, E., Ulrich, M., Doring, G., Botzenhart, K., and Gott, P. (2000). Trefoil factor family domain peptides in the human respiratory tract. J. Pathol. 190, 133–142.

Eissa, A., and Diamandis, E.P. (2008). Human tissue kallikreins as promiscuous modulators of homeostatic skin barrier functions. Biol. Chem. 389, 669–680.

Eissa, A., Amodeo, V., Smith, C.R., and Diamandis, E.P. (2011). Kallikrein-related peptidase-8 (KLK8) is an active serine protease in human epidermis and sweat and is involved in a skin barrier proteolytic cascade. J. Biol. Chem. 286, 687–706.

Emami, N., and Diamandis, E.P. (2007). New insights into the functional mechanisms and clinical applications of the kallikrein-related peptidase family. Mol. Oncol. 1, 269–287.

Emami, N., and Diamandis, E.P. (2008). Human kallikrein-related peptidase 14 (KLK14) is a new activator component of the KLK proteolytic cascade. Possible function in seminal plasma and skin. J. Biol. Chem. 283, 3031–3041.

Emami, N., and Diamandis, E.P. (2010). Potential role of multiple members of the kallikrein-related peptidase family of serine proteases in activating latent TGF beta 1 in semen. Biol. Chem. 391, 85–95.

Evans, C.M., and Koo, J.S. (2009). Airway mucus: the good, the bad, the sticky. Pharmacol. Ther. 121, 332–348.

Evans, C.M., Kim, K., Tuvim, M.J., and Dickey, B.F. (2009). Mucus hypersecretion in asthma: causes and effects. Curr. Opin. Pulm. Med. 15, 4–11.

Forteza, R., Casalino-Matsuda, S.M., Monzon, M.E., Fries, E., Rugg, M.S., Milner, C.M., and Day, A.J. (2007). TSG-6 potentiates the antitissue kallikrein activity of inter-alpha-inhibitor through bikunin release. Am. J. Respir. Cell. Mol. Biol. 36, 20–31.

Fujisawa, T., Velichko, S., Thai, P., Hung, L.Y., Huang, F., and Wu, R. (2009). Regulation of airway MUC5AC expression by IL-1beta and IL-17A; the NF-kappaB paradigm. J. Immunol. 183, 6236–6243.

Ghio, P., Cappia, S., Selvaggi, G., Novello, S., Lausi, P., Zecchina, G., Papotti, M., Borasio, P., and Scagliotti, G.V. (2006). Prognostic role of protease-activated receptors 1 and 4 in resected stage IB non-small-cell lung cancer. Clin. Lung Cancer 7, 395–400.

Goettig, P., Magdolen, V., and Brandstetter, H. (2010). Natural and synthetic inhibitors of kallikrein-related peptidases (KLKs). Biochimie 92, 1546–1567.

Gooptu, B., Ekeowa, U.I., and Lomas, D.A. (2009). Mechanisms of emphysema in alpha1-antitrypsin deficiency: molecular and cellular insights. Eur. Respir. J. 34, 475–488.

Guillon-Munos, A., Oikonomopoulou, K., Michel, N., Smith, C.R., Petit-Courty, A., Canepa, S., Reverdiau, P., Heuze-Vourc'h, N., Diamandis, E.P., and Courty, Y. (2011). The kallikrein-related peptidase 12 hydrolyzes matricellular proteins of the CCN family and modifies interactions of CCN1 and CCN5 with growth factors. J. Biol. Chem. 286, 25505–25518.

Henkhaus, R.S., Gerner, E.W., and Ignatenko, N.A. (2008). Kallikrein 6 is a mediator of K-RAS-dependent migration of colon carcinoma cells. Biol. Chem. 389, 757–764.

Herr, C., Shaykhiev, R., and Bals, R. (2007). The role of cathelicidin and defensins in pulmonary inflammatory diseases. Expert Opin. Biol. Ther. 7, 1449–1461.

Hershberger, P.A., Vasquez, A.C., Kanterewicz, B., Land, S., Siegfried, J.M., and Nichols, M. (2005). Regulation of endogenous gene expression in human non-small cell lung cancer cells by estrogen receptor ligands. Cancer Res. 65, 1598–1605.

Heuze-Vourc'h, N., Planque, C., Guyetant, S., Coco, C., Brillet, B., Blechet, C., Parent, C., Briollais, L., Reverdiau, P., Jourdan, M.L., and Courty, Y. (2009). High kallikrein-related peptidase 6 in non-small cell lung cancer cells: an indicator of tumor proliferation and poor prognosis. J. Cell. Mol. Med. 3, 4014–4022.

Holbourn, K.P., Acharya, K.R., and Perbal, B. (2008). The CCN family of proteins: structure-function relationships. Trends Biochem. Sci. 10, 461–473.

Hollenberg, M.D., Oikonomopoulou, K., Hansen, K.K., Saifeddine, M., Ramachandran, R., and Diamandis, E.P. (2008). Kallikreins and proteinase-mediated signaling: proteinase-activated receptors (PARs) and the pathophysiology of inflammatory diseases and cancer. Biol. Chem. 389, 643–651.

Hollingsworth, M.A., and Swanson, B.J. (2004). Mucins in cancer: protection and control of the cell surface. Nat. Rev. Cancer 4, 45–60.

Ishibashi, H., Suzuki, T., Suzuki, S., Niikawa, H., Lu, L., Miki, Y., Moriya, T., Hayashi, S., Handa, M., Kondo, T., and Sasano, H. (2005). Progesterone receptor in non-small cell lung cancer--a potent prognostic factor and possible target for endocrine therapy. Cancer Res. 65, 6450–6458.

Jun, J.I. and Lau, L.F. (2011). Taking aim at the extracellular matrix: CCN proteins as emerging therapeutic targets. Nat. Rev. Drug Discov. 10, 945–963.

Kim, J.H., Choi, D.S., Lee, O.H., Oh, S.H., Lippman, S.M., and Lee, H.Y. (2011a). Antiangiogenic antitumor activities of IGFBP-3 are mediated by IGF-independent suppression of Erk1/2 activation and Egr-1-mediated transcriptional events. Blood 118, 2622–2631.

Kim, T.M., Yim, S.H., Lee, J.S., Kwon, M.S., Ryu, J.W., Kang, H.M., Fiegler, H., Carter, N.P., and Chung, Y.J. (2005). Genome-wide screening of genomic alterations and their clinicopathologic implications in non-small cell lung cancers. Clin. Cancer Res. 11, 8235–8242.

Kim, W.Y., Kim, M.J., Moon, H., Yuan, P., Kim, J.S., Woo, J.K., Zhang, G., Suh, Y.A., Feng, L., Behrens, C., van Pelt, C.S., Kang, H., Lee, J.J., Hong, W.K., Wistuba, I.I., and Lee, H.Y. (2011b). Differential impacts of insulin-like growth factor-binding protein-3 (IGFBP-3) in epithelial IGF-induced lung cancer development. Endocrinology 152, 2164–2173.

Kular, L., Pakradouni, J., Kitabgi, P., Laurent, M., and Martinerie, C. (2010). The CCN family: a new class of inflammation modulators? Biochimie 93, 377–388.

Landvik, N.E., Hart, K., Skaug, V., Stangeland, L.B., Haugen, A., and Zienolddiny, S. (2009). A specific interleukin-1B haplotype correlates with high levels of IL1B mRNA in the lung and increased risk of non-small cell lung cancer. Carcinogenesis 30, 1186–1192.

Lawrence, M.G., Lai, J., and Clements, J.A. (2010). Kallikreins on steroids: structure, function, and hormonal regulation of prostate-specific antigen and the extended kallikrein locus. Endocr. Rev. 31, 407–446.

Lovgren, J., Rajakoski, K., Karp, M., Lundwall, A., and Lilja, H. (1997). Activation of the zymogen form of prostate-specific antigen by human glandular kallikrein 2. Biochem. Biophys. Res. Commun. 238, 549–555.

Luo, L.Y., Grass, L., Howarth, D.J., Thibault, P., Ong, H., and Diamandis, E.P. (2001). Immunofluorometric assay of human kallikrein 10 and its identification in biological fluids and tissues. Clin. Chem. 47, 237–246.

Maeda, H., Yonou, H., Yano, K., Ishii, G., Saito, S., and Ochiai, A. (2009). Prostate-specific antigen enhances bioavailability of insulin-like growth factor by degrading insulin-like growth factor binding protein 5. Biochem. Biophys. Res. Commun. 381, 311–316.

Marquez-Garban, D.C., Mah, V., Alavi, M., Maresh, E.L., Chen, H.W., Bagryanova, L., Horvath, S., Chia, D., Garon, E., Goodglick, L., and Pietras, R.J. (2011). Progesterone and estrogen receptor expression and activity in human non-small cell lung cancer. Steroids 76, 910–920.

Meyer-Hoffert, U., Wu, Z., Kantyka, T., Fischer, J., Latendorf, T., Hansmann, B., Bartels, J., He, Y., Glaser, R., and Schroder, J.M. (2010). Isolation of SPINK6 in human skin: selective inhibitor of kallikrein-related peptidases. J. Biol. Chem. 285, 32174–32181.

Michael, I.P., Pampalakis, G., Mikolajczyk, S.D., Malm, J., Sotiropoulou, G., and Diamandis, E.P. (2006). Human tissue kallikrein 5 is a member of a proteolytic cascade pathway involved in seminal clot liquefaction and potentially in prostate cancer progression. J. Biol. Chem. 281, 12743–12750.

Miki, Y., Suzuki, T., Abe, K., Suzuki, S., Niikawa, H., Iida, S., Hata, S., Akahira, J., Mori, K., Evans, D.B., Kondo, T., Yamada-Okabe, H., and Sasano, H. (2010). Intratumoral localization of aromatase and interaction between stromal and parenchymal cells in the non-small cell lung carcinoma microenvironment. Cancer Res. 70, 6659–6669.

Mikkonen, L., Pihlajamaa, P., Sahu, B., Zhang, F.P., and Janne, O.A. (2009). Androgen receptor and androgen-dependent gene expression in lung. Mol. Cell. Endocrinol. 317, 14–24.

Moriyuki, K., Sekiguchi, F., Matsubara, K., Nishikawa, H., and Kawabata, A. (2009). Proteinase-activated receptor-2-triggered prostaglandin E(2) release, but not cyclooxygenase-2 upregulation, requires activation of the phosphatidylinositol 3-kinase / Akt / nuclear factor-kappaB pathway in human alveolar epithelial cells. J. Pharmacol. Sci. 111, 269–275.

Netea, M.G., Simon, A., van de Veerdonk, F., Kullberg, B.J., van der Meer, J.W., and Joosten, L.A. (2010). IL-1beta processing in host defense: beyond the inflammasomes. PLoS Pathog. 6, e1000661.

Nicholas, B., Skipp, P., Mould, R., Rennard, S., Davies, D.E., O'Connor, C.D., and Djukanovic, R. (2006). Shotgun proteomic analysis of human-induced sputum. Proteomics 6, 4390–4401.

Niikawa, H., Suzuki, T., Miki, Y., Suzuki, S., Nagasaki, S., Akahira, J., Honma, S., Evans, D. B., Hayashi, S., Kondo, T., and Sasano, H. (2008). Intratumoral estrogens and estrogen receptors in human non-small cell lung carcinoma. Clin. Cancer Res. 14, 4417–4426.

Nylander-Lundqvist, E., and Egelrud, T. (1997). Formation of active IL-1 beta from pro-IL-1 beta catalyzed by stratum corneum chymotryptic enzyme *in vitro*. Acta Derm. Venereol. 77, 203–206.

Oikonomopoulou, K., Hansen, K.K., Saifeddine, M., Vergnolle, N., Tea, I., Blaber, M., Blaber, S.I., Scarisbrick, I., Diamandis, E.P., and Hollenberg, M.D. (2006). Kallikrein-mediated cell signalling: targeting proteinase-activated receptors (PARs). Biol. Chem. 387, 817–824.

Oikonomopoulou, K., Batruch, I., Smith, C.R., Soosaipillai, A., Diamandis, E.P., and Hollenberg, M.D. (2010a). Functional proteomics of kallikrein-related peptidases in ovarian cancer ascites fluid. Biol. Chem. 391, 381–390.

Oikonomopoulou, K., Diamandis, E.P., and Hollenberg, M.D. (2010b). Kallikrein-related peptidases: proteolysis and signaling in cancer, the new frontier. Biol. Chem. 391, 299–310.

Olsson, A.Y., Bjartell, A., Lilja, H., and Lundwall, A. (2005). Expression of prostate-specific antigen (PSA) and human glandular kallikrein 2 (hK2) in ileum and other extraprostatic tissues. Int. J. Cancer 113, 290–297.

Paliouras, M., Borgoño, C., and Diamandis, E.P. (2007). Human tissue kallikreins: the cancer biomarker family. Cancer Lett. 249, 61–79.

Paliouras, M., and Diamandis, E.P. (2008a). An AKT activity threshold regulates androgen-dependent and androgen-independent PSA expression in prostate cancer cell lines. Biol. Chem. 389, 773–780.

Paliouras, M., and Diamandis, E.P. (2008b). Intracellular signaling pathways regulate hormone-dependent kallikrein gene expression. Tumour Biol. 29, 63–75.

Pallis, A.G., Karamouzis, M.V., Konstantinopoulos, P.A., and Papavassiliou, A.G. (2010). Molecular networks in respiratory epithelium carcinomas. Cancer Lett. 295, 1–6.

Pampalakis, G., and Sotiropoulou, G. (2007). Tissue kallikrein proteolytic cascade pathways in normal physiology and cancer. Biochim. Biophys. Acta 1776, 22–31.

Peters, T., and Henry, P.J. (2009). Protease-activated receptors and prostaglandins in inflammatory lung disease. Br. J. Pharmacol. 158, 1017–1033.

Petraki, C.D., Karavana, V.N., Skoufogiannis, P.T., Little, S.P., Howarth, D.J., Yousef, G.M., and Diamandis, E.P. (2001). The spectrum of human kallikrein 6 (zyme/protease M/neurosin) expression in human tissues as assessed by immunohistochemistry. J. Histochem. Cytochem. 49, 1431–1441.

Petraki, C.D., Karavana, V.N., Luo, L.Y., and Diamandis, E.P. (2002). Human kallikrein 10 expression in normal tissues by immunohistochemistry. J. Histochem. Cytochem. 50, 1247–1261.

Petraki, C.D., Karavana, V.N., and Diamandis, E.P. (2003). Human kallikrein 13 expression in normal tissues: an immunohistochemical study. J. Histochem. Cytochem. 51, 493–501.

Petraki, C.D., Papanastasiou, P.A., Karavana, V.N., and Diamandis, E.P. (2006). Cellular distribution of human tissue kallikreins: immunohistochemical localization. Biol. Chem. 387, 653–663.

Planque, C., de Monte, M., Guyetant, S., Rollin, J., Desmazes, C., Panel, V., Lemarie, E., and Courty, Y. (2005). KLK5 and KLK7, two members of the human tissue kallikrein family, are differentially expressed in lung cancer. Biochem. Biophys. Res. Commun. 329, 1260–1266.

Planque, C., Ainciburu, M., Heuze-Vourc'h, N., Regina, S., de Monte, M., and Courty, Y. (2006). Expression of the human kallikrein genes 10 (KLK10) and 11 (KLK11) in cancerous and non-cancerous lung tissues. Biol. Chem. 387, 783–788.

Planque, C., Blechet, C., Ayadi-Kaddour, A., Heuze-Vourc'h, N., Dumont, P., Guyetant, S., Diamandis, E.P., El Mezni, F., and Courty, Y. (2008a). Quantitative RT-PCR analysis and immunohisto-chemical localization of the kallikrein-related peptidases 13 and 14 in lung. Biol. Chem. 389, 781–786.

Planque, C., Li, L., Zheng, Y., Soosaipillai, A., Reckamp, K., Chia, D., Diamandis, E.P., and Goodglick, L. (2008b). A multiparametric serum kallikrein panel for diagnosis of non-small cell lung carcinoma. Clin. Cancer Res. 14, 1355–1362.

Planque, C., Choi, Y. H., Guyetant, S., Heuze-Vourc'h, N., Briollais, L., and Courty, Y. (2010). Alternative splicing variant of kallikrein-related peptidase 8 as an independent predictor of unfavorable prognosis in lung cancer. Clin. Chem. 56, 987–997.

Poblete, M.T., Garces, G., Figueroa, C.D., and Bhoola, K.D. (1993). Localization of immunoreactive tissue kallikrein in the seromucous glands of the human and guinea-pig respiratory tree. Histochem. J. 25, 834–839.

Proud, D., and Vio, C.P. (1993). Localization of immunoreactive tissue kallikrein in human trachea. Am. J. Respir. Cell.Mol. Biol. 8, 16–19.

Radiloff, D.R., Wakeman, T.P., Feng, J., Schilling, S., Seto, E. and Wang, X.F. (2011). Trefoil factor 1 acts to suppress senescence induced by oncogene activation during the cellular transformation process. Proc. Natl. Acad. Sci. USA 108, 6591–6596.

Rehault, S., Monget, P., Mazerbourg, S., Tremblay, R., Gutman, N., Gauthier, F., and Moreau, T. (2001). Insulin-like growth factor binding proteins (IGFBPs) as potential physiological substrates for human kallikreins hK2 and hK3. Eur. J. Biochem. 268, 2960–2968.

Sanders, H.R., and Albitar, M. (2010). Somatic mutations of signaling genes in non-small-cell lung cancer. Cancer Genet. Cytogenet. 203, 7–15.

Sano, A., Sangai, T., Maeda, H., Nakamura, M., Hasebe, T., and Ochiai, A. (2007). Kallikrein 11 expressed in human breast cancer cells releases insulin-like growth factor through degradation of IGFBP-3. Int. J. Oncol. 30, 1493–1498.

Santibanez, J.F., Quintanilla, M., and Bernabeu, C. (2011). TGF-beta/TGF-beta receptor system and its role in physiological and pathological conditions. Clin. Sci. (Lond) 121, 233–251.

Sasaki, H., Kawano, O., Endo, K., Suzuki, E., Haneda, H., Yukiue, H., Kobayashi, Y., Yano, M., and Fujii, Y. (2006). Decreased kallikrein 11 messenger RNA expression in lung cancer. Clin Lung Cancer 8, 45–48.

Seiz, L., Kotzsch, M., Grebenchtchikov, N.I., Geurts-Moespot, A.J., Fuessel, S., Goettig, P., Gkazepis, A., Wirth, M. P., Schmitt, M., Lossnitzer, A., Sweep, F.C., and Magdolen, V. (2010). Polyclonal antibodies against kallikrein-related peptidase 4 (KLK4): immunohistochemical assessment of KLK4 expression in healthy tissues and prostate cancer. Biol. Chem. 391, 391–401.

Shaw, J.L., and Diamandis, E.P. (2007). Distribution of 15 human kallikreins in tissues and biological fluids. Clin. Chem. 53, 1423–1432.

Shaw, J.L., and Diamandis, E.P. (2008). A potential role for tissue kallikrein-related peptidases in human cervico-vaginal physiology. Biol. Chem. 389, 681–688.

Shaw, J.L., Petraki, C., Watson, C., Bocking, A., and Diamandis, E.P. (2008). Role of tissue kallikrein-related peptidases in cervical mucus remodeling and host defense. Biol. Chem. 389, 1513–1522.

Sher, Y.P., Chou, C.C., Chou, R.H., Wu, H.M., Wayne Chang, W.S., Chen, C.H., Yang, P.C., Wu, C.W., Yu, C.L., and Peck, K. (2006). Human kallikrein 8 protease confers a favorable clinical outcome in non-small cell lung cancer by suppressing tumor cell invasiveness. Cancer Res. 66, 11763–11770.

Siegfried, J.M., Hershberger, P.A., and Stabile, L.P. (2009). Estrogen receptor signaling in lung cancer. Semin. Oncol. 36, 524–531.

Singh, J., Naran, A., Misso, N.L., Rigby, P.J., Thompson, P.J., and Bhoola, K.D. (2008). Expression of kallikrein-related peptidases (KRP/hK5, 7, 6, 8) in subtypes of human lung carcinoma. Int. Immunopharmacol. 8, 300–306.

Sotiropoulou, G., Pampalakis, G., and Diamandis, E.P. (2009). Functional roles of human kallikrein-related peptidases. J. Biol. Chem. 284, 32989–32994.

Sotiropoulou, G., and Pampalakis, G. (2010). Kallikrein-related peptidases: bridges between immune functions and extracellular matrix degradation. Biol. Chem. 391, 321–331.

Stabile, L.P., Davis, A.L., Gubish, C.T., Hopkins, T.M., Luketich, J.D., Christie, N., Finkelstein, S., and Siegfried, J.M. (2002). Human non-small cell lung tumors and cells derived from normal lung express both estrogen receptor alpha and beta and show biological responses to estrogen. Cancer Res. 62, 2141–2150.

Stefansson, K., Brattsand, M., Roosterman, D., Kempkes, C., Bocheva, G., Steinhoff, M., and Egelrud, T. (2008). Activation of proteinase-activated receptor-2 by human kallikrein-related peptidases. J. Invest. Dermatol. 128, 18–25.

Tecle, T., Tripathi, S., and Hartshorn, K. L. (2010). Review: Defensins and cathelicidins in lung immunity. Innate Immun. 16, 151–159.

Thai, P., Loukoianov, A., Wachi, S., and Wu, R. (2008). Regulation of airway mucin gene expression. Annu. Rev. Physiol. 70, 405–429.

Thornton, D.J., Rousseau, K., and McGuckin, M.A. (2008). Structure and function of the polymeric mucins in airways mucus. Annu. Rev. Physiol. 70, 459–486.

Toonkel, R.L., Borczuk, A.C., and Powell, C.A. (2010). TGF-beta signaling pathway in lung adenocarcinoma invasion. J. Thorac. Oncol. 5, 153–157.

van de Veerdonk, F.L., Netea, M.G., Dinarello, C.A., and Joosten, L.A. (2011). Inflammasome activation and IL-1β and IL-18 processing during infection. Trends Immunol. 32, 110–116.

Verma, M.K., Miki, Y., and Sasano, H. (2011). Aromatase in human lung carcinoma. Steroids 76, 759–764.

von Haussen, J., Koczulla, R., Shaykhiev, R., Herr, C., Pinkenburg, O., Reimer, D., Wiewrodt, R., Biesterfeld, S., Aigner, A., Czubayko, F., and Bals, R. (2008). The host defence peptide LL-37/hCAP-18 is a growth factor for lung cancer cells. Lung Cancer 59, 12–23.

Waugh, D.J., and Wilson, C. (2008). The interleukin-8 pathway in cancer. Clin. Cancer Res. 14, 6735–6741.

White, N.M., Chow, T.F., Mejia-Guerrero, S., Diamandis, M., Rofael, Y., Faragalla, H., Mankaruous, M., Gabril, M., Girgis, A., and Yousef, G.M. (2010). Three dysregulated miRNAs control kallikrein 10 expression and cell proliferation in ovarian cancer. Br. J. Cancer 102, 1244–1253.

Wu, W.K., Wang, G., Coffelt, S.B., Betancourt, A.M., Lee, C.W., Fan, D., Wu, K., Yu, J., Sung, J.J., and Cho, C.H. (2010). Emerging roles of the host defense peptide LL-37 in human cancer and its potential therapeutic applications. Int. J. Cancer 127, 1741–1747.

Xu, N., Wang, Y.S., Pan, W.B., Xiao, B., Wen, Y.J., Chen, X.C., Chen, L.J., Deng, H.X., You, J., Kan, B., Fu, A.F., Li, D., Zhao, X., and Wei, Y.Q. (2008). Human alpha-defensin-1 inhibits growth of human lung adenocarcinoma xenograft in nude mice. Mol. Cancer Ther. 7, 1588–1597.

Yagyu, T., Kobayashi, H., Matsuzaki, H., Wakahara, K., Kondo, T., Kurita, N., Sekino, H., and Inagaki, K. (2006). Enhanced spontaneous metastasis in bikunin-deficient mice. Int. J. Cancer 118, 2322–2328.

Yamasaki, K., Schauber, J., Coda, A., Lin, H., Dorschner, R.A., Schechter, N.M., Bonnart, C., Descargues, P., Hovnanian, A., and Gallo, R.L. (2006). Kallikrein-mediated proteolysis regulates the antimicrobial effects of cathelicidins in skin. FASEB J. 20, 2068–2080.

Yano, T., Haro, A., Shikada, Y., Maruyama, R., and Maehara, Y. (2011). Non-small cell lung cancer in never smokers as a representative ‚non-smoking-associated lung cancer': epidemiology and clinical features. Int. J. Clin. Oncol. 1, 287–293.

Yao, C., Karabasil, M.R., Purwanti, N., Li, X., Akamatsu, T., Kanamori, N., and Hosoi, K. (2006). Tissue kallikrein mK13 is a candidate processing enzyme for the precursor of interleukin-1beta in the submandibular gland of mice. J. Biol. Chem. 281, 7968–7976.

Yoon, H., Laxmikanthan, G., Lee, J., Blaber, S. I., Rodriguez, A., Kogot, J.M., Scarisbrick, I. A., and Blaber, M. (2007). Activation profiles and regulatory cascades of the human kallikrein-related peptidases. J. Biol. Chem. 282, 31852–31864.

Yoon, H., Blaber, S.I., Evans, D.M., Trim, J., Juliano, M.A., Scarisbrick, I.A., and Blaber, M. (2008). Activation profiles of human kallikrein-related peptidases by proteases of the thrombostasis axis. Protein Sci. 17, 1998–2007.

Zarghami, N., Levesque, M., D'Costa, M., Angelopoulou, K., and Diamandis, E.P. (1997). Frequency of expression of prostate-specific antigen mRNA in lung tumors. Am. J. Clin. Pathol. 108, 184–190.

Zhang, Y., Song, H., Miao, Y., Wang, R., and Chen, L. (2009). Frequent transcriptional inactivation of Kallikrein 10 gene by CpG island hypermethylation in non-small cell lung cancer. Cancer Sci. 101, 934–940.

Zuo, G.W., Kohls, C.D., He, B.C., Chen, L., Zhang, W., Shi, Q., Zhang, B.Q., Kang, Q., Luo, J., Luo, X., Wagner, E. R., Kim, S.H., Restegar, F., Haydon, R.C., Deng, Z.L., Luu, H. H., He, T.C., and Luo, Q. (2010). The CCN proteins: important signaling mediators in stem cell differentiation and tumorigenesis. Histol. Histopathol. 25, 795–806.

Maroulio Talieri, Marina Devetzi, and Manfred Schmitt

2 Clinical Relevance of Kallikrein-related Peptidases in Gastric and Colorectal Cancer

2.1 Introduction

The human kallikrein-related peptidase (KLK) family of enzymes (Lundwall *et al.*, 2006), widely expressed throughout the human body, is now being studied for its clinical utility in several human diseases. KLKs are hormonally regulated enzymes. Consequently, their expression in cancerous tissue was first studied in steroid hormone-regulated cancer diseases such as that of the breast, ovary, or prostate (Avgeris *et al.*, 2010; Borgoño and Diamandis 2004; Clements *et al.*, 2004; Lawrence *et al.*, 2010; Mavridis and Scorilas, 2010; Paliouras *et al.*, 2007; Sotiropoulou *et al.*, 2009; Yousef *et al.*, 2005). Regarding the clinical utility of KLKs in gastrointestinal cancers, clinically relevant studies have been conducted only recently, to explore the putative role of these novel cancer biomarkers for progression and/or metastasis of gastric cancer (GC) and colorectal cancer (CRC).

Surprisingly, only seven of the 15 KLKs (KLK5, 6, 7, 10, 11, 13, and 14) have been studied for their clinical utility in GC or CRC, although relevant antibodies/ELISA tests and gene probes are available for such kind of investigations. In gastric cancer, only four of those are of clinical relevance, namely KLK6, 10, 11, and 13 (Scorilas *et al.*, 2012); only five KLKs (KLK5-7, 10, and 14) are clinically relevant for CRC. Interestingly, in CRC, overexpression of these KLKs is always associated with a poor prognosis, which is not the case for GC. Overexpression of two of the KLKs, KLK6 and KLK10, is associated with a poor clinical outcome, whereas KLK11 and KL13 upregulation is associated with a favorable prognosis.

2.2 Features of gastric and colorectal cancers

Gastric cancer or stomach cancer (GC) is a malignancy of the upper gastrointestinal tract. Advanced gastric cancer is generally refractory to chemotherapy, which leads to a poor prognosis, but if diagnosed at an early stage, it is a curable disease. Therefore it is highly important to identify cancer biomarkers associated with early-stage gastric cancer. Where it was once the second most common cancer worldwide, stomach cancer has dropped to fourth place, behind cancers of the lung, breast, and colorectum, although there are significant geographic variations in its incidence and outcome (Crew and Neugut, 2006). GC is particularly common in Asia. Estimated new cases and deaths from GC, for instance for the United States of America in 2011, are 21,520 new cases and 10,340 deaths (American Cancer Society, 2011). Metastasis occurs in 80–90% of individuals with GC, with a six-month survival rate of 65% in those diagnosed in early stages and <15% of those diagnosed in late stages. GC rates are about twice as high in males as in females. Gastric adenocarcinoma has the highest incidence (90%) among GCs originating from the glandular epithelium of the gastric mucosa.

Clinical staging and types of GC are very similar to colorectal cancer (CRC), which is by far the most common malignancy of the gastrointestinal tract, representing the third most commonly diagnosed cancer in males and the second in females, with an estimated 141,210 new cases and 49,380 deaths in 2011 in the United States of America (Jemal *et al.*, 2011; American Cancer Society, 2011). Rates are substantially higher in males than in females. The incidence of CRC varies greatly between different regions of the world. Much of this can be attributed to differences in diet, particularly the consumption of red and processed meat, alcohol, smoking, excessive bodyweight, and low physical activity (Chan and Giovannucci, 2010; Cho *et al.*, 2012; Wu *et al.*, 2006). CRC incidence and mortality rates have been decreasing over the past two decades in both males and females, especially in economically developed countries in which CRC has become a preventable disease. Currently, endoscopy remains the most important screening procedure with therapeutic value for CRC, because prognostic and/or predictive markers for this type of malignancy are scarce or lack specificity and sensitivity (Bohanes *et al.*, 2011; He *et al.*, 2011; Winder *et al.*, 2010).

2.3 Established biomarkers in gastric and colorectal cancer

Tumorigenesis is a multi-step process that progresses on the basis of genetic mutations. In GC etiology, in addition to environmental factors, genetic factors may play an important role (Resende *et al.*, 2010, 2011). The rationale behind this is that accumulation of genetic and epigenetic alterations in GC may play crucial roles in the process of cellular tumorigenesis and immortalization, e.g. involving the *DNMT3A*, *PSCA*, *VEGF*, and *XRCC1* genes. Other genes were reported to be associated with gastric carcinogenesis through oncogenic activation (*MYC*, *SEMA5A*, *BCL2L12*, *RBP2*, *BUBR1*, *GSK3b*, *CD133*, *DSC2*, *P-cadherin*, *CDH17*, *CD168*, *CD44*, *MMP-7* and *MMP-11*, and a subset of miRNAs) or tumor suppressor gene inactivation mechanisms (*KLF6*, *RELN*, *PTCH1A*, *CLDN11*, *SFRP5*, *TFF1*, *PDX1*, *BCL2L10*, *XRCC*, *psiTPTE-HERV*, *HAI-2*, *GRIK2*, and *RUNX3*).

Also, genetic variants of *IL-10*, *IL-17*, *MUC1*, *MUC6*, *DNMT3B*, *SMAD4*, and *SERPINE1* have been reported to modify the risk of developing GC. Remarkably, none of these biomarkers has entered clinical practice yet (Scorilas and Mavridis, 2012). Apart from that, the oncoprotein HER2 and the cyclooxygenase COX-2 evolved as molecular targets for therapy. Furthermore, environmental factors such as Helicobacter pylori infection, consumption of processed or salted/nitrated foods, and cigarette smoking have been found to be associated with a risk of developing GC (Crew and Neugut 2006; Resende *et al.*, 2010; 2011; Shikata *et al.*, 2008; Vogiatzi *et al.*, 2007).

Prognosis of GC is traditionally assessed by TNM status (tumor size, nodal status, metastasis), tumor location, nuclear grading, and histological subtype, but not by the use of tissue-associated biomarkers, although molecular biomarkers have also been widely investigated for their potential for predicting the outcome. Cancer biomarkers with prognostic and/or predictive potential that are currently under investigation encompass the thymidylate synthase and the excision repair cross-complementing ERCC1 gene expression in GC patients treated with either 5-fluorouracil or cisplatin (Kim *et al.*, 2010). HER2 was also identified as a prognostic marker in cases of differentiated gastric cancer. Furthermore, p53 protein expression and mutation, urokinase-type plasminogen activator (uPA) and its inhibitor PAI-1, xanthine oxidoreductase, claudin-4, vascular endothelial growth factor, interleukin-8, and cyclin E have all been associated with poor survival of GC patients. Combined expression of MYC, EGFR, and FGFR2 is predictive of poor survival in cisplatin/5-fluorouracil-treated, advanced-stage GC patients (Kim *et al.*, 2010).

CRC develops in a series of genetic steps, corresponding to the histological progression from normal colonic epithelium to adenomatous dysplasia through microinvasion, adenocarcinoma and, finally, metastasis. Over 90% of CRC contain two or more genetic alterations, generally arising from pre-existing adenomas (Anwar *et al.*, 2004). Colorectal carcinogenesis mostly involves adenoma to carcinoma alterations with chromosomal **in**stability (CIN) on chromosome 5q. CIN colorectal tumors (80–85% of CRC) manifest themselves with a high frequency of allelic imbalance, chromosomal amplifications, and translocations. Every step of CRC initiation and progression involves well-defined genetic alterations in tumor suppressor genes (e.g. APC, p53), oncogenes (Morán *et al.*, 2010), and mitotic spindle proteins such as BUB1 and AURKA (Ross, 2011).

The second pathway of colorectal carcinogenesis, the "mutator" pathway, characterized by microsatellite instability, is present in most tumors that occur in hereditary non-polyposis colorectal cancer and in approximately 15–20% of sporadic CRC (Anwar *et al.*, 2004). Microsatellite mutations are present in the genes implicated in colorectal carcinogenesis, such as *TGFβRII, IGF2R, BAX, APC, β-catenin, axin, MMP-3*, and *BCL-10*. This accumulation of mutations is caused by a primary defect in mismatch repair genes, including *hMLH1, hMSH2, and MSH6* (Ross, 2011).

Besides investigating the molecular pathways leading to CRC tumorigenesis it is of eminent clinical importance to investigate potential prognostic and/or predictive cancer biomarkers connected to the clinical outcome of CRC patients (George and Kopetz, 2011). Conventional clinical and pathological risk factors such as tumor stage and residual tumor mass provide limited prognostic information and do not predict response to adjuvant chemotherapy (Kelley and Venook, 2011). The gene chip-based Oncotype Dx Colon Cancer test (Genomic Health, Redwood City, CA, USA) assesses mRNA expression of twelve target genes in CRC tumor tissue specimens, in order to create a disease recurrence score which predicts CRC recurrence in stage II tumors treated by surgery.

Likewise, by the same technology, the ColoPrint mRNA expression test (Agendia BV, Amsterdam, The Netherlands) assesses a set of 18 prognosis-related genes (Salazar *et al.*, 2011). Other genes under investigation for their prognostic value in stage II/III CRC are *Ki-67, p21, p27, PTEN, BRAF, PIK3CA, EGFR, HER2, BCL2, E-cadherin, β-catenin, TGF-β, VEGF, uPA, MMPs/TIMPs, thymidylate synthetase*, and *telomerase* (Ross, 2011). Concerning therapy resistance, indicators of the efficacy of 5-fluorouracil, irinotecan, or oxaliplatin (e.g. for *DPD, MTHFR, thymidilate synthetase, topoisomerase I, UGT1A1, ERCC1, DNA polymerase β, glutathione S-transferase π1*) have not achieved standard-of-practice usefulness. Likewise, for markers PIK3CA, p53, PTEN, and the Fc γ recep-tor, more clinical evidence has to be presented before they can be recommended as markers of resistance to antibody-based cetuximab or panitumumab immunotherapy (Ross, 2011). Otherwise, for metastatic CRC, mutation analyses for the GTPase *KRAS* and the Raf kinase *bRAF* have already entered clinical practice as biomarkers for pre-dicting resistance to EGFR-directed antibody therapy (Chibaudel *et al.*, 2012; Hamf-jord *et al.*, 2011).

2.4 KLKs: novel biomarkers in gastric and colorectal cancer

To fully understand the physiologic and pathophysiologic functions of KLKs, explicit knowledge of their expression patterns is indispensable. At least for the physiologi-cal situation, Shaw and Diamandis (2007) have studied global protein and mRNA level expression of all of the 15 KLKs in various human tissues and biological fluids. However, such a comprehensive investigation is missing for the malignant state. Regarding gastric and colorectal cancer, we reviewed the scientific literature for pub-lished data on the clinical relevance of *KLK* mRNA and/or KLK protein expression (**Tab. 2.1, 2.2, and 2.3**). So far, clinical utility of these biomarkers has been demon-strated only for KLK5, 6, 7, 10, 11, 13, and 14.

Nevertheless, as a consequence of these rather scattered unicentric investiga-tions, for future studies, harmonization of tools (antibodies, ELISA tests, qRT-PCR-assays) is highly recommended, and subgroup analyses of the study populations (early stage *vs.* advanced stage of GC/CRC, nodal-positive versus nodal-negative, low versus high nuclear grade, metastasis-free versus metastasized) and precise knowl-edge about treatment modalities (neoadjuvant, adjuvant, palliative) are all prereq-uisites. Only then, after multicenter validation of the findings, meta-analyses can be performed to demonstrate the robustness of individual KLKs as cancer biomark-ers to predict the course of the GC/CRC disease and/or response to systemic cancer therapy.

Tab. 2.1 Gene and protein expression of KLKs in normal and cancerous gastric tissue

	Normal tissue[a]		Cancerous tissue		Reference
	mRNA[b]	Protein[c]	mRNA[b]	Protein[c]	
KLK1	moderate	moderate	not determined	elevated	Sawant et al., 2001
KLK2	absent	absent	not determined	not determined	
KLK3	low	low	not determined	absent	Milne et al., 2007
KLK4	low	absent	not determined	not determined	
KLK5	low	absent	not determined	not determined	
KLK6	high	low	elevated	elevated	Kim et al., 2011 Nagahara et al., 2005
KLK7	high	absent	not determined	not determined	
KLK8	high	absent	not determined	not determined	
KLK9	high	low	not determined	not determined	
KLK10	moderate	absent	elevated	elevated	Feng et a., 2006 Huang et al., 2007 Li et al., 2011
KLK11	high	moderate	decreased	decreased	Wen et al., 2011
KLK12	high	high	not determined	not determined	
KLK13	low	low	decreased	not determined	Konstantoudakis et al., 2010
KLK14	moderate	low	not determined	not determined	
KLK15	low	absent	not determined	not determined	

a Data compiled from Shaw and Diamandis (2007)
b Data collected by real-time polymerase chain reaction (RT-PCR), quantitative real-time polymerase chain reaction (qRT-PCR), or in silico mRNA expression analysis
c Data collected by immunohistochemistry, ELISA, or western blotting

2.4.1 Review of the clinical relevance of KLK expression in gastric cancer

Nine of the KLK mRNAs (KLK1, 6–12, and 14) are moderately/highly expressed in normal gastric tissue, whereas the other six KLK mRNAs (KLK2-5, 13, and 15) are not (KLK2) or only slightly (KLK3-5, 13, and 15) expressed (**Tab. 2.1**). This expression pattern is also reflected at the protein level (ELISA) for the majority of the KLKs; only KLK6-9 did not match. Aside form this, seven KLKs are not expressed at the protein level (KLK2, 4, 5, 7, 8, 10, and 15) in the normal stomach tissue. Surprisingly, for these KLKs, only KLK2 is not expressed at the mRNA or protein level. The other six KLKs, although not detected at the protein level, are moderately/highly (KLK7, 8, and 10) or at least slightly expressed (KLK4, 5, and 15) at the mRNA level.

In GC, four studies concerned themselves with KLK mRNA detection, namely for KLK6 and KLK10, which are increased in GC tumor tissue compared to healthy gastric tissue, and for KLK11 and KLK13, the levels of which are decreased, compared to the healthy counterparts. For five of the KLKs (KLK1, 3, 6, 10, and 11) protein expression

Tab. 2.2 Gene and protein expression of KLKs in normal and cancerous colorectal tissue

	Normal tissue[a]		Cancerous tissue		Reference
	mRNA[b]	Protein[c]	mRNA[b]	Protein[c]	
KLK1	high	moderate	decreased	not determined	Yousef *et al.*, 2004
KLK2	absent	absent	not determined	not determined	
KLK3	low	low	not determined	not determined	
KLK4	moderate	absent	not determined	increased	Gratio *et al.*, 2010
KLK5	low	absent	not determined	no change	Talieri *et al.*, 2009a
KLK6	low	absent	increased	increased	Agesen *et al.*, 2012
					Kim *et al.*, 2011
					Ogawa *et ul.*, 2005
					Petraki *et al.*, 2012
					Talieri *et al.*, 2009a, 2011
					Yousef *et al.*, 2004
KLK7	low	absent	increased	increased	Inoue *et al.*, 2010
					Talieri *et al.*, 2009a
					Talieri *et al.*, 2009b
KLK8	low	absent	increased	increased	Talieri *et al.*, 2009a
					Yousef *et al.*, 2004
KLK9	absent	moderate	not determined	not determined	
KLK10	absent	absent	increased	increased	Feng *et al.*, 2006
					Talieri *et al.*, 2009a
					Talieri *et al.*, 2011
					Yousef *et al.*, 2004
KLK11	low	low	not determined	increased	Talieri *et al.*, 2009a
KLK12	low	high	not determined	not determined	
KLK13	low	absent	not determined	increased	Talieri *et al.*, 2009a
KLK14	low	absent	not determined	no change	Gratio *et al.*, 2011
					Talieri *et al.*, 2009a
KLK15	high	absent	not determined	increased	Talieri *et al.*, 2009a

a Data compiled from Shaw and Diamandis (2007)
b Data collected by real-time polymerase chain reaction (RT-PCR), quantitative real-time polymerase chain reaction (qRT-PCR), or *in silico* mRNA expression analysis
c Data collected by immunohistochemistry, ELISA, or western blotting

was determined by western blotting and/or immunohistochemistry. KLK3 (PSA) was absent in GC tissue, KLK1, 6, and 10 were elevated, and KLK11 was decreased. The clinical impact of gene and/or protein expression **(Tab. 2.3)** was investigated for four KLKs (KLK6, 10, 11, and 13). Thus, clinically relevant data in GC is missing for eleven of the fifteen KLKs (KLK1-5, 7–9, 11, 12, 14, and 15).

KLK1. Protein expression of KLK1 (also known as tissue kallikrein) in GC tumor tissue and normal gastric mucosa counterparts was assessed by immunohistochemistry by Sawant *et al.* (2001), which is the only publication so far concerning KLK1 expression in GC. They reported enhanced immunoreactivity of KLK1 in the tumor

Tab. 2.3 Clinical impact of KLK gene/protein expression in primary tumor tissue in gastric or colorectal cancer

	Type of cancer	Clinical impact of tumor tissue mRNA and/or protein KLK expression[a]	Reference
KLK5	Colorectal	Upregulation → poor prognosis	Talieri et al., 2009a
KLK6	Gastric	Upregulation → poor prognosis	Nagahara et al., 2005
	Colorectal	Upregulation → poor prognosis	Agesen et al., 2012
			Kim et al., 2011
			Ogawa et al., 2005
			Petraki et al., 2012
			Talieri et al., 2009a
KLK7	Colorectal	Upregulation → poor prognosis	Inoue et al., 2010
			Talieri et al., 2009a
			Talieri et al., 2009b
KLK8	Colorectal	Increase in KLK8 protein not related to prognosis	Talieri et al., 2009a
KLK10	Gastric	Upregulation → advanced clinical stage	Feng et al., 2006
			Huang et al., 2007
			Li et al., 2011
	Colorectal	Upregulation → poor prognosis	Feng et al., 2006
			Talieri et al., 2011
KLK11	Gastric	Downregulation → poor prognosis	Wen et al., 2011
	Colorectal	Increase in KLK11 protein not related to prognosis	Talieri et al., 2009a
KLK13	Gastric	Upregulation → favorable prognosis	Konstantoudakis et al., 2010
	Colorectal	Upregulation → poor prognosis	Talieri et al., 2009a
KLK14	Colorectal	Upregulation → poor prognosis	Talieri et al., 2009a
KLK15	Colorectal	Increase in KLK15 protein not related to prognosis	Talieri et al., 2009a

a So far, clinically relevant data regarding the course of the cancer disease were obtained for KLK6, 10, 11, and 13 (gastric cancer) and KLK5-8, 10, 11, 13, and 14 (colorectal cancer) only.

tissue, compared to normal mucosa tissue, which is suggestive of a critical role for KLK1 in GC. No data was presented regarding *KLK1* mRNA expression. The impact of KLK1 expression on disease-free or overall survival of GC patients was not yet determined.

KLK3. Protein, but not mRNA expression, of KLK3 (also known as PSA, prostate specific antigen) was assessed with immunohistochemistry by Milne et al. (2007) in GC tumor tissues. No KLK3 protein expression was observed, but instead a copy number gain for *KLK3* was seen, by employing methylation-specific multiplex ligation-dependent probe amplification. Since no KLK3 protein expression was observed, disease-free or overall survival of GC patients could not be correlated with KLK3 protein expression levels. *KLK3* mRNA was not determined yet.

KLK6. Two reports have been published so far, investigating the clinical value of KLK6 expression in GC patients (Kim *et al.*, 2011; Nagahara *et al.*, 2005). According to the authors, *KLK6* mRNA expression is significantly elevated in cancerous tissue over non-cancerous tissue, which was confirmed by immunohistochemistry. Nagahara *et al.* (2005) also showed that tumor tissue *KLK6* mRNA overexpression is a marker of poor overall survival in GC patients, a finding which still has to be validated by others. Data for disease-free survival was not reported. Clinical relevance of KLK6 protein GC tumor tissue overexpression was not assessed in this study or by anyone else.

KLK10. Three reports are available regarding expression of KLK10 in tumor tissues of GC patients (Feng *et al.*, 2006; Huang *et al.*, 2007; and Li *et al.*, 2011). Feng *et al.* (2006) showed that in primary GC KLK10 protein expression is elevated compared to normal gastric mucosa. They also showed that tumor cell *KLK10* mRNA expression is elevated in tumor tissues over the corresponding normal mucosa and that an increase in tumor tissue *KLK10* mRNA correlates with clinical stage, Lauren type, and depth of tumor invasion. This finding was confirmed by Li *et al.* (2011) who also described upregulation of *KLK10* mRNA in GC tumor tissues. Huang *et al.* (2007), assessing *KLK10* mRNA expression by *in situ* hybridization of normal and diseased tissue specimens, collected by routine upper gastrointestinal endoscopy, reported that *KLK10* mRNA expression is located in normal mucosa cells but is low or absent in tumor cells. They also reported that downregulation of tumor tissue *KLK10* mRNA correlates with *KLK10* exon 3 CpG island hypermethylation. Taken together, these findings need to be validated, especially when considering that the study presented by Huang *et al.* (2007) investigated early stage endoscopy biopsies, whereas the studies by Feng *et al.* (2006) and Li *et al.* (2011) examined advanced-stage tumor biopsies, obtained during primary surgery. Survival analyses were not presented.

KLK11. Only one published study concerned itself with assessment of *KLK11* expression in GC tumor tissues (Wen *et al.*, 2011). In this study, by use of real-time PCR, western blot, and immunohistochemistry, *KLK11* mRNA expression and KLK11 protein expression were identified as a novel cancer biomarker for GC patients. Different from KLK6 and KLK10 expression, KLK11 gene and protein expression is significantly decreased in GC tumor tissues, compared to normal gastric mucosa. In addition, *KLK11* expression is even lower in poorly differentiated cancer specimens than in well-differentiated ones. Survival analysis based on immunohistochemistry showed that a lack of KLK11 protein expression is associated with poor disease-free and overall survival. Survival analysis based on *KLK11* mRNA expression was not performed.

KLK13. Similar to KLK11, in GC, *KLK13* mRNA expression is decreased in tumor tissue specimens, compared to their paired non-malignant tissues (Konstantoudakis *et al.*, 2010). Indeed, life table analyses implied that *KLK13*-positive GC patients are characterized by a low risk of disease recurrence, associated with improved overall survival. KLK13 protein expression was not assessed. Hence, KLK13 protein-based survival analysis was not performed.

2.4.2 Review of the clinical relevance of KLK expression in colorectal cancer

Three of the *KLK* mRNAs (*KLK1, 4,* and *15*) are moderately/highly expressed at the gene level in normal colon tissue, whereas the other twelve *KLK* mRNAs (*KLK2, 3, 5-*and *14*) are not (*KLK2, 9,* and *10*) or only slightly (*KLK3, 5–8,* and *11–14*) expressed (**Tab. 2.2**). This expression pattern is also reflected for the majority of KLKs at the protein level (ELISA), where only KLK4, 9, 12, and 15 did not match. No KLK4 or KLK15 protein is expressed in the normal colon, although fair amounts of *KLK* mRNA were expressed. The opposite is true for KLK9 and KLK12. In CRC, studies regarding *KLK* mRNA expression were reported for five *KLK* mRNAs (*KLK1, 6–8,* and *10*). Of these, *KLK6-8,* and *10* mRNA levels were increased compared to the healthy counterparts and *KLK1* mRNA was decreased.

In CRC, for ten of the KLKs (KLK4-8, 10, 11, and 13–15) protein expression was determined by ELISA, western blotting, and/or immunohistochemistry. KLK5 and KLK14 protein expression was not different, compared to the healthy mucosa counterpart. For the other KLKs (KLK4, 6–8, 10, 11, 13, and 15) protein expression was increased. The clinical impact of gene and/or protein expression (**Tab. 2.3**) was investigated for six KLKs (KLK5-7, 10, 13, 14). Hence, clinical relevant data in CRC is missing for nine of the fifteen KLKs (KLK1-4, 8, 9, 11, 12, and 15).

KLK1. Yousef *et al.* (2004), studying *in silico KLK* genes' expression, reported downregulation of *KLK1* in CRC tumor tissue, compared to normal colon mucosa tissue. No further investigations have been published concerning KLK1 expression in CRC tumor tissues at the gene or protein level. Consequently, life-table analyses were not performed.

KLK4. Gratio *et al.* (2010) demonstrated KLK4 protein expression in CRC tumor tissues and absence of immunoreactivity in normal colon mucosa tissues. They also showed its ability to induce PAR1 (protease-activated receptor 1; thrombin receptor) signalling in colon cancer cells. KLK4 protein expression data was not correlated with the course of the CRC disease. Also, *KLK4* mRNA expression has so far not been examined in CRC tumor tissues.

KLK5. As of yet, only one report concerning the clinical relevance of KLK5 in CRC has been published (Talieri *et al.*, 2009a). In this publication, KLK5 protein in tumor tissue extracts and in extracts of the normal mucosa was determined by ELISA. Presence of about equal amounts of KLK5 protein in normal mucosa tissue and CRC tumor tissue was demonstrated. High KLK5 protein present in tumor tissue extracts is an indicator of shorter disease-free and overall survival. *KLK5* mRNA expression was not determined.

KLK6. Yousef *et al.* (2004), by employing *in silico* analysis, demonstrated overexpression of *KLK6* mRNA in CRC tumor tissues, compared to normal colon mucosa in which *KLK6* mRNA was absent. Following this investigation, five other, independent reports have been published which focus on the clinical utility of KLK6 expression in CRC tumor tissue. Soon after Yousef *et al.* (2004), Ogawa *et al.* (2005), by using real-time PCR, reported that *KLK6* mRNA expression is indeed significantly elevated in CRC tumor tissue, compared to normal mucosa tissue. This elevation was confirmed also at the protein level by the authors, using immunohistochemical analysis, which showed intense staining for KLK6 in tumor cells, but absence of staining in normal epithelial cells. High *KLK6* mRNA values were associated with shorter overall survival, whereas disease-free survival was not assessed. Taking a different approach by using ELISA, Talieri *et al.* (2009a) investigated the clinical impact of KLK6 protein for predicting the course of disease of CRC patients. KLK6 protein was strongly increased in CRC tumor tissue extracts, compared to normal mucosa tissue extracts. High tumor tissue KLK6 protein values indicated shorter disease-free and overall survival.

Soon after, Kim *et al.* (2011) confirmed the finding that *KLK6* mRNA and KLK6 protein are significantly up-regulated in CRC tumor tissue, but not in the normal mucosa or in pre-malignant dysplastic tissues. They also reported a correlation between KLK6 protein expression (immunohistochemistry) and disease-free and overall survival, confirming the results obtained by Talieri *et al.* (2009a). Applying immunohistochemistry as well, a direct correlation between positive KLK6 protein values and shorter disease-free/overall survival was also demonstrated by Petraki *et al.* (2012). It is worth mentioning, however, that in this study only about one-third of CRC patients expressed KLK6 protein in their tumor tissues and that most of these patients were in an advanced stage of the disease. In contrast with the previous studies, the authors claimed that in CRC KLK6 protein expression is decreased, compared to normal colonic mucosa tissue. Thus, diverse functions of KLK6 at different stages of tumor initiation and progression might be conceivable.

Agesen *et al.* (2012) have developed a 13-gene classifier (ColoGuideEx), each providing independent prognostic information specific to predicting the course of stage II CRC disease. An increase in mRNA expression of eight of the genes, including KLK6, is associated with poor disease-free survival of stage II, but not stage III CRC patients. The authors remark that now, with ColoGuideEx, individualized prognostic stratification of CRC patients in order to guide postoperative patient management is warranted, but that the performance of this classifier has to be compared to other, independent test formats such as ColoPrint and Oncotype DX Colon.

KLK7. Three reports were published demonstrating the clinical utility of KLK7 as a biomarker for predicting the course of the CRC disease (Inoue *et al.*, 2010; Talieri *et al.*, 2009a and b). Both groups reported increased KLK7 protein (Talieri *et al.*, 2009a and b), using ELISA/immunohistochemistry or *KLK7* mRNA expression (Inoue *et al.*, 2010; Talieri *et al.*, 2009b) in CRC tumor tissues, compared to the normal mucosa. Both groups agree that elevation of KLK7 protein or *KLK7* mRNA is associated with shorter disease-free and overall survival of CRC patients, especially in less differentiated tumors in a more advanced stage (Talieri *et al.*, 2009b).

KLK8. Yousef *et al.* (2004) showed, by *in silico* analysis, upregulation of *KLK8* in CRC tumor tissues over normal mucosa tissues. Talieri *et al.* (2009a) investigated, by ELISA, the clinical impact of KLK8 protein for predicting disease-free and overall survival in CRC patients. Although KLK8 protein was increased in CRC tumor tissue extracts compared to normal mucosa tissue extracts, such elevation was not associated with the course of the CRC disease. *KLK8* mRNA expression was not determined.

KLK10. Yousef *et al.* (2004), by *in silico* analysis, already predicted upregulation of *KLK10* in CRC tumor tissues over normal mucosa tissues, data which was confirmed by Feng *et al.* (2006), using mRNA expression analysis plus western blotting and IHC for KLK10 expression in CRC tumor tissues. Since in this report elevated *KLK10* mRNA expression was especially associated with advanced stages of CRC, the authors speculated that *KLK10* mRNA expression might be a predictor of poor clinical outcome. In fact, in two other publications by Talieri *et al.* (2009a; 2011) it was shown that both KLK10 protein and *KLK10* mRNA overexpression are associated with disease-free and overall survival in CRC, including early and advanced stage patients.

KLK11. Talieri *et al.* (2009a) investigated, by ELISA, the clinical impact of KLK11 protein for predicting disease-free and overall survival in CRC patients. KLK11 protein was elevated in tumor tissue extracts, compared to extracts of the normal mucosa. In their analysis, KLK11 only failed to exhibit statistical significance as an indicator for predicting the course of the CRC disease. *KLK11* mRNA expression was not determined

KLK13. So far, only one report concerning the clinical relevance of KLK13 in CRC was published (Talieri *et al.*, 2009a). In this publication, KLK13 protein was determined, by ELISA, in tumor tissue extracts and in the extracts of the normal mucosa. KLK13 protein expression in CRC tumor tissue extracts was enhanced over normal mucosa tissue extracts. Furthermore, elevated tumor tissue KLK13 protein expression was significantly associated with shorter disease-free and overall survival. *KLK13* mRNA expression was not determined.

KLK14. Presence of about equal amounts of KLK14 protein in normal mucosa-containing tissue and CRC tumor tissue was demonstrated by assessing tissue extracts with ELISA (Talieri *et al.*, 2009a). In another report (Gratio *et al.*, 2011), localization of KLK14 protein expression in normal mucosa tissue and tumor tissue was visualized by KLK14-directed immunohistochemistry. The authors showed that in the normal mucosa tissue, stromal cells were positive for KLK14, which is different from the tumor compartment, in which the tumor cells exhibited KLK14 staining. The clinical relevance of KLK14 protein overexpression in CRC tumor tissues was exemplified in the publication by Talieri *et al.* (2009a). Elevated tumor tissue KLK14 protein expression was significantly associated with shorter disease-free and overall survival. *KLK14* mRNA expression was not determined.

KLK15. Talieri *et al.* (2009a) investigated, by ELISA, the clinical impact of KLK15 protein for predicting disease-free and overall survival in CRC patients. In their analysis, KLK15 protein was increased, compared to normal mucosa tissue, but such an elevation was not associated with the course of the CRC. *KLK15* mRNA expression was not determined.

2.5 Proteolytic activity of KLKs in gastric/colorectal cancers

A prominent member of the KLK family that is upregulated in gastric and colorectal cancers and is associated with tumor invasion and metastasis is KLK6 (Kim *et al.*, 2011; Nagahara *et al.*, 2005; Ogawa *et al.*, 2005; Petraki *et al.*, 2012). In this respect, it is worthwhile to mention that KLK6 can degrade the extracellular matrix proteins fibrin(ogen), laminin, fibronectin, and collagen type I, as well as collagen type IV representing a major part of the basement membrane (Bernett *et al.*, 2002; Magklara *et al.*, 2003). These findings point to the fact that KLK6 is important in pericellular proteolysis and tissue remodeling. Upregulation of KLK7, another member of the KLK family, for example in colorectal and ovarian cancer or intracranial tumors, is also associated with enhanced tumor invasion capacity (Dong *et al.*, 2003; Inoue, 2010; Prezas *et al.*, 2006; Shan *et al.*, 2006; Talieri, 2009a and b). Likewise, overexpression of KLK4-6 increased the malignant, invasive phenotype of colorectal cancer and ovarian cancer cells (Dong *et al.*, 2003; Prezas *et al.*, 2006; Shan *et al.*, 2006).

There is no published study on GC or CRC which examines the effect of KLKs on members of the cadherin family. Cadherins, a class of calcium-dependent transmembrane glycoproteins, ensure that cells within tissues are bound together. For instance, a specific member of the cadherins, E-cadherin, forms a complex with cytoplasmic proteins, termed catenins. In most human gastrointestinal cancers, E-cadherin/catenin complexes are disturbed, and this underscores their pivotal role in the progression of these tumors (Debruyne *et al.*, 1999). Another member, P-cadherin, is important for maintaining cellular localization and tissue integrity. Sun *et al.* (2011), however, found that P-cadherin expression was higher in liver metastases of CRC patients than in the corresponding primary tumor, suggesting a prominent role of P-cadherin in the regulation of CRC metastasis. Although as of yet not investigated for gastrointestinal cancers, it is worthwhile to mention that Klucky *et al.* (2007) reported that ectopic KLK6 expression impaired E-cadherin protein expression in cell membranes of mouse keratinocytes associated with enhanced cell proliferation and translocation of β-catenin to the nucleus. Also, Chou *et al.* (2011) stated that overexpression of KLK13 by demethylation of its gene resulted in an increase in lung cancer's cell migratory and invasive properties, accompanied by an increase in N-cadherin. Whether such findings may also apply to GC/CRC is a topic of future investigations.

Proteinase-activated receptors (PAR) are a family of four G-protein-coupled cell surface receptors. PARs are activated by various proteinases (Hollenberg *et al.*, 2008, Oikonomopoulou *et al.*, 2006). After activation, PARs trigger a cascade of downstream events which lead to signal transduction, resulting in, for example. various pathophysiological events at the gene transcription, cell proliferation, and tissue repair level (Darmoul *et al.*, 2004a and b; Ossovskaya and Bunnett, 2004). Recent studies have implicated PAR1, PAR4 (thrombin receptors), and PAR2 (trypsin receptor) in human CRC growth (Darmoul *et al.*, 2004a and b). Gratio *et al.* (2010) analyzed *KLK4* gene expression in CRC and explored the activation process of PAR1 and PAR2 in colon tumor cells. They reported aberrant KLK4 immunoreactivity in colonic adenocarcinoma, in comparison to the mild dysplastic mucosa contiguous to cancerous lesions, or to the "normal-appearing" mucosa from CRC patients, and showed loss of PAR1 and PAR2 by KLK4 attack. Furthermore, significant ERK1/2 phosphorylation was triggered by KLK4. Gratio *et al.* (2011) also reported strong KLK14 immunostaining in human CRC tumor tissue specimens, which was absent in the normal mucosa. Furthermore, they demonstrated that KLK14 may stimulate internalization of PAR2, accompanied by a rise in intracellular calcium and an increase in ERK1/2 phosphorylation.

2.6 Effect of KLK expression on cell regulation and metabolic pathways

In vitro studies indicate that KLKs might participate in neoplastic progression by either promoting or inhibiting tumor cell proliferation, e.g. by affecting growth hormones. In this context, in hormone-regulated cancer tissues, such as that of the breast or prostate, IGFBP (insulin-like growth factor binding protein) is a physiological target for KLK2 and KLK3 (Réhault *et al.*, 2001; Sano *et al.*, 2007). No reports for GC/CRC regarding the action of KLK2 or KLK3 on IGFBP have been issued so far (Avgeris *et al.*, 2010; Milne *et al.*, 2007).

KLK4 is strongly expressed by colon cancer tissues, in contrast to normal colon mucosa (Gratio *et al.*, 2011). Interestingly, KLK4 can activate single-chain pro-uPA, which is also increased in GC/CRC patients (Beyer *et al.*, 2006; Halamkova *et al.*, 2011; Schmitt *et al.*, 2010), leading to the activation of the plasminogen system (Beaufort *et al.*, 2010), supporting pericellular degradation of the extracellular matrix and release and/or activation of growth factors.

KLK6 is elevated in both CRC and GC and seems to play a role in GC/CRC progression (Nagahara *et al.*, 2005; Ogawa *et al.*, 2005). For CRC, Ignatenko *et al.* (2004) reported the upregulation of *KLK6* mRNA in human colon cancer cells that were stably transfected with a mutant *K-ras* allele (K-RAS[G12V]). *K-ras* is a proto-oncogene, encoding a membrane-bound G protein that is responsible for extra- and intracellular mitogenic signals in various pathways. *K-ras* mutations may contribute to colorectal adenoma development (Fearon, 2011). K-ras-dependent KLK6 expression and secretion in CRC was examined by Henkhaus *et al.* (2008a). Using pharmacological inhibitors of pathways downstream of K-ras, they showed that PI3K and p42/44MAPK pathways play important roles in the induction of *KLK6* in mutant K-ras-expressing colon cancer cells. Increased *KLK6* expression enhanced colon cancer cell migration through the extracellular matrix. Inhibition of KLK6 activity using specific KLK6 antibodies or siRNA treatment of the cancer cells expressing mutant K-ras, resulted in a reduction of their invasive capacity, supporting an oncogenic role of KLK6 in CRC. Such findings have not yet been reported for GC.

On the other hand, KLK6 expression and secretion by colon cancer cells is mediated by caveolin-1 (CAV-1) (Henkhaus *et al.*, 2008b). Caveolins are integral plasma membrane proteins present in the caveolae and involved in receptor-independent endocytosis, which function as lipid rafts and scaffolding domains within the plasma membrane in almost all cell types (Stan, 2005). CAV-1 is modulating mitogenic signaling pathways and plays a central role in integrin signaling (e.g. PI3K, AKT) and KLK6 secretion (Kim *et al.*, 2006). CAV-1 and KLK6 co-localize to plasma membrane lipid rafts (Henkhaus *et al.*, 2008b). Whether this is also true for GC remains to be elucidated.

2.7 Conclusion

In GC/CRC, KLK gene and/or protein expression have been examined to some extent. Yet, the data presented so far is limited, which implies that much more work is required in order to support their clinical utility. Furthermore, the pathophysiological role of KLKs in GC/CRC is presently still poorly understood, also owing to the fact that in relation to these cancer diseases their role in biological pathways has been examined only for a few KLKs.

Bibliography

Agesen, T.H., Sveen, A., Merok, M.A., Lind, G.E., Nesbakken, A., Skotheim, R.I., and Lothe, R.A. (2012). ColoGuideEx, a robust gene classifier specific for stage II colorectal cancer prognosis. Gut, [Epub ahead of print].

American Cancer Society: Cancer facts and figures 2011. Accessed November 16, 2011, at http://www.cancer.gov.

Anwar, S., Frayling, I.M., Scott, N.A., and Carlson, G.L. (2004). Systematic review of genetic influences on the prognosis of colorectal cancer. Br. J. Surg. 91, 1275–1291.

Avgeris, M., Mavridis, K., and Scorilas, A. (2010). Kallikrein-related peptidase genes as promising biomarkers for prognosis and monitoring of human malignancies. Biol. Chem. 391, 505–511.

Beaufort, N., Plaza, K., Utzschneider, D., Schwarz, A., Burkhart, J.M., Creutzburg, S., Debela, M., Schmitt, M., Ries, C., and Magdolen, V. (2010). Interdependence of kallikrein-related peptidases in proteolytic networks. Biol. Chem. 391, 581–587.

Bernett, M.J., Blaber, S.I., Scarisbrick, I.A., Dhanarajan, P., Thompson, S.M., and Blaber, M. (2002). Crystal structure and biochemical characterization of human kallikrein 6 reveals that a trypsin-like kallikrein is expressed in the central nervous system. J. Biol. Chem. 277, 24562–24570.

Beyer, B.C., Heiss, M.M., Simon, E.H., Gruetzner, K.U., Babic, R., Jauch, K.W., Schildberg, F.W., and Allgayer, H. (2006). Urokinase system expression in gastric carcinoma, prognostic impact in an independent patient series and first evidence of predictive value in preoperative biopsy and intestinal metaplasia specimens. Cancer 106, 1026–1035.

Bohanes, P., LaBonte, M.J., Winder, T., and Lenz, H.J. (2011). Predictive molecular classifiers in colorectal cancer. Semin. Oncol. 38, 576–587.

Borgoño, C.A., and Diamandis, E.P. (2004). The emerging roles of human kallikreins in cancer. Nat. Rev. Cancer 4, 876–890.

Chan, A.T., and Giovannucci, E.L. (2010). Primary prevention of colorectal cancer. Gastroenterology. 138, 2029–2043.

Chibaudel, B., Tournigand, C., André, T., and de Gramont, A. (2012). Therapeutic strategy in unresectable metastatic colorectal cancer. Ther. Adv. Med. Oncol. 4, 75–89.

Cho, E., Lee, J.E., Rimm, E.B., Fuchs, C.S., and Giovannucci, E.L. (2012). Alcohol consumption and the risk of colon cancer by family history of colorectal cancer. Am. J. Clin. Nutr. 95, 413–419.

Chou, R.H., Lin, S.C., Wen, H.C., Wu, C.W., and Chang, W.S. (2011). Epigenetic activation of human kallikrein 13 enhances malignancy of lung adenocarcinoma by promoting N-cadherin expression and laminin degradation. Biochem. Biophys. Res. Commun. 409, 442–447.

Clements, J.A., Willemsen, N.M., Myers, S.A., and Dong, Y. (2004). The tissue kallikrein family of serine proteases: functional roles in human disease and potential as clinical biomarkers. Crit. Rev. Clin. Lab. Sci. 41, 265–312.

Crew K.D., and Neugut A.I. (2006). Epidemiology of gastric cancer. World J. Gastroenterol. 12, 354–362.

Darmoul, D., Gratio, V., Devaud, H., and Laburthe, M. (2004a). Protease-activated receptor 2 in colon cancer trypsin-induced MAPK phosphorylation and cell proliferation are mediated by epidermal growth factor receptor transactivation. J. Biol. Chem. 279, 20927–20934.

Darmoul, D., Gratio V., Devaud, H., Peiretti, F., and Laburthe, M. (2004b). Activation of proteinase-activated receptor 1 promotes human colon cancer cell proliferation through epidermal growth factor receptor transactivation. Mol. Cancer Res. 2, 514–522.

Debruyne, P., Vermeulen, S., and Mareel, M. (1999). The role of the E-cadherin/catenin complex in gastrointestinal cancer. Acta Gastroenterol. Belg. 62, 393–402.

Dong, Y., Kaushal, A., Brattsand, M., Nicklin, J., and Clements, J.A. (2003). Differential splicing of KLK5 and KLK7 in epithelial ovarian cancer produces novel variants with potential as cancer biomarkers. Clin. Cancer Res. 9, 1710–1720.

Fearon, E.R. (2011). Molecular genetics of colorectal cancer. Annu. Rev. Pathol. 6, 479–507.

Feng, B., Xu, W.B., Zheng, M.H., Ma, J.J., Cai, Q., Zhang, Y., Ji, J., Lu, A.G., Qu, Y., Li, J.W., Wang, M.L., Hu, W.G., Liu, B.Y., and Zhu, Z.G. (2006). Clinical significance of human kallikrein 10 gene expression in colorectal cancer and gastric cancer. J. Gastroenterol. Hepatol. 21, 1596–1603.

George, B., and Kopetz, S. (2011). Predictive and prognostic markers in colorectal cancer. Curr. Oncol. Rep. 13, 206–215.

Gratio, V., Beaufort, N., Seiz, L., Maier, J., Virca, G.D., Debela, M., Grebenchtchikov, N., Magdolen, V., and Darmoul, D. (2010). Kallikrein-related peptidase 4, a new activator of the aberrantly expressed protease-activated receptor 1 in colon cancer cells. Am. J. Pathol. 176, 1452–1461.

Gratio, V., Loriot, C., Virca, G.D., Oikonomopoulou, K., Walker, F., Diamandis, E.P., Hollenberg, M.D., and Darmoul, D. (2011). Kallikrein-related peptidase 14 acts on proteinase-activated receptor 2 to induce signaling pathway in colon cancer cells. Am. J. Pathol. 179, 2625–2636.

Halamkova, J., Kiss, I., Pavlovsky, Z., Jarkovsky, J., Tomasek, J., Tucek, S., Hanakova, L., Moulis, M., Cech, Z., Zavrelova, J., and Penka, M. (2011). Clinical relevance of uPA, uPAR, PAI 1 and PAI 2 tissue expression and plasma PAI 1 level in colorectal carcinoma patients. Hepatogastroenterology 58, 1918–1925.

Hamfjord, J., Stangeland, A.M., Skrede, M.L., Tveit, K.M., Ikdahl, T., and Kure, E.H. (2011). Wobble-enhanced ARMS method for detection of KRAS and BRAF mutations. Diagn. Mol. Pathol. 20, 158–165.

He, J., and Efron, J.E. (2011). Screening for colorectal cancer. Adv. Surg. 45, 31–44.

Henkhaus, R.S., Gerner, E.W., and Ignatenko, N.A. (2008a). Kallikrein 6 is a mediator of K-RAS-dependent migration of colon carcinoma cells. Biol. Chem. 389, 757–764.

Henkhaus, R.S., Roy, U.K., Cavallo-Medved, D., Sloane, B.F., Gerner, E.W., and Ignatenko, N.A. (2008b). Caveolin-1-mediated expression and secretion of kallikrein 6 in colon cancer cells. Neoplasia. 10, 140–148.

Hollenberg, M.D., Oikonomopoulou, K., Hansen, K.K., Saifeddine, M., Ramachandran, R., and Diamandis, E.P. (2008). Kallikreins and proteinase-mediated signaling, proteinase-activated receptors (PARs) and the pathophysiology of inflammatory diseases and cancer. Biol. Chem. 389, 643–651.

Huang, W., Zhong, J., Wu, L.Y., Yu, L.F., Tian, X.L., Zhang, Y.F., and Li, B. (2007). Downregulation and CpG island hypermethylation of NES1/hK10 gene in the pathogenesis of human gastric cancer. Cancer Lett. 251, 78–85.

Ignatenko, N.A., Babbar, N., Mehta, D., Casero, R.A. Jr., and Gerner, E.W. (2004). Suppression of polyamine catabolism by activated Ki-ras in human colon cancer cells. Mol. Carcinog. 39, 91–102.

Inoue, Y., Yokobori, T., Yokoe, T., Toiyama, Y., Miki, C., Mimori, K., Mori, M., and Kusunoki, M. (2010). Clinical significance of human kallikrein7 gene expression in colorectal cancer. Ann. Surg. Oncol. 17, 3037–3042.

Jemal, A., Bray, F., Center, M.M., Ferlay, J., Ward, E., and Forman, D. (2011). Global cancer statistics. CA Cancer J. Clin. 61, 69–90.

Kelley, R.K., and Venook, A.P. (2011). Prognostic and predictive markers in stage II colon cancer: is there a role for gene expression profiling? Clin. Colorectal Cancer. 10, 73–80.

Kim, H.A., Kim K.H., and Lee, R.A. (2006). Expression of caveolin-1 is correlated with Akt-1 in colorectal cancer tissues. Exp. Mol. Pathol. 80, 165–170.

Kim, H.K., Choi, I.J., Kim, C.G., Kim, H.S., Oshima, A., Yamada, Y., Arao, T., Nishio, K., Michalowski, A., and Green, J.E. (2010). Three-gene predictor of clinical outcome for gastric cancer patients treated with chemotherapy. Pharmacogenomics J. 12, 119–127.

Kim, J.T., Song, E.Y., Chung, K.S., Kang, M.A., Kim, J.W., Kim, S.J., Yeom, Y.I., Kim, J.H., Kim, K.H., and Lee, H.G. (2011). Up-regulation and clinical significance of serine protease kallikrein 6 in colon cancer. Cancer 117, 2608 -2619.

Klucky, B., Mueller, R., Vogt, I., Teurich, S., Hartenstein, B., Breuhahn, K., Flechtenmacher, C., Angel, P., and Hess, J. (2007). Kallikrein 6 induces E-cadherin shedding and promotes cell proliferation, migration, and invasion. Cancer Res. 67, 8198–8206.

Konstantoudakis, G., Florou, D., Mavridis, K., Papadopoulos, I.N., and Scorilas, A. (2010). Kallikrein-related peptidase 13 (KLK13) gene expressional status contributes significantly in the prognosis of primary gastric carcinomas. Clin. Biochem. 43, 1205–1211.

Lawrence, M.G., Lai, J., and Clements, J.A. (2010). Kallikreins on steroids, structure, function and hormonal regulation of prostate-specific antigen and the extended kallikrein locus. Endocr. Rev. 31, 407–446.

Li, M., Zhao, Z.W., Zhang, Y., and Xin, Y. (2011). Over-expression of Ephb4 is associated with carcino-genesis of gastric cancer. Dig. Dis. Sci. 56, 698–706.

Lundwall, A., Band, V., Blaber, M., Clements, J., Courty, Y., Diamandis, E.P., Fritz, H., Lilja, H., Malm, J., Maltais, L.J., Olsson, A.Y., Petraki, C., Scorilas, A., Sotiropoulou, G., Stenman, U.H., Stephan, C., Talieri, M., and Yousef, G.M. (2006). A comprehensive nomenclature for serine proteases with homology to tissue kallikreins. Biol. Chem. 387, 637–641.

Magklara, A., Mellati, A.A., Wasney, G.A., Little, S.P., Sotiropoulou, G., Becker, G.W., and Diamandis, E.P. (2003). Characterization of the enzymatic activity of human kallikrein 6, autoactivation, substrate specificity and regulation by inhibitions. Biochem. Biophys. Res. Commun. 307, 948–955.

Mavridis, K., and Scorilas, A. (2010). Prognostic value and biological role of the kallikrein-related peptidases in human malignancies. Future Oncol. 6, 269–285.

Milne, A.N., Sitarz, R., Carvalho, R., Polak, M.M., Ligtenberg, M., Pauwels, P., Offerhaus, G.J., and Weterman, M.A. (2007). Molecular analysis of primary gastric cancer, corresponding xenografts, and 2 novel gastric carcinoma cell lines reveals novel alterations in gastric carcinogenesis. Hum. Pathol. 38, 903–913.

Morán, A., Ortega, P., de Juan, C., Fernández-Marcelo, T., Frías, C., Sánchez-Pernaute, A., Torres, A.J., Díaz-Rubio, E., Iniesta, P., and Benito M. (2010). Differential colorectal carcinogenesis, molecular basis and clinical relevance. World J. Gastrointest. Oncol. 2, 151–158.

Nagahara, H., Mimori, K., Utsunomiya, T., Barnard, G.F., Ohira, M., Hirakawa, K., and Mori, M. (2005). Clinicopathologic and biological significance of KLK6 overexpression in human gastric cancer. Clin. Cancer Res. 11, 6800–6806.

Ogawa, K., Utsunomiya, T., Mimori, K., Tanaka, F., Inoue, H., Nagahara, H., Murayama, S., and Mori, M. (2005). Clinical significance of human kallikrein gene 6 messenger RNA expression in colorectal cancer. Clin. Cancer Res. 11, 2889–2893.

Oikonomopoulou, K., Hansen, K.K., Saifeddine, M., Tea, I., Blaber, M., Blaber, S.I., Scarisbrick, I., Andrade-Gordon, P., Cotrell, G.S., Bunnett, N.W., Diamandis, E.P., and Hollenberg, M.D. (2006). Proteinase-activated receptors: targets for kallikrein signaling. J. Biol. Chem. 281, 32095–32112.

Ossovskaya, V.S., and Bunnett, N.W. (2004). Protease-activated receptors: contribution to physiology and disease. Physiol. Rev. 84, 579–621.

Paliouras, M., Borgoño, C.A., and Diamandis, E.P. (2007). Human tissue kallikreins, the cancer biomarker family. Cancer Lett. 249, 61–79.

Petraki, C., Dubinski, W., Scorilas, A., Saleh, C., Pasic, M.D., Komborozos, V., Khalil, B., Gabril, M.Y., Streutker, C., Diamandis, E.P., and Yousef, G.M. (2012). Evaluation and prognostic significance of human tissue kallikrein-related peptidase 6 (KLK6) in colorectal cancer. Pathol. Res. Pract. 208, 104–108.

Prezas, P., Scorilas, A., Yfanti, C., Viktorov, P., Agnanti, N., Diamandis, E., and Talieri, M. (2006). The role of human tissue kallikrein 7 and 8 in intracranial malignancies. Biol. Chem. 387, 1607–1612.

Réhault, S., Monget, P., Mazerbourg, S., Tremblay, R., Gutman, N., Gauthier, F., and Moreau, T. (2001). Insulin-like growth factor binding proteins (IGFBPs) as potential physiological substrates for human kallikreins hK2 and hK3. Eur. J. Biochem. 268, 2960–2968.

Resende, C., Ristimäki, A., and Machado, J.C. (2010). Genetic and epigenetic alteration in gastric carcinogenesis. Helicobacter 15 (Suppl. 1), 34–39.

Resende, C., Thiel, A., Machado, J.C., and Ristimäki, A. (2011). Gastric cancer, basic aspects. Helicobacter 16 (Suppl. 1), 38–44.

Ross, J.S. (2011). Biomarker-based selection of therapy for colorectal cancer. Biomark. Med. 5, 319–332.

Salazar, R., Roepman, P., Capella, G., Moreno, V., Simon, I., Dreezen, C., Lopez-Doriga, A., Santos, C., Marijnen, C., Westerga, J., Bruin, S., Kerr, D., Kuppen, P., van de Velde, C., Morreau, H., van Velthuysen, L., Glas, A.M., van't Veer, L.J., and Tollenaar, R. (2011). Gene expression signature to improve prognosis prediction of stage II and III colorectal cancer. J. Clin. Oncol. 29, 17–24.

Sano, A., Sangai, T., Maeda, H., Nakamura, M., Hasebe, T., and Ochiai, A. (2007). Kallikrein 11 expressed in human breast cancer cells releases insulin-like growth factor through degradation of IGFBP-3. Int. J. Oncol. 30, 1493–1498.

Sawant, S., Snyman, C., and Bhoola, K. (2001). Comparison of tissue kallikrein and kinin receptor expression in gastric ulcers and neoplasms. Int. Immunopharmacol. 1, 2063–2080.

Schmitt, M., Mengele, K., Napieralski, R., Gkazepis, A., Magdolen, V., Reuning, U., Sweep, F., Brünner, N., Foekens, J., and Harbeck, N. (2010). Clinical utility of Level-of-Evidence-1 disease forecast cancer biomarkers urokinase-type plasminogen activator (uPA) and its inhibitor PAI-1. Expert Review Mol. Diagn. 10,1051–1067.

Scorilas, A., and Mavridis, K. (2012). Kallikrein-related peptidases (KLKs) as novel potential biomarkers in gastric cancer: an open yet challenging road lies ahead. J. Surg. Oncol. 105, 223–224.

Shan, S.J., Scorilas, A., Katsaros, D., Rigault de la Longrais I., Massobrio, M., and Diamandis, E.P. (2006). Unfavorable prognostic value of human kallikrein 7 quantified by ELISA in ovarian cancer cytosols. Clin. Chem. 52, 1879–1886.

Shaw, J.L., and Diamandis, E.P. (2007). Distribution of 15 human kallikreins in tissues and biological fluids. Clin. Chem. 53, 1423–1432.

Shikata, K., Doy, Y., Yonemoto, K., Arima, H., Ninomiya, T., Kubo, M., Tanizaki, Y., Matsumoto, T., Iida, M., and Kiyohara, Y. (2008). Population-based prospective study of the combined influence of cigarette smoking and *Helicobacter pylori* infection on gastric cancer incidence, the Hisayama study. Am. J. Epidemiol. 12, 1409–1415.

Sotiropoulou, G., Pampalakis, G., and Diamandis, E.P. (2009). Functional roles of human kallikrein-related peptidases. J. Biol. Chem. 284, 32989–32994.

Stan, R.V. (2005). Structure of caveolae. Biochim. Biophys. Acta. 1746, 334 -348.

Sun, L., Hu, H., Peng, L., Zhou, Z., Zhao, X., Pan, J., Sun, L., Yang, Z., and Ran, Y. (2011). P-cadherin promotes liver metastasis and is associated with poor prognosis in colon cancer. Am. J. Pathol. 179, 380–390.

Talieri, M., Li, L., Zheng, Y., Alexopoulou, D.K., Soosaipillai, A., Scorilas, A., Xynopoulos, D., and Diamandis, E.P. (2009a). The use of kallikrein-related peptidases as adjuvant prognostic markers in colorectal cancer. Br. J. Cancer. 100, 1659–1665.

Talieri, M., Mathioudaki, K., Prezas, P., Alexopoulou, D.K., Diamandis, E.P., Xynopoulos, D., Ardavanis, A., Arnogiannaki, N., and Scorilas, A. (2009b). Clinical significance of Kallikrein-related peptidase 7 (KLK7) in colorectal cancer. Thromb. Haemost. 101, 741–747.

Talieri, M., Alexopoulou, D.K., Scorilas, A., Kypraios, D., Arnogiannaki, N., Devetzi, M., Patsavela, M., and Xynopoulos, D. (2011). Expression analysis and clinical evaluation of kallikrein-related peptidase 10 (KLK10) in colorectal cancer. Tumour Biol. 32, 737–744.

Vogiatzi, P., Vindigni, C., Roviello, F., Renieri, A., and Giordano, A. (2007). Deciphering the underlying genetic and epigenetic events leading to gastric carcinogenesis. J. Cell. Physiol. 211, 287–295.

Wen, Y.G., Wang, Q., Zhou, C.Z., Yan, D.W., Qiu, G.Q., Yang, C., Tang, H.M., and Peng, Z.H. (2011). Identification and validation of kallikrein-related peptidase 11 as a novel prognostic marker of gastric cancer based on immunohistochemistry. J. Surg. Oncol. 104, 516–524.

Winder, T., and Lenz, H.J. (2010). Molecular predictive and prognostic markers in colon cancer. Cancer Treat. Rev. 36, 550–556.

Wu, K., Giovannucci, E., Byrne, C., Platz, E.A., Fuchs, C., Willett, W.C., and Sinha, R. (2006). Meat mutagens and risk of distal colon adenoma in a cohort of U.S. men. Cancer Epidemiol. Biomarkers Prev. 15, 1120–1125.

Yousef, G.M., Borgoño, C.A., Popalis, C., Yacoub, G.M., Polymaris, M.E., Soosaipillai, A., and Diamandis, E.P. (2004). *In-silico* analysis of kallikrein gene expression in pancreatic and colon cancers. Anticancer Res. 24, 43–52.

Yousef, G.M., Obiezu, C.V., Luo, L.Y., Magklara, A., Borgoño, C.A., Kishi, T., Memari, N., Michael, P., Sidiropoulos, M., Kurlender, L., Economopoulou, K., Kapadia, C., Komatsu, N., Petraki, C., Elliott, M., Scorilas, A., Katsaros, D., Levesque, M.A., and Diamandis, E.P. (2005). Human tissue kallikreins, from gene structure to function and clinical applications. Adv. Clin. Chem. 39, 11–79.

Rong Jiang, Zonggao Shi, Jeffrey Johnson, and M. Sharon Stack

3 Pathophysiology of Kallikrein-related Peptidases in Head and Neck Cancer

3.1 Introduction

The term head and neck squamous cell carcinoma (HNSCC) refers to a diverse group of cancers that originate from the upper aerodigestive tract (including lip, oral cavity, nasal cavity, nasopharynx, and larynx), the paranasal sinuses, and the salivary glands **(Fig. 3.1)**. As these anatomical sites are important for functions such as respiration, speech, swallowing, smell and taste, treatment of head and neck cancers can have considerable functional consequences and severely impair quality of life. Among the histopathological types, approximately 90% of head and neck cancers are squamous cell carcinoma. HNSCC is one of the most common tumors, with at least 500,000 new cases annually worldwide, the majority of which are located in the oral cavity, larynx, and pharynx (Parkin *et al.*, 2005). Squamous cell carcinoma of the oral cavity (OSCC) is the most commonly diagnosed malignancy of the oral cavity, including cancers that affect the tongue, floor of the mouth, buccal mucosa, lips, palate, and gingiva. OSCC results in over 200,000 deaths annually and is one of the top eight most frequently diagnosed cancers worldwide (Jemal *et al.*, 2010; Rusthoven *et al.*, 2008; Sano and Myers, 2007). In the United States of America, an estimated 36,540 new cases were detected in 2010. However, despite advances in surgery, chemother-

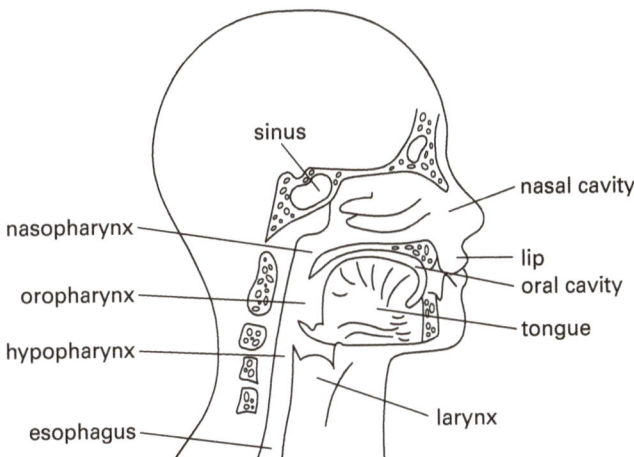

Fig. 3.1 **Anatomical sites of HNSCC incidence.** HNSCC arises from the upper aerodigestive tract and includes the oral and nasal cavities, lip, sinuses, pharynx (nasopharynx, oropharynx, hypopharynx), and larynx. Within the oral cavity, tumors may involve the tongue, alveolar ridge, retromolar trigone, floor of the mouth, and hard palate (Marur and Forastiere, 2008).

apy, and radiation treatment, only 60% of these individuals will survive for five years (Jemal *et al.*, 2010).

Understanding the mechanism underlying the carcinogenesis and progression of head and neck cancers remains a challenge. The most prominent risk factors for HNSCC identified by epidemiologic studies are tobacco and alcohol consumption, accounting for about 75% of cases, and their effects are multiplicative (Conway *et al.*, 2009). More recent studies have emphasized the role of human papilloma virus (HPV), especially type 16, as a contributor to the etiology of HNSCC in the oropharynx (Fakhry and Gillison, 2006). The overall incidence of HNSCC has experienced a slow decline during the past decade, which is attributed to a decrease in prevalence of the more traditional risk factors, such as smoking. Conversely, the increased incidence of cancers of the oral tongue and particularly oropharyngeal cancers are potentially related to the fact that oral and oropharyngeal HPV infections are becoming increasingly common (Leemans *et al.*, 2011). Paradoxically, patients with HPV-positive head and neck cancers have a much better overall survival compared to their HPV-negative counterparts (Fakhry *et al.*, 2008). In addition to the above-mentioned exogenous risk factors, genetic predispositions have also been shown to be important. Certain inherited disorders, such as Fanconi anemia, predispose to HNSCC (Hopkins *et al.*, 2008; Kutler *et al.*, 2003). Other risk factors identified in case-control studies include gender (men more common than women), long term passive smoking (Conway *et al.*, 2009), low body mass index, and sexual behavior (Heck *et al.*, 2010).

The high mortality of HNSCC can primarily be attributed to the fact that tumors are often diagnosed at a late stage, after the primary tumor has metastasized. Indeed, the oral cavity is associated with a rich lymphatic network and metastasis to regional lymph nodes in the neck is an early event in HNSCC progression. This is critical to survival, and patients with carcinomas that have spread to regional lymph nodes have a poor prognosis, with a five-year survival rate approximately half that of patients without lymphatic spread (Fu *et al.*, 1998). Treatment of late stage OSCC is also associated with a high morbidity and severe functional impairment (Rusthoven *et al.*, 2008; Sano and Myers, 2007). Invasion and metastasis of OSCC require multiple cellular events, including alterations in cytoskeleton, cell-cell adhesive contacts and basement membrane attachments, extracellular matrix proteolysis, and migration (Liotta *et al.*, 1983). Thus, while early detection is key to reducing the morbidity and mortality of OSCC, a more detailed understanding of the pathobiological processes that control both local invasion and distant metastasis could significantly improve diagnostic and prognostic indicators and identify novel therapeutic targets.

Evaluation of the prognosis for patients with OSCC is challenging, due to the variety of anatomical sites and subsites, which give rise to tumors with diverse histopathology, but is largely determined by clinical stage, i.e. depending on the status of tumor, node, and metastasis (TNM) (Patel and Shah, 2005). Clinical staging of OSCC is based on results from physical examination, imaging, cytology of lymph nodes, and the detailed histopathology after surgery. The tumor, or T, category is based on the size of the primary tumor. N, or nodal status, is based on the level of lymph nodes in the neck to which the tumor has metastasized, and M is based on the presence of distant metastases. For oral cancer in general, only approximately 34% of patients are diagnosed with early stage (stage I, II) disease and have an 83% 5-year survival rate. In contrast, 46% of patients are diagnosed with locally advanced disease (stage III, IVa, IVb), with a 54% 5-year survival, while the 14% of patients that present with distant metastatic disease (stage IVc) only have a 32% 5-year survival rate (Jemal *et al.*, 2010). Obviously, early stage tumors treated with surgery and/or radiotherapy have a more favorable prognosis, compared to late stage tumors.

The main approaches to treatment of advanced HNSCC cancers is surgery together with postoperative chemotherapy (combination therapies using taxanes, 5-fluoroura-cil, platinum compounds, and/or methotrexate) and radiotherapy (Brockstein, 2011). Recent studies have supported the use of molecular targeted therapies in HNSCC, particularly cetuximab, a monoclonal antibody directed against the epidermal growth factor receptor, providing a potentially less toxic alternative to the current standard of care (Frampton, 2010; Leemans *et al.*, 2011). Additional data is needed, in order to determine whether this approach will result in improved overall survival, which has not markedly improved in recent decades. Further elucidation of additional molecular targets involved in HNSCC progression and metastasis may lead to novel therapeutic strategies and improved individualization of current therapies.

3.2 A murine orthotopic xenograft model using urinary-type plasminogen activator receptor (uPAR) overexpressing OSCC cells mimics aggressive human OSCC

Metastatic spread of cancer cells requires a coordinated synergy between cell adhesion, proteolysis, and motility. A key molecular player in cancer metastasis is urinary-type plasminogen activator (uPA) receptor (uPAR, CD87) and has been well studied (Blasi and Carmeliet, 2002; Shi and Stack, 2007). This molecule was originally identified as a surface-anchored binding protein for uPA, a serine protease involved in the activation of plasminogen to the broad spectrum protease plasmin, and the presence of uPAR at the cell membrane facilitates efficient and focused pericellular proteolytic activity (Blasi and Carmeliet, 2002; D'Alessio and Blasi, 2009; Shi and Stack, 2007). The uPAR itself is a highly glycosylated protein (apparent molecular weight 55–60 kD), expressed by both epithelial and stromal cells, and is anchored to the outer layer of the plasma membrane via a glycosylphosphatidylinositol linkage at the C-terminus (Blasi and Carmeliet, 2002). In addition to its role in the regulation of pericellular proteolysis, uPAR has been implicated in quite a number of non-proteolytic activities that may also contribute to metastasis (Jo *et al.*, 2009; Shi *et al.*, 2011; Waltz *et al.*, 1997; Wei *et al.*, 1994), providing a functional link between adhesive function and proteinase regulation. As such, this molecule has been implicated in many physiological and pathological processes, including fibrinolysis, matrix remodeling, cell proliferation, invasion, and migration (Dass *et al.*, 2008).

Studies of human OSCC using cDNA microarray analyses to evaluate genetic changes associated with primary oral tumors, together with immunohistochemical detection of corresponding protein expression, have identified uPA as one of 25 genes that are part of an "OSCC gene signature" for the molecular classification of oral tumors (Nagata *et al.*, 2003; Ziober *et al.*, 2006). Together with its receptor (uPAR), tumors that bear high levels of uPA are usually more invasive and exhibit enhanced lymph node metastasis (Lindberg *et al.*, 2006; Nozaki *et al.*, 1998; Yasuda *et al.*, 1997) and more frequent tumor relapse (Hundsdorfer *et al.*, 2005). We have recently identified a matrix-induced physical interaction between uPAR and α3β1 integrin that initiates a Src/MEK/ERK-dependent signaling pathway, culminating in enhanced invasive activity *in vitro* (Ghosh *et al.*, 2000; 2006). To evaluate the potential contribution of these molecular interactions to tumor growth and invasion *in vivo*, cells with modified uPAR levels were injected submucosally into the anterior tongue in an orthotopic xenograft model of OSCC. In this model, downregulation of uPAR expression resulted in tumors that exhibited features of well-differentiated squamous cell carcinoma, with well-defined borders and numerous keratin pearls, some dystrophic calcification, low mitotic index, and mild nuclear pleomorphism (Ghosh *et al.*, 2010; Pettus *et al.*, 2009). In contrast, tumors generated from cells that overexpress uPAR had morphologic characteristics consistent with poorly differentiated SCC, including poorly circumscribed, ill-defined borders with infiltrative cords of tumor cells, with both peri-neural and vascular invasion (Ghosh *et al.*, 2010; Pettus *et al.*, 2009). In human OSCC, these features are associated with aggressive tumors and carry a poor prognosis.

Because the orthotopic murine xenograft model of uPAR-overexpressing OSCC cells was consistent with clinical studies of human OSCC, showing that tumors with high levels of uPA or uPAR behave more aggressively and have more frequent tumor relapse (Hundsdorfer *et al.*, 2005; Lindberg *et al.*, 2006; Nozaki *et al.*, 1998; Yasuda *et al.*, 1997), we utilized this model as a discovery platform to identify additional genes associated with a more aggressive OSCC phenotype, through comparative cDNA microarray analysis (Ghosh *et al.*, 2010). This approach identified 148 differentially expressed genes, involved in a variety of biological processes, by functional annotation clustering, including developmental processes, inflammation, proliferation, and adhesion. Interestingly, expression of a number of proteinases (serine and metalloproteinases), proteinase inhibitors, and extracellular matrix proteins was modified, including four members of the KLK family (KLK5, 7, 8, and 10) (Ghosh *et al.*, 2010; Pettus *et al.*, 2009).

3.3 Expression of KLKs in OSCC

Proteolytic enzymes have attracted considerable attention in HNSCC research, because of their mechanistic involvement in cancer progression. Among them, the human kallikrein-related peptidases (KLK) are a single family of 15 highly conserved trypsin- or chymotrypsin-like serine proteases, encoded by the largest continuous cluster of protease-encoding genes (*KLK1-15*) in the genome (Lundwall *et al.*, 2006). Aberrant *KLK* expression patterns have been reported in many malignancies, including those of the breast, prostate, and ovary, and have been widely implicated as cancer biomarkers (Borgoño *et al.*, 2004; Clements *et al.*, 2004; Paliouras *et al.*, 2007; Pampalakis and Sotiropoulou, 2007). While the physiological roles and natural endogenous substrates for most KLKs have yet to be defined, expression of multiple KLKs in a single tissue compartment suggests their participation in proteolytic cascades (Borgoño *et al.*, 2004; Clements *et al.*, 2004; Paliouras *et al.*, 2007; Pampalakis and Sotiropoulou, 2007).

Based on cDNA microarray data showing altered expression of *KLK5, 7, 8,* and *10* in aggressive uPAR-overexpressing oral tumors (Ghosh *et al.*, 2010), our group evaluated both murine and human OSCC for KLK expression. In murine tumors, immunohistochemical staining of serial sections for expression of KLK5, 7, 8 and 10 showed a significantly higher percentage of cells staining positive in poorly differentiated tumors (70–90%) relative to well-differentiated tumors (30–40%), whereas control sections did not stain positively for KLK3 (Pettus *et al.*, 2009). In normal murine tongue, only the lower stratum superficiale exhibited positive KLK expression. Similarly, in human tongue OSCC, 65–100% of tumors analyzed were positive for this panel of KLKs and negative for KLK3, with the same tumor expressing multiple KLKs (Pettus *et al.*, 2009). This data was validated in cell line models of OSCC, using quantitative real-time polymerase chain reaction (qRT-PCR), showing higher expression levels of *KLK5, 7, 8,* and *10* in OSCC cell lines relative to normal oral mucosal cells and pre-malignant oral keratinocytes (Jiang *et al.*, 2011). Expression of multiple KLKs is also detected at other OSCC sites, including the cheek, lip, and gingiva (**Fig. 3.2**). While the role of these KLKs in OSCC remains under active investigation, these *in vivo* data support biochemical studies, which suggest the potential for participation in zymogen activation cascades (Clements *et al.*, 2004; Pampalakis and Sotiropoulou, 2007). While additional data on the role of KLKs in OSCC or HNSCC is very limited, results of a recent study report that 85% of a cohort of 80 cases of moderately- to poorly-differentiated oral carcinomas stain positively for KLK4 and KLK7, with particularly intense staining observed at the infiltration front, suggesting a potential role in localized invasion (Zhao *et al.*, 2011). Furthermore, strong staining positively correlated with poor survival, suggesting that the analysis of KLK expression in OSCC may have prognostic significance.

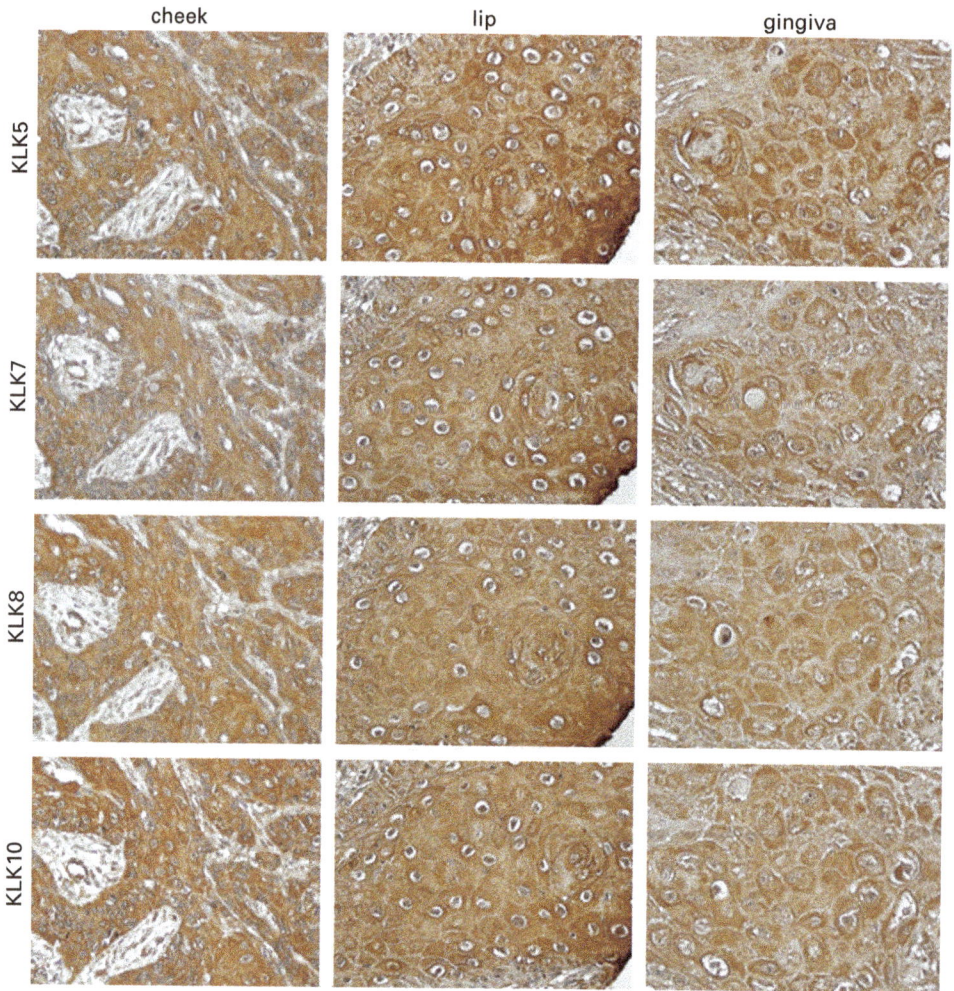

Fig. 3.2 **Immunohistochemical analysis of KLK expression in OSCC.** Serial sections of OSCC lesions from the cheek, lip, and gingiva, as indicated, were immunostained with antibodies against KLK5 (rabbit anti-human KLK5, Abcam ab28565, 1:20 dilution), KLK7 (Abcam ab28309, 1:20 dilution), KLK8 (Abcam ab28310, 1:20 dilution), or KLK10 (Abcam ab28300, 1:20 dilution), as indicated, followed by a biotinylated secondary antibody, avidin-POX, and DAB chromogen as the substrate (Pettus *et al.*, 2009).

3.4 Potential functional role of KLK5 in regulating cell-cell junctional integrity in OSCC

While the biological substrates of most KLKs are currently unknown, a major function of KLK5 and KLK7 is skin desquamation through cleavage of desmoglein 1 (Dsg1), desmocollin 1 (Dsc1) and corneodesmosin (Cds) (Borgoño *et al.*, 2007; Caubet *et al.*, 2004; Descargues *et al.*, 2006) and expression of KLK5 and KLK7 in the stratum superficiale of normal tongue supports a similar role in mucosal tissue (Pettus *et al.*, 2009). KLK5 and KLK7 activity as well as the process of desquamation are regulated by the proteinase inhibitor designated LEKTI, encoded by the *SPINK5* gene (see Chapters 6 and 13, Volume 1). Genetic defects in SPINK5 cause Netherton syndrome, which is characterized by a detached stratum corneum (Descargues *et al.*, 2005) and defective epidermal barrier function. Interestingly, transcriptomic analysis of OSCC relative to normal tongue identified *SPINK5* as a gene significantly downregulated in tongue tumors (Ye *et al.*, 2008), suggesting that KLK5 and KLK7 activity in OSCC is unregulated.

As the processes of desquamation and metastasis exhibit analogous functional requirements for loss of cell-cell cohesion, the effect of KLK5 expression on the integrity of the major desmosomal cadherin component Dsg1 was evaluated in normal oral keratinocytes and OSCC cells. OSCC cells exhibited proteolytic processing of Dsg1, which was lost upon siRNA knockdown of *KLK5* expression (Jiang *et al.*, 2011), while the functionally related cell-cell adhesion molecule E-cadherin was unaffected. Conversely, normal oral keratinocytes engineered to overexpress KLK5 exhibited a proteolytically processed form of Dsg1 relative to control cells (Jiang *et al.*, 2011). Other studies also provided evidence for the direct cleavage of Dsg1 by KLK5 in corneocytes and identified a number of putative cleavage sites located within the first three extracellular cadherin repeat domains, designated ECI, ECII, and ECIII, respectively. These sites also mediate heterophilic binding to desmocollins (Borgoño *et al.*, 2007). Among these sites, two reside within calcium binding motifs, responsible for the calcium-dependent conformation needed for proper Dsg1 function. Thus, based on the location of the cleavage sites, KLK5 action may destabilize Dsg1 structure and abolish its

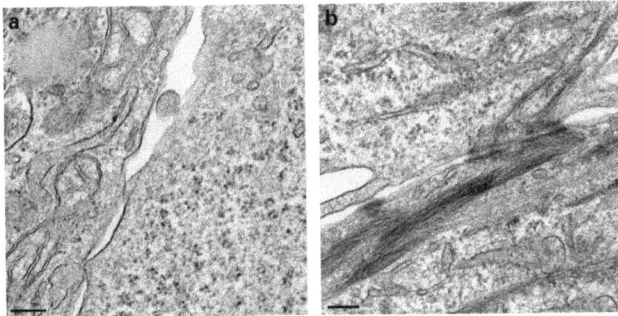

Fig. 3.3 **Modulation of KLK5 expression alters desmosome density in OSCC cells. (a)** The parental OSCC cell line SCC25 or **(b)** a variant line with reduced KLK5 expression, designated SCC25-KLK5-KD (Jiang *et al.*, 2011), were cultured on coverslips, fixed using 2% glutaraldehyde plus 2% paraform-aldehyde in 0.1 M cacodylate buffer, and further processed for transmission electron microscopy. Ultrathin sections were cut and collected on celloidine-coated 100-mesh nickel grids. Observations were made with a JEOL 1400 transmission electron microscope. Electron-dense desmosomal cadherin plaques are rare in the cell-cell junctions of SCC25 cells (panel A), while SCC25-KLK5KD cells have abundant desmosomes (panel B). Magnification 20,000 x.

adhesion to desmocollins on neighboring cells. Further, extracellular Dsg1 proteolytic fragments produced by KLK5 may further destabilize intercellular adhesion via a competition with full length Dsg1 and desmocollin 1 for binding sites. In support of this hypothesis, ultrastructural analysis of OSCC cells showed a striking, 3-fold increase in desmosomal plaques in cells with reduced KLK5 expression, relative to parental cells (**Fig. 3.3**) (Jiang *et al.*, 2011).

To assess the functional consequences of altered desmosomal integrity, cell-cell aggregation dynamics were evaluated. Parental OSCC cells aggregated slowly, while *KLK5* knockdown significantly enhanced aggregation kinetics. Conversely, normal oral mucosal cells rapidly form robust aggregates. However, introduction of *KLK5* expression significantly delays aggregation kinetics. Normal oral mucosal cells also usually retain sheet-like epithelial integrity. However, introduction of *KLK5* expression leads to a significant loss of epithelial cohesivity, while OSCC monolayers with silenced *KLK5* expression retain epithelial integrity (Jiang *et al.*, 2011). Interestingly, modulation of *KLK5* expression in OSCC and normal oral epithelium also alters barrier function, as evaluated using FITC-Dextran to examine the permeability of cell monolayers. Silencing *KLK5* expression in OSCC cells restores barrier function (**Fig. 3.4a**), while introduction of *KLK5* to normal oral mucosal cells increases monolayer perme-

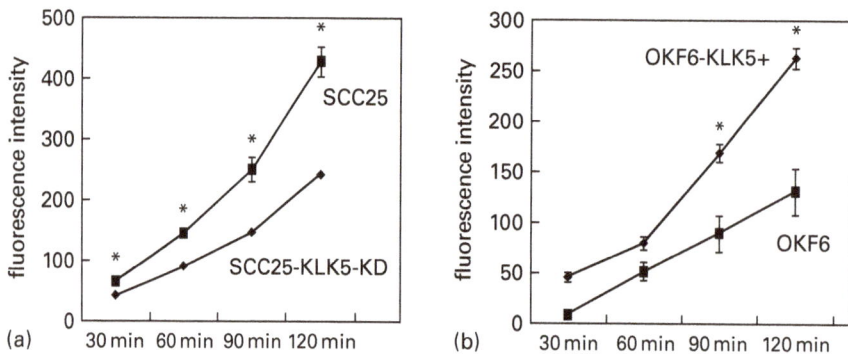

Fig. 3.4 **Modulation of KLK5 expression alters epidermal barrier function.** The OSCC cell lines **(a)** SCC25 and a variant line with reduced *KLK5* expression, designated SCC25-KLK5-KD (Jiang *et al.*, 2011), and the normal oral mucosal cell lines **(b)** OKF6 and a variant line with elevated *KLK5* expression, designated OKF6-KLK5+ (Jiang *et al.*, 2011) were evaluated for epithelial barrier function using fluorescein isothiocyanate (FITC)-Dextran. Cells were cultured in triplicate to confluence on a cell culture insert of 6.5 mm diameter with 0.4 µm pore size. FITC-Dextran (70 kDa, 250 µg/ml in PBS) was applied to the apical chamber and Hank's buffered salt solution to the bottom chamber. Monolayers were incubated at 37 °C, aliquots (200 µl) were taken from the bottom chambers at 30, 60, 90, and 120 min, and fluorescence intensity was measured using a SpectraMax Gemini (Molecular Devices, Sunnyvale, CA, USA) fluorimetric plate reader with excitation at 490 nm and emission at 520 nm. Results are representative of triplicate cultures from three independent experiments. (*) designates a statistically significant difference in fluorescence intensity ($p < 0.05$). Note that downregulation of *KLK5* expression improves epithelial barrier function in OSCC cells **(a)**, while a gain of *KLK5* expression leads to a loss of barrier function in normal oral mucosal cells **(b)**.

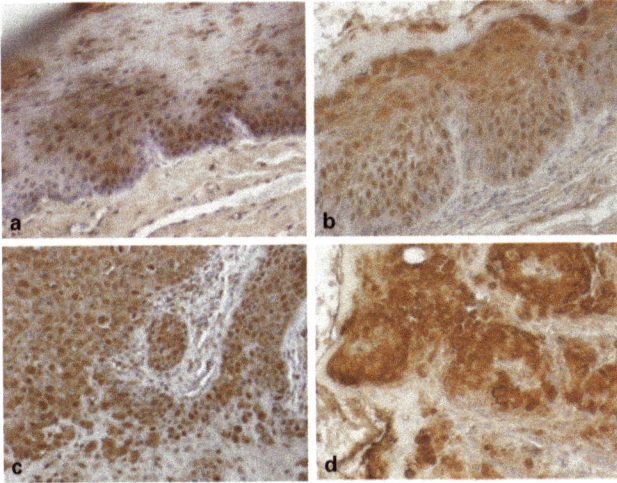

Fig. 3.5 **Immunohistochemical analysis of KLK5 expression in pre-malignant OSCC.** Sections of oral lesions representative of **(a)** mild dysplasia, **(b)** moderate dysplasia, **(c)** severe dysplasia, or **(d)** carcinoma *in situ* were immunostained with antibodies against KLK5 (rabbit anti-human KLK5, Abcam, Cambridge, MA, USA; 1:20 dilution), followed by a biotinylated secondary antibody, avidin-POX, and DAB chromogen as the substrate (Pettus *et al.*, 2009).

ability and thereby disrupts barrier function (**Fig. 3.4b**). Together, this data indicates that *KLK5*-mediated Dsg1 processing functionally regulates cell-cell adhesion and dispersal, as well as epidermal barrier function, suggesting that KLK5 may potentiate metastatic dissemination of OSCC by contributing to dissolution of Dsg1-containing cell-cell contacts. This is consistent with immunohistochemical analysis of human OSCC, showing reduced Dsg1 staining relative to normal oral mucosa (Jiang *et al.*, 2011). Furthermore, our preliminary data suggests an increase in KLK5 staining during oral cancer progression from mild and moderate dysplasia to severe dysplasia and carcinoma *in situ* (**Fig. 3.5**), indicating a potential role for KLK5 in OSCC progression.

While the role of other KLKs in the regulation of junctional integrity has not been investigated in OSCC, KLK7 has been shown to cleave other cell-cell junctional components, including corneodesmosin and desmocollin 1 (Caubet *et al.*, 2004; Descargues *et al.*, 2006). Furthermore, a hyperkeratosis phenotype was observed in *KLK8* knockout mice (Kishibe *et al.*, 2007), implicating a role for KLK8 in normal skin structure and function. Together with data showing upregulated KLK expression and loss of Dsg-1 staining in tongue OSCC (Jiang *et al.*, 2011; Pettus *et al.*, 2009) as well as downregulation of SPINK5 in human tongue tumors relative to normal tongue (Ye *et al.*, 2008), it is interesting to speculate that uncontrolled KLK activities may be involved in cleavage of desmosomal cadherin components in order to release cell-cell junctions and thereby potentiate OSCC metastasis.

3.5 Conclusions and future directions

The study of KLK family proteases in HNSCC is emerging as an important new area of investigation. While expression of KLK4, 5, 7, 8, and 10 has been reported in HNSCC cells and tissues, the role of KLKs in head and neck cancer initiation, progression, and metastasis remains unclear. Furthermore, the potential diagnostic and/or prognostic significance of expression of specific KLKs or groups of KLKs also requires additional investigation. Based on current data, it is interesting to speculate that one or more KLKs may also represent a potential therapeutic target(s).

Identification of KLK substrates in the oral cavity will likely yield key mechanistic insight into the role of these proteases in HNSCC. Although it is evident that KLK5 participates in Dsg1 ectodomain shedding, the potential involvement of other downstream protease(s) zymogens that could be activated by KLK5 cannot be ruled out using the current approaches. It has been shown that pro-KLK5 undergoes auto-activation and may process downstream KLK zymogens such as pro-KLK7, thereby functioning as an initiator of potential KLK cascades (Brattsand *et al.*, 2005; Michael *et al.*, 2006). Furthermore, the transmembrane serine protease matriptase can function in pro-KLK activation, suggesting the involvement of enzymes further upstream in a regulatory cascade (Sales *et al.*, 2010).

In addition to modifying the physical structure of the epidermis via Dsg1 cleavage, KLK5 has been implicated in the processing of additional substrates, which may regulate the immune barrier in the oral cavity. For example, KLK5 has been shown to activate protease-activated receptor-2 (PAR2) (Briot *et al.*, 2009; Oikonomopoulou *et al.*, 2006; Ramsay *et al.*, 2008; Stefansson *et al.*, 2008), a G-protein-coupled cell surface receptor that is activated by proteolytic cleavage within the extracellular N-terminus, generating a 'tethered ligand' that binds the receptor in order to initiate cell signaling (Ishihara *et al.*, 1997; Nystedt *et al.*, 1995; Vu *et al.*, 1991; Xu *et al.*, 1998). Activation of PAR2 initiates a number of cellular processes associated with cancer progression, including proliferation, adhesion, cytoskeletal reorganization, filopodial protrusion, and migration (Ramsay *et al.*, 2008). PARs are expressed in a variety of HNSCC tissues and their expression is correlated with advanced stage, lymph node metastasis, and poor prognosis (Zhang *et al.*, 2004). It merits further exploration to determine whether KLK5 may potentiate OSCC progression via processing of PAR2 and activation of a variety of G-protein-coupled receptor signaling pathways.

In keratinocytes, PAR2 activation has also been shown to mediate expression of anti-microbial peptides, such as defensins and cathelicidins (Lee *et al.*, 2010). These naturally occurring polypeptides are secreted by epithelia and neutrophils and play a crucial role in the protection of mucosal epithelia from invading microorganisms, but have also been shown to participate in cancer progression (Herr *et al.*, 2007). In humans, the cathelicidin gene product, hCAP18, is processed to the active anti-microbial peptide LL37, which exhibits anti-microbial activity against a broad spectrum of pathogens. The hCAP18 processing in skin is blocked by LEKTI, and recent data demonstrates that KLK5 and KLK7 catalyze cathelicidin activation (Yamasaki *et al.*, 2006). Additional studies show that the precursor of defensin-1a (also called human neutrophil peptide-1, HNP-1) is processed by KLK5 (Shaw *et al.*, 2008). In the oral cavity, defensin-1a is synthesized in the salivary glands and is also released into the gingival crevicular fluid by neutrophils (Dale *et al.*, 2006). Defensin-1a is increased in saliva of OSCC patients relative to healthy volunteers (Mizukawa *et al.*, 2001) and staining for defensin-1a is abundant in tongue OSCC, relative to non-tumor tissue (Lundy *et al.*, 2004). Defensin-1a is produced as an early humoral mediator of the innate immune response and may thereby play a role in host response to tumor invasion. Based on this data, additional studies are needed to determine whether KLK5-mediated inactivation of defensin-1a may potentiate OSCC metastasis.

Although the role of specific KLKs in HNSCC remains elusive, emerging data supports a potential link between KLK expression and HNSCC progression through multiple pathways. KLK5-catalyzed processing of desmosomal cadherin ectodomains may contribute to a loss of tissue cohesion and oral tumor progression. At the same time, KLK5 may also process other substrates, including PAR2 or defensin-1a, to potentiate metastasis. Studies designed to determine the role of other KLKs in the processing of substrates in the oral cavity that modulate HNSCC initiation, progression, and metastasis will be of future interest.

Acknowledgement

This work in part was supported by Research Grant CA085870 (to M.S.S.) from the National Cancer Institute, National Institutes of Health, USA.

Bibliography

Blasi, F., and Carmeliet, P. (2002). uPAR: a versatile signalling orchestrator. Nat. Rev. Mol. Cell. Biol. 3, 932–943.

Borgoño, C.A., Michael, I.P., and Diamandis, E.P. (2004). Human tissue kallikreins: physiologic roles and applications in cancer. Mol. Cancer Res. 2, 257–280.

Borgoño, C.A., Michael, I.P., Komatsu, N., Jayakumar, A., Kapadia, R., Clayman, G.L., Sotiropoulou, G., and Diamandis, E.P. (2007). A potential role for multiple tissue kallikrein serine proteases in epidermal desquamation. J. Biol. Chem. 282, 3640–3652.

Brattsand, M., Stefansson, K., Lundh, C., Haasum, Y., and Egelrud, T. (2005). A proteolytic cascade of kallikreins in the stratum corneum. J. Invest. Dermatol. 124, 198–203.

Briot, A., Deraison, C., Lacroix, M., Bonnart, C., Robin, A., Besson, C., Dubus, P., and Hovnanian, A. (2009). Kallikrein 5 induces atopic dermatitis-like lesions through PAR2-mediated thymic stromal lymphopoietin expression in Netherton syndrome. J. Exp. Med. 206, 1135–1147.

Brockstein, B.E. (2011). Management of recurrent head and neck cancer: recent progress and future directions. Drugs 71, 1551–1559.

Caubet, C., Jonca, N., Brattsand, M., Guerrin, M., Bernard, D., Schmidt, R., Egelrud, T., Simon, M., and Serre, G. (2004). Degradation of corneodesmosome proteins by two serine proteases of the kallikrein family, SCTE/KLK5/hK5 and SCCE/KLK7/hK7. J. Invest. Dermatol. 122, 1235–1244.

Clements, J.A., Willemsen, N.M., Myers, S.A., and Dong, Y. (2004). The tissue kallikrein family of serine proteases: functional roles in human disease and potential as clinical biomarkers. Crit. Rev. Clin. Lab. Sci. 41, 265–312.

Conway, D.I., Hashibe, M., Boffetta, P., Wunsch-Filho, V., Muscat, J., La Vecchia, C., and Winn, D.M. (2009). Enhancing epidemiologic research on head and neck cancer: INHANCE – The international head and neck cancer epidemiology consortium. Oral Oncol. 45, 743–746.

D'Alessio, S., and Blasi, F. (2009). The urokinase receptor as an entertainer of signal transduction. Front. Biosci. 14, 4575–4587.

Dale, B.A., Tao, R., Kimball, J.R., and Jurevic, R.J. (2006). Oral antimicrobial peptides and biological control of caries. BMC Oral Health 6 Suppl 1, S13.

Dass, K., Ahmad, A., Azmi, A.S., Sarkar, S.H., and Sarkar, F.H. (2008). Evolving role of uPA/uPAR system in human cancers. Cancer Treat. Rev. 34, 122–136.

Descargues, P., Deraison, C., Bonnart, C., Kreft, M., Kishibe, M., Ishida-Yamamoto, A., Elias, P., Barrandon, Y., Zambruno, G., Sonnenberg, A., Hovnanian, A. (2005). Spink5-deficient mice mimic Netherton syndrome through degradation of desmoglein 1 by epidermal protease hyperactivity. Nat. Genet. 37, 56–65.

Descargues, P., Deraison, C., Prost, C., Fraitag, S., Mazereeuw-Hautier, J., D'Alessio, M., Ishida-Yamamoto, A., Bodemer, C., Zambruno, G., and Hovnanian, A. (2006). Corneodesmosomal cadherins are preferential targets of stratum corneum trypsin- and chymotrypsin-like hyperactivity in Netherton syndrome. J. Invest. Dermatol. 126, 1622–1632.

Fakhry, C., and Gillison, M.L. (2006). Clinical implications of human papillomavirus in head and neck cancers. J. Clin. Oncol. 24, 2606–2611.

Fakhry, C., Westra, W.H., Li, S., Cmelak, A., Ridge, J.A., Pinto, H., Forastiere, A., and Gillison, M.L. (2008). Improved survival of patients with human papillomavirus-positive head and neck squamous cell carcinoma in a prospective clinical trial. J. Natl. Cancer Inst. 100, 261–269.

Frampton, J.E. (2010). Cetuximab: a review of its use in squamous cell carcinoma of the head and neck. Drugs 70, 1987–2010.

Fu, K.F., Silverman, S., and Kramer, A.M. (1998). Spread of tumor, staging, and survival. In Oral Cancer, 4th edition, S. Silverman, ed. (Ontario, B. C. Decker), pp. 67–74.

Ghosh, S., Brown, R., Jones, J.C., Ellerbroek, S.M., and Stack, M.S. (2000). Urinary-type plasminogen activator (uPA) expression and uPA receptor localization are regulated by α3ß1 integrin in oral keratinocytes. J. Biol. Chem. 275, 23869–23876.

Ghosh, S., Johnson, J.J., Sen, R., Mukhopadhyay, S., Liu, Y., Zhang, F., Wei, Y., Chapman, H.A., and Stack, M.S. (2006). Functional relevance of urinary-type plasminogen activator receptor-α3ß1 integrin association in proteinase regulatory pathways. J. Biol. Chem. 281, 13021–13029.

Ghosh, S., Koblinski, J., Johnson, J., Liu, Y., Ericsson, A., Davis, J.W., Shi, Z., Ravosa, M.J., Crawford, S., Frazier, S., Stack, M.S. (2010). Urinary-type plasminogen activator receptor/α3ß1 integrin signaling, altered gene expression, and oral tumor progression. Mol. Cancer Res. 8, 145–158.

Heck, J.E., Berthiller, J., Vaccarella, S., Winn, D.M., Smith, E.M., Shan'gina, O., Schwartz, S.M., Purdue, M.P., Pilarska, A., Eluf-Neto, J., et al., and Hashibe, M. (2010). Sexual behaviours and the risk of head and neck cancers: a pooled analysis in the International Head and Neck Cancer Epidemiology (INHANCE) consortium. Int. J. Epidemiol. 39, 166–181.

Herr, C., Shaykhiev, R., and Bals, R. (2007). The role of cathelicidin and defensins in pulmonary inflammatory diseases. Expert Opin. Biol. Ther. 7, 1449–1461.

Hopkins, J., Cescon, D.W., Tse, D., Bradbury, P., Xu, W., Ma, C., Wheatley-Price, P., Waldron, J., Goldstein, D., Meyer, F., Bairati, I., Liu, G. (2008). Genetic polymorphisms and head and neck cancer outcomes: a review. Cancer Epidemiol. Biomarkers Prev. 17, 490–499.

Hundsdorfer, B., Zeilhofer, H.F., Bock, K.P., Dettmar, P., Schmitt, M., Kolk, A., Pautke, C., and Horch, H.H. (2005). Tumour-associated urokinase-type plasminogen activator (uPA) and its inhibitor PAI-1 in normal and neoplastic tissues of patients with squamous cell cancer of the oral cavity – clinical relevance and prognostic value. J. Craniomaxillofac. Surg. 33, 191–196.

Ishihara, H., Connolly, A.J., Zeng, D., Kahn, M.L., Zheng, Y.W., Timmons, C., Tram, T., and Coughlin, S.R. (1997). Protease-activated receptor 3 is a second thrombin receptor in humans. Nature 386, 502–506.

Jemal, A., Siegel, R., Xu, J., and Ward, E. (2010). Cancer statistics, 2010. CA Cancer J. Clin. 60, 277–300.

Jiang, R., Shi, Z., Johnson, J.J., Liu, Y., and Stack, M.S. (2011). Kallikrein-5 promotes cleavage of desmoglein-1 and loss of cell-cell cohesion in oral squamous cell carcinoma. J. Biol. Chem. 286, 9127–9135.

Jo, M., Takimoto, S., Montel, V., and Gonias, S.L. (2009). The urokinase receptor promotes cancer metastasis independently of urokinase-type plasminogen activator in mice. Am. J. Pathol. 175, 190–200.

Kishibe, M., Bando, Y., Terayama, R., Namikawa, K., Takahashi, H., Hashimoto, Y., Ishida-Yamamoto, A., Jiang, Y.P., Mitrovic, B., Perez, D., Iizuka, H., and Yoshida, S. (2007). Kallikrein 8 is involved in skin desquamation in cooperation with other kallikreins. J. Biol. Chem. 282, 5834–5841.

Kutler, D.I., Auerbach, A.D., Satagopan, J., Giampietro, P.F., Batish, S.D., Huvos, A.G., Goberdhan, A., Shah, J.P., and Singh, B. (2003). High incidence of head and neck squamous cell carcinoma in patients with Fanconi anemia. Arch. Otolaryngol. Head Neck Surg. 129, 106–112.

Lee, S.E., Kim, J.M., Jeong, S.K., Jeon, J.E., Yoon, H.J., Jeong, M.K., and Lee, S.H. (2010). Protease-activated receptor-2 mediates the expression of inflammatory cytokines, antimicrobial peptides, and matrix metalloproteinases in keratinocytes in response to Propionibacterium acnes. Arch. Dermatol. Res. 302, 745–756.

Leemans, C.R., Braakhuis, B.J., and Brakenhoff, R.H. (2011). The molecular biology of head and neck cancer. Nat. Rev. Cancer 11, 9–22.

Lindberg, P., Larsson, A., and Nielsen, B.S. (2006). Expression of plasminogen activator inhibitor-1, urokinase receptor and laminin gamma-2 chain is an early coordinated event in incipient oral squamous cell carcinoma. Int. J. Cancer 118, 2948–2956.

Liotta, L.A., Rao, C.N., and Barsky, S.H. (1983). Tumor invasion and the extracellular matrix. Lab. Invest. 49, 636–649.

Lundwall, A., Band, V., Blaber, M., Clements, J.A., Courty, Y., Diamandis, E.P., Fritz, H., Lilja, H., Malm, J., Maltais, L.J., Olsson, A.Y., Petraki, C., Scorilas, A., Sotiropoulou, G., Stenman, U.H., Stephan, C., Talieri, M., and Yousef, G.M. (2006). A comprehensive nomenclature for serine proteases with homology to tissue kallikreins. Biol. Chem. 387, 637–641.

Lundy, F.T., Orr, D.F., Gallagher, J.R., Maxwell, P., Shaw, C., Napier, S.S., Gerald Cowan, C., Lamey, P.J., and Marley, J.J. (2004). Identification and overexpression of human neutrophil alpha-defensins (human neutrophil peptides 1, 2 and 3) in squamous cell carcinomas of the human tongue. Oral Oncol. 40, 139–144.

Marur, S., and Forastiere, A.A. (2008). Head and neck cancer: changing epidemiology, diagnosis, and treatment. Mayo Clin. Proc. 83, 489–501.

Michael, I.P., Pampalakis, G., Mikolajczyk, S.D., Malm, J., Sotiropoulou, G., and Diamandis, E.P. (2006). Human tissue kallikrein 5 is a member of a proteolytic cascade pathway involved in seminal clot liquefaction and potentially in prostate cancer progression. J. Biol. Chem. 281, 12743–12750.

Mizukawa, N., Sawaki, K., Nagatsuka, H., Kamio, M., Yamachika, E., Fukunaga, J., Ueno, T., Takagi, S., and Sugahara, T. (2001). Human – and ß-defensin immunoreactivity in oral mucoepidermoid carcinomas. Anticancer Res. 21, 2171–2174.

Nagata, M., Fujita, H., Ida, H., Hoshina, H., Inoue, T., Seki, Y., Ohnishi, M., Ohyama, T., Shingaki, S., Kaji, M., Saku, T., and Takagi, R. (2003). Identification of potential biomarkers of lymph node metastasis in oral squamous cell carcinoma by cDNA microarray analysis. Int. J. Cancer 106, 683–689.

Nozaki, S., Endo, Y., Kawashiri, S., Nakagawa, K., Yamamoto, E., Yonemura, Y., and Sasaki, T. (1998). Immunohistochemical localization of a urokinase-type plasminogen activator system in squamous cell carcinoma of the oral cavity: association with mode of invasion and lymph node metastasis. Oral Oncol. 34, 58–62.

Nystedt, S., Emilsson, K., Larsson, A.K., Strombeck, B., and Sundelin, J. (1995). Molecular cloning and functional expression of the gene encoding the human proteinase activated receptor 2. Eur. J. Biochem. 232, 84–89.

Oikonomopoulou, K., Hansen, K.K., Saifeddine, M., Vergnolle, N., Tea, I., Blaber, M., Blaber, S.I., Scarisbrick, I., Diamandis, E.P., and Hollenberg, M.D. (2006). Kallikrein-mediated cell signalling: targeting proteinase-activated receptors (PARs). Biol. Chem. 387, 817–824.

Paliouras, M., Borgoño, C., and Diamandis, E.P. (2007). Human tissue kallikreins: the cancer biomarker family. Cancer Lett. 249, 61–79.

Pampalakis, G., and Sotiropoulou, G. (2007). Tissue kallikrein proteolytic cascade pathways in normal physiology and cancer. Biochim. Biophys. Acta 1776, 22–31.

Parkin, D.M., Bray, F., Ferlay, J., and Pisani, P. (2005). Global cancer statistics, 2002. CA Cancer J. Clin. 55, 74–108.

Patel, S.G., and Shah, J.P. (2005). TNM staging of cancers of the head and neck: striving for uniformity among diversity. CA Cancer J. Clin. 55, 242–258.

Pettus, J.R., Johnson, J.J., Shi, Z., Davis, J.W., Koblinski, J., Ghosh, S., Liu, Y., Ravosa, M.J., Frazier, S., and Stack, M.S. (2009). Multiple kallikrein (KLK 5, 7, 8, and 10) expression in squamous cell carcinoma of the oral cavity. Histol. Histopathol. 24, 197–207.

Ramsay, A.J., Reid, J.C., Adams, M.N., Samaratunga, H., Dong, Y., Clements, J.A., and Hooper, J.D. (2008). Prostatic trypsin-like kallikrein-related peptidases (KLKs) and other prostate-expressed tryptic proteinases as regulators of signalling via proteinase-activated receptors (PARs). Biol. Chem. 389, 653–668.

Rusthoven, K., Ballonoff, A., Raben, D., and Chen, C. (2008). Poor prognosis in patients with stage I and II oral tongue squamous cell carcinoma. Cancer 112, 345–351.

Sales, K.U., Masedunskas, A., Bey, A.L., Rasmussen, A.L., Weigert, R., List, K., Szabo, R., Overbeek, P.A., and Bugge, T.H. (2010). Matriptase initiates activation of epidermal pro-kallikrein and disease onset in a mouse model of Netherton syndrome. Nat. Genet. 42, 676–683.

Sano, D., and Myers, J.N. (2007). Metastasis of squamous cell carcinoma of the oral tongue. Cancer Metastasis Rev. 26, 645–662.

Shaw, J.L., Petraki, C., Watson, C., Bocking, A., and Diamandis, E.P. (2008). Role of tissue kallikrein-related peptidases in cervical mucus remodeling and host defense. Biol. Chem. 389, 1513–1522.

Shi, Z., and Stack, M.S. (2007). Urinary-type plasminogen activator (uPA) and its receptor (uPAR) in squamous cell carcinoma of the oral cavity. Biochem. J. 407, 153–159.

Shi, Z., Liu, Y., Johnson, J.J., and Stack, M.S. (2011). Urinary-type plasminogen activator receptor (uPAR) modulates oral cancer cell behavior with alteration in p130cas. Mol. Cell Biochem. 357, 151–161.

Stefansson, K., Brattsand, M., Roosterman, D., Kempkes, C., Bocheva, G., Steinhoff, M., and Egelrud, T. (2008). Activation of proteinase-activated receptor-2 by human kallikrein-related peptidases. J. Invest. Dermatol. 128, 18–25.

Vu, T.K., Hung, D.T., Wheaton, V.I., and Coughlin, S.R. (1991). Molecular cloning of a functional thrombin receptor reveals a novel proteolytic mechanism of receptor activation. Cell 64, 1057–1068.

Waltz, D.A., Natkin, L.R., Fujita, R.M., Wei, Y., and Chapman, H.A. (1997). Plasmin and plasminogen activator inhibitor type 1 promote cellular motility by regulating the interaction between the urokinase receptor and vitronectin. J. Clin. Invest. 100, 58–67.

Wei, Y., Waltz, D.A., Rao, N., Drummond, R.J., Rosenberg, S., and Chapman, H.A. (1994). Identification of the urokinase receptor as an adhesion receptor for vitronectin. J. Biol. Chem. 269, 32380–32388.

Xu, W.F., Andersen, H., Whitmore, T.E., Presnell, S.R., Yee, D.P., Ching, A., Gilbert, T., Davie, E.W., and Foster, D.C. (1998). Cloning and characterization of human protease-activated receptor 4. Proc. Natl. Acad. Sci. USA 95, 6642–6646.

Yamasaki, K., Schauber, J., Coda, A., Lin, H., Dorschner, R.A., Schechter, N.M., Bonnart, C., Descargues, P., Hovnanian, A., and Gallo, R.L. (2006). Kallikrein-mediated proteolysis regulates the antimicrobial effects of cathelicidins in skin. FASEB J. 20, 2068–2080.

Yasuda, T., Sakata, Y., Kitamura, K., Morita, M., and Ishida, T. (1997). Localization of plasminogen activators and their inhibitor in squamous cell carcinomas of the head and neck. Head Neck 19, 611–616.

Ye, H., Yu, T., Temam, S., Ziober, B.L., Wang, J., Schwartz, J.L., Mao, L., Wong, D.T., and Zhou, X. (2008). Transcriptomic dissection of tongue squamous cell carcinoma. BMC Genomics 9, 69.

Zhang, X., Hunt, J.L., Landsittel, D.P., Muller, S., Adler-Storthz, K., Ferris, R.L., Shin, D.M., and Chen, Z.G. (2004). Correlation of protease-activated receptor-1 with differentiation markers in squamous cell carcinoma of the head and neck and its implication in lymph node metastasis. Clin. Cancer Res. 10, 8451–8459.

Zhao, H., Dong, Y., Quan, J., Smith, R., Lam, A., Weinstein, S., Clements, J., Johnson, N.W., and Gao, J. (2011). Correlation of the expression of human kallikrein-related peptidases 4 and 7 with the prognosis in oral squamous cell carcinoma. Head Neck 33, 566–572.

Ziober, A.F., Patel, K.R., Alawi, F., Gimotty, P., Weber, R.S., Feldman, M.M., Chalian, A.A., Weinstein, G.S., Hunt, J., and Ziober, B.L. (2006). Identification of a gene signature for rapid screening of oral squamous cell carcinoma. Clin. Cancer Res. 12, 5960–5971.

Hannu Koistinen, and Ulf-Håkan Stenman

4 PSA (Prostate-Specific Antigen) and other Kallikrein-related Peptidases in Prostate Cancer

4.1 Introduction

Prostate cancer is a major cause of cancer death in developed countries (Jemal *et al.*, 2011). Presently, the majority of prostate cancers is detected at an early stage and carries a favorable prognosis. Some, however, are diagnosed at an extracapsular stage, when no curative therapy is available. Another problem is overdiagnosis, i.e. detection of indolent cancers that would not surface during the remaining lifetime. Laboratory tests discriminating between cancers that need to be cured and indolent tumors are not available. Therefore, it is important to develop better prognostic methods and treatments. As the prostate produces several kallikrein-related peptidases (KLK), including prostate specific antigen (PSA or KLK3) and KLK2 (also known as human kallikrein 2, hK2), KLKs are potentially useful for prostate cancer diagnosis.

The functions of KLKs in the normal prostate and prostate cancer are not well known, although several roles have been suggested. Generally, increased proteolytic activity is associated with several forms of cancer, in which proteases play a major role in invasion and formation of metastases (Turk, 2006). However, tumor suppressive properties of proteases have also been described (Lopez-Otin and Matrisian, 2007). The hypothesized roles of KLKs in prostate cancer include both activities that promote and inhibit cancer growth and metastasis. As it may be possible to control prostate cancer growth by modulating the proteolytic activities of KLKs, these proteases have evolved as potential targets for prostate cancer treatment. We here review the clinical use of KLKs with respect to prostate cancer and their potential biological roles in prostate cancer.

4.2 The role of KLKs in prostate cancer diagnosis, prognosis, and monitoring

4.2.1 PSA

The clinical use of PSA (KLK3) has been extensively and critically reviewed in several recent articles, and readers interested in a more detailed account are referred to these (De Angelis *et al.*, 2007; Lawrence *et al.*, 2010; Lilja *et al.*, 2008; Shariat *et al.*, 2011; Ulmert *et al.*, 2009). PSA is expressed in differentiated luminal epithelial cells of the prostate, from which it is secreted into the seminal fluid. It is the most abundant KLK expressed in the prostate, with estimated levels in extracellular fluid of the prostate up to 60 mg/L (~2 µM), or in prostate tissue 10 mg/g of tissue (Denmeade *et al.*, 2001; Shaw and Diamandis, 2007). PSA is a major constituent of seminal fluid, but from the normal prostate only a minor part leaks out into the circulation. However, when tissue architecture is deranged as in prostate cancer, and the tumor loses contact with the prostatic ducts, PSA is secreted directly into the extracellular fluid and the circulation (Stenman, 1997). Although PSA expression is higher in the normal than in the malignant prostatic tissue (Abrahamsson *et al.*, 1988; Paju *et al.*, 2007), its leakage into the circulation, together with its prostate specificity, makes it a very sensitive marker for prostate cancer.

PSA was detected in the serum of prostate cancer patients in 1980 (Papsidero *et al.*, 1980). Since then, PSA has become the most useful and most widely used tumor marker (Stenman *et al.*, 2005; Ulmert *et al.*, 2009). Determination of serum PSA is widely used to detect prostate cancer at an early preclinical stage (Catalona *et al.*, 1991; Lilja *et al.*, 2008; Stamey *et al.*, 1987; Stenman *et al.*, 2005). This is possible because the PSA concentration in the serum usually increases 5–10 years before the cancer gives rise to symptoms (Lilja *et al.*, 2011; Stenman *et al.*, 1994). The US Food and Drug Administration (FDA) approved the PSA test for the detection of prostate cancer in 1994. While high PSA levels have been found to be associated with large tumors and advanced tumor stage, PSA is not cancer-specific, as its serum levels are also increased in nonmalignant conditions such as benign prostatic hyperplasia and prostatitis (Stenman, 1997).

The serum concentrations of PSA can vary from <0.1 to more than 10,000 µg/L in men with advanced prostate cancer. Men are generally referred to a biopsy examination if their PSA levels exceed 4 µg/L, although lower cut-off levels have been recommended as well (Lilja *et al.*, 2008). When serum PSA is over 4 µg/L, cancer is detected in 27–44% of apparently healthy men over 50 years of age (Lilja *et al.*, 2008). The use of a single threshold for PSA has been criticized, as the concentrations increase with age and are also dependent on ethnicity and body mass index (Shariat *et al.*, 2011; Ulmert *et al.*, 2009). About three quarters of cancers are clinically organ-confined and potentially curable when the PSA levels are in the range of 4–10 µg/L (Catalona *et al.*, 1991; 1994).

In different studies, the specificity and sensitivity of PSA for prostate cancer detection varies, due to the biopsy techniques used to detect cancer, but also due to a lack of standardization of the PSA assays in early studies (Bangma *et al.*, 2010; Lilja *et al.*, 2008). Previous participation in PSA screening may have a dramatic effect on the performance of statistical models predicting biopsy outcome (Vickers *et al.*, 2010). Because benign prostatic hyperplasia that causes increased plasma PSA levels becomes more common with age, the specificity of a PSA test can be improved by using age-specific reference values (Oesterling *et al.*, 1993). Furthermore, as serum PSA reflects prostate volume, the cancer specificity of PSA can be improved by correcting the value for prostate volume, also called PSA density (Benson *et al.*, 1992;

Finne *et al.*, 2004; Veneziano *et al.*, 1990). PSA velocity, i.e. the increase of PSA over time, can be used to detect tumor development. However, these methods are not widely used (Ulmert *et al.*, 2009; Vickers *et al.*, 2011a).

While total PSA is not a specific marker for prostate cancer it is a very reliable indicator of disease recurrence (Lilja *et al.*, 2008). Increasing serum concentrations reveal the presence of residual cancer cells that were not removed by primary therapy. With very few exceptions, these produce measurable amounts of PSA long before the tumor is detectable via other means. After successful radical prostatectomy, the PSA concentrations become undectable within weeks, whereas they decrease more slowly after radiotherapy and often remain detectable due to the presence of benign residual prostatic tissue. During antiandrogen treatment, PSA levels decline if the tumor is hormone-responsive. Increasing PSA concentrations reliably reveal recurrent disease and the need for additional therapy.

Screening for prostate cancer by PSA determination is controversial (Vickers and Lilja, 2009). The use of opportunistic screening with PSA has dramatically increased prostate cancer incidence, leading to detection of cancers that would not have surfaced clinically without screening (Stenman *et al.*, 2005). Recently, two large screening studies have been published. In the European Randomized Study of Screening for Prostate Cancer (ERSPC), screening reduced mortality by about 20–30% (Roobol *et al.*, 2009; Schroder *et al.*, 2009), while in the US trial, the Prostate, Lung, Colorectal, and Ovarian Cancer Screening Trial (PLCO), no reduction in mortality was observed (Andriole *et al.*, 2009). Both studies showed that screening leads to an overdiagnosis in about 50% of cases (Draisma *et al.*, 2003). This leads to overtreatment, which causes unnecessary costs, reduced quality of life, and even death due to complications of surgery. The risk of overdiagnosis increases with age, and in 1994 we predicted that screening for prostate cancer will not reduce mortality in men over 65 years of age (Stenman *et al.*, 1994). This has recently been confirmed (Bill-Axelson *et al.*, 2011). In men under 50 years of age, an elevated serum PSA is a reliable predictor of development of clinically relevant prostate cancer within 20 to 30 years (Lilja *et al.*, 2011; Vickers and Lilja, 2009).

Different molecular forms of PSA

In the circulation, enzymatically active PSA rapidly forms complexes with serine protease inhibitors (serpins), including α1-antichymotrypsin (ACT, serpinA3) (Lilja *et al.*, 1991; Stenman *et al.*, 1991) and α1-protease inhibitor (API, serpinA1) (Zhang *et al.*, 1997), and α2-macroglobulin (A2M) (Zhang *et al.*, 2000) (**Fig. 4.1**). All of these protease inhibitors are produced in the liver and are present in the blood circulation in vast excess over PSA (Hortin *et al.*, 2008; Stenman *et al.*, 1991). PSA-ACT is the major immunoreactive form of PSA in the circulation, while the PSA-A2M complex is not detectable by conventional immunoassays but can be detected after denaturation of the sample at high pH (Zhang *et al.*, 2000). About 5–40% of circulating PSA is not

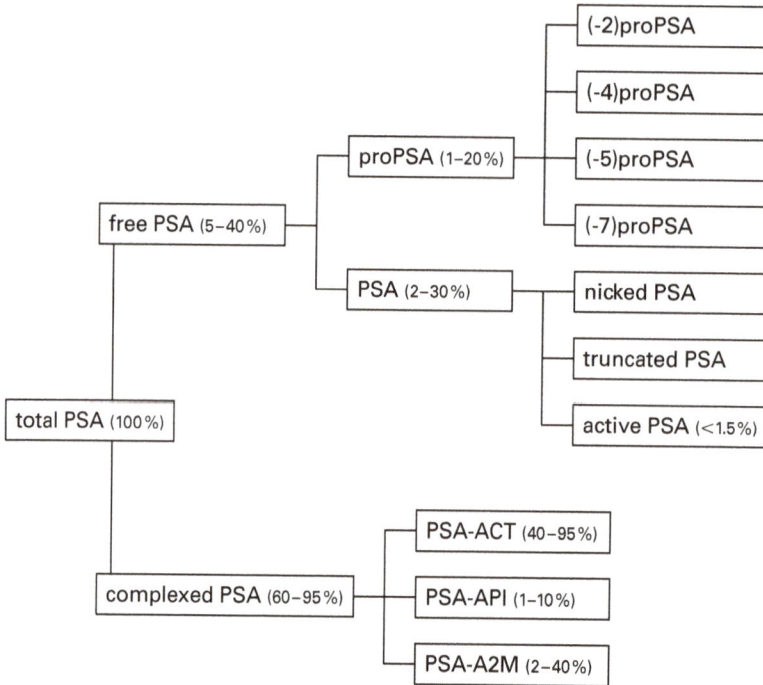

Fig. 4.1 **Different molecular forms of PSA found in blood circulation.** Numbers in parenthesis refer to the amount of given forms as a percentage of total PSA in blood circulation and are only sugges-tive. Nicked PSA refers to internally cleaved forms of PSA, while truncated PSA refers to PSA-lacking amino acids from N-terminus. (-7)proPSA is full-length proPSA with a 7 amino acid propeptide. Adapted from Zhu (2009).

complexed with inhibitors. These uncomplexed forms include active PSA, different forms of proPSA, truncated and internally cleaved forms, including so called benign PSA (BPSA), which is internally cleaved at Lys 182 (Bangma *et al.*, 2010; Mikolajczyk *et al.*, 2000). Differentially glycosylated forms have also been found (Tabares *et al.*, 2006). Specific determination of these forms is potentially useful for improving the diagnostic accuracy of PSA.

In patients with prostate cancer, the proportion of PSA-ACT in serum has been shown to be elevated, whereas that of free PSA is reduced (Stenman *et al.*, 1991). PSA-ACT alone has been found a more accurate marker of prostate cancer than either total PSA or free PSA by itself (Stenman *et al.*, 1991). Free and total PSA are easier to measure than PSA-ACT. Therefore, an assay of the proportion of free PSA in relation to total PSA (%fPSA) is routinely used. A low %fPSA indicates increased risk of prostate cancer among men with elevated serum PSA (Morote *et al.*, 2002; Partin *et al.*, 1996). This is also true when the total PSA levels are in the range of 2–4 μg/L. When total PSA is in this range, a low %fPSA is a strong predictor of later diagnosis of prostate cancer (Finne *et al.*, 2008).

Several isoforms of PSA have been studied in various cohorts of the ERSPC study (Bangma *et al.*, 2010). The use of free PSA in addition to total PSA has been found to reduce the number of negative sextant biopsies by 30%, at a PSA cut-off level of 3 μg/L at initial screening, at the cost of losing 10% of detectable cancers which, however, are predominantly histologically well-differentiated (Bangma *et al.*, 2010). Recently, determining the ratio of a proform of PSA, containing only two of the amino acids of the propeptide [(-2)proPSA], to free PSA has been shown to increase the predictive value and specificity for prostate cancer, compared to total PSA and %fPSA. However, it showed limited additional value in identifying aggressive prostate cancer (Jansen *et al.*, 2010).

4.2.2 KLK2

Among the KLKs, KLK2 (hK2) has attracted most interest, after PSA, due to its prostate specificity and relatively high expression levels (Shaw and Diamandis, 2007; Young *et al.*, 1996). Contrary to PSA, KLK2 expression is higher in prostate cancer than in the benign prostate. Furthermore, high KLK2 expression is associated with an adverse prognosis (Darson *et al.*, 1997). However, the concentrations of KLK2 are much lower than those of PSA, and KLK2 alone is not as good a marker for prostate cancer as PSA is (Shariat *et al.*, 2011). Several studies have shown that, when used together with different forms of PSA, KLK2 provides additional prognostic information and increases specificity in prostate cancer screening (Becker *et al.*, 2000; Vickers *et al.*, 2011b). A panel comprising KLK2, and total, free and intact PSA, has been shown to reduce the number of unnecessary biopsies in men with elevated total PSA levels (Vickers *et al.*, 2011b).

4.2.3 Other KLKs

In addition to PSA and KLK2, several other KLKs have been determined in the normal prostate and in prostate cancer. Shaw and Diamandis (2007) reported that all 15 KLKs are expressed in the prostate at the mRNA level. In tissue extracts they found, in addition to KLK2 and PSA, relatively high protein levels of KLK11, lesser amounts of KLK1 and KLK9, and even less KLK4, 5, and 13–15. Comprehensive data on KLK expression in serum is restricted to KLK2 and PSA. The other KLKs have mostly been studied at the tissue level. These studies have shown that expression of KLK5 and KLK7 is reduced in prostate cancer (Jamaspishvili *et al.*, 2011; Xuan *et al.*, 2008; Yousef *et al.*, 2002), while the expression of KLK2, 4, 11, 12, and 14 is increased (Diamandis *et al.*, 2002; Memari *et al.*, 2007; Yousef *et al.*, 2003; Xi *et al.*, 2004), compared to the normal prostate or hyperplastic prostate. It is noteworthy that in prostate cancer KLK11 serum levels are increased (Diamandis *et al.*, 2002). Regarding KLK6, 10, 13, and 15, the changes in expression are still matter of debate (Petraki *et al.*, 2003; Rabien *et al.*, 2010; Yousef *et al.*, 2001). While there is currently not enough information to be able to elucidate whether these other KLKs could be clinically useful, presently available data suggests that, as markers, they are not as good as PSA. However, they may be valuable as additional markers, together with PSA (Paliouras *et al.*, 2007). Several KLKs can possibly be used together with other parameters to build models that predict the risk of prostate cancer, like those established for different forms of PSA and prostate volume (Finne *et al.*, 2004; Stephan *et al.*, 2006).

4.2.4 Splicing and polymorphic variants

Alternative splicing is usual in cancers (Venables *et al.*, 2009). Splicing variants of *KLKs* have been described and some of those are associated with prostate cancer (Kurlender *et al.*, 2005; Whitbread *et al.*, 2010). Alternative splicing may take place due to differences in splicing machinery or genetic variation that create novel splicing sites. Genetic variation in the *KLK* locus is common (Parikh *et al.*, 2010). Several studies, including genome-wide association studies (GWAS), have proposed that single nucleotide polymorphisms (SNP) in *KLK2*, *PSA*, and other *KLKs* are associated with prostate cancer (Eeles *et al.*, 2008; Klein *et al.*, 2010; Lose *et al.*, 2011). However, these results are conflicting and are dependent on the selection of the control population, as some of these SNPs are associated with increased PSA levels, which may affect tumor incidence by increasing the chance to being diagnosed with prostate cancer on the basis of elevated PSA (Ahn *et al.*, 2008). Still, some of the SNPs may be associated with prostate cancer (Kote-Jarai *et al.*, 2011). Combining SNPs in *KLK* genes with protein levels of KLKs has provided some improvement in prostate cancer prediction (Johansson *et al.*, 2012; Klein *et al.*, 2010). However, it remains to be established whether genotyping, in addition to PSA measurement, would give any clinically useful information. SNPs and alternative splicing of *KLKs* may affect the levels and functions of KLKs, but their potential functional roles in prostate cancer remains to be established. SNPs in the *KLK* locus are described in more detail in Chapter 2, Volume 1.

4.3 Potential functional roles of KLKs in prostate cancer

Several of the fifteen KLKs are expressed in the prostate and may play a functional role in prostate cancer development (Borgoño and Diamandis, 2004; Lawrence *et al.*, 2010; Shaw and Diamandis, 2007). These KLKs are able to activate each other and are thought to form a "KLK cascade", or "KLK activome" (Pampalakis and Sotiropoulou, 2007; Sotiropoulou *et al.*, 2009). Other proteases produced in the prostate, like trypsin, may also activate KLKs (Paju *et al.*, 2000). Interestingly, several of the prostatic KLKs are regulated by androgens, although other mechanisms are also involved (Lawrence *et al.*, 2010). The PSA concentrations in serum start to increase well before, some-times even decades before the development of otherwise detectable tumors (Stenman *et al.*, 1994; Lilja *et al.*, 2011). This raises the question as to whether PSA initiates or facilitates cancer development. The PSA concentration in extracellular fluid of the prostate is high, up to 2 μM, and most of it is enzymatically active (Denmeade *et al.*, 2001). However, when active PSA, like some other KLKs, reaches the circulation, they are rapidly inactivated by serpins and α2-macroglobulin. Therefore, KLKs may have a functional role in the prostate, but systemic effects are unlikely. The physiological function of PSA, and perhaps also several other prostatic KLKs, is to promote sperm motility by dissolving the seminal clot formed after ejaculation by cleaving semeno-gelins (Lilja, 1985). However, several other functions of KLKs that may be relevant for the development of prostate cancer have been suggested (reviewed in Borgoño and Diamandis, 2004; Lawrence *et al.*, 2010; Williams *et al.*, 2007). These include both promoting and inhibiting activities on tumor growth and metastasis formation. KLKs cleave several protein substrates, at least *in vitro* (a comprehensive list can be found in Lawrence *et al.*, 2010). Several of these may be relevant to prostate cancer devel-opment. However, the hypothesized functions based on *in vitro* cleavage need to be interpreted with caution.

We here review the functional properties of KLKs with respect to prostate cancer, and with special reference to PSA, which is by far the most abundant and intensively studied KLK. The functions are discussed according to the *Hallmarks of Cancer* pre-sented by Hanahan and Weinberg (2011). Some of the *Hallmarks of Cancer* are not likely to be relevant to KLKs and are thus not considered here. In addition to those presented here, several other cancer relevant functions, which may be also important in prostate cancer, have been ascribed to KLKs in non-prostate cancer models. These are not covered here, but are reviewed in Borgoño and Diamandis (2004), Lawrence *et al.* (2010), and Sotiropoulou *et al.* (2009). It is worth mentioning that sometimes the results are very much dependent on the model system used and perhaps on cancer type, like in the case of KLK6, which stimulates proliferation, migration, and invasion of mouse keratinocyte cell lines (Klucky *et al.*, 2007), while it has a tumor protective role in aggressive MDA-MB-231 breast cancer cells, including reduced proliferation rate and cell motility (Pampalakis *et al.*, 2009).

4.3.1 Sustaining proliferative signaling and evading growth suppressors

In order for cancer cells to survive, it is essential that they sustain chronic prolifera-tion and evade growth-suppressing signals. Several KLKs have been associated with increased cell proliferation. KLK4-transfections to prostate cancer cells have been shown to both decrease and increase proliferation (Klokk *et al.*, 2007; Veveris-Lowe *et al.*, 2005). Williams *et al.* (2011) created a stable LNCaP cell line, in which PSA production was knocked down using RNA interference. These cells showed reduced growth both *in vitro* and *in vivo*, compared to the control cells. They also expressed PSA ectopically in PSA-negative prostate cancer cells, and showed that active PSA, but not an inactive variant of PSA, promotes cell growth. This is in contrast to the study by Niu and coworkers (2008), which showed that stable transfection of PSA into androgen receptor (AR)-positive cells promote cell growth, irrespective of the activity state of PSA. No effect was seen when two AR-negative cell lines were studied. In contrast to these transfection studies, Bindukumar and coworkers (2005) found that subcutaneous administration of PSA in mice reduced xenograft tumor growth.

KLKs may mediate their stimulatory effect on cell proliferation via different mechanisms, most of which are related to their enzymatic activity. Several KLKs, like KLK2 and PSA, and other prostatic proteases cleave insulin-like growth factor (IGF)-binding proteins (IGFBPs) (Cohen *et al.*, 1992; Koistinen *et al.*, 2002; Lawrence *et al.*, 2010; Rehault *et al.*, 2001). IGF-I is a potent growth factor, which may form complexes with several IGFBPs, which restrict binding to its receptor, IGFR, and thereby inhibit the bioavailability of IGF-I (Jogie-Brahim *et al.*, 2009). However, binding of IGF-I to its binding proteins also prolongs the half-life of IGF-I and may facilitate its delivery to target tissues, where it is released by proteases, e.g. the KLKs, that cleave IGFBPs, especially IGFBP-3, the major IGFBP in blood circulation and in several tissues (Jogie-Brahim *et al.*, 2009). Increased IGF-I activity is implicated in several cancers, including prostate cancer (Samani *et al.*, 2007). Thus, the cleavage of IGFBPs by KLKs could contribute to prostate cancer growth. IGFBP-3 and its fragments have been described to also have IGF-independent activities, such as induction of apoptosis (Ingermann *et al.*, 2010). Furthermore, the role of the IGF-system in the development of prostate cancer is complex, and the results from various studies are controversial (Samani *et al.*, 2007).

PSA is also able to cleave the latent form of transforming growth factor-β2 (TGF-β2), leading to its activation, but not to release of TGF-β2 stored in the extracellular matrix (ECM) (Dallas *et al.*, 2005). TGF-β2 promotes the growth of malignant cells, but it also has tumor-suppressing functions. Recently, PSA was found to cleave galectin-3, releasing functionally active monovalent lectin from galectin-3 homodimer (Saraswati *et al.*, 2011). For some of its functions, galectin-3 needs to be in the intact dimeric form. In prostate cancer, galectin-3 expression is downregulated and it is cleaved (Merseburger *et al.*, 2008; Wang *et al.*, 2009). However, knock-down of galectin-3 reduces tumor cell migration, invasion, and proliferation as well as tumor growth, in immunocompromized mice (Wang *et al.*, 2009). Perhaps this discrepancy could be explained by findings showing that nuclear galectin-3 has an effect on invasion and tumor growth opposite to cytoplasmic galectin-3, the latter promoting tumor growth (Califice *et al.*, 2004).

The PSA-α2-macroglobulin complex binds to the cell surface protein GRP78, which leads to activation of signaling molecules that stimulate DNA and protein synthesis and may thus promote prostate cancer growth (Misra *et al.*, 2011). PSA has also been reported to enhance ARA70-induced AR transactivation by modulating the p53 pathway, resulting in increased proliferation (Niu *et al.*, 2008). Several KLKs have been found to activate protease activated receptors (PAR) (Lawrence *et al.*, 2010; Oikonomopoulou *et al.*, 2010) (see Chapter 15, Volume 1). PARs form a family of four G-protein-coupled receptors, which are activated by serine proteases that cleave their N-terminus. Activation leads to a wide array of responses, which in many cases have been found to promote cancer cell growth. KLK2 activates PAR-2, while KLK4 activates PAR-1 and PAR-2 in prostate cancer cells and at least PAR-1 in stromal cells (Mize *et al.*, 2008; Ramsay *et al.*, 2008). KLK4 and PAR-2 are co-expressed in primary prostate cancer and bone metastases (Ramsay *et al.*, 2008). The activation of stromal PAR-1 by KLK4 induces the release of several cytokines, including IL-6, which induces prostate cancer cell proliferation (Wang *et al.*, 2010). IL-6 has been suggested to play a pathogenic role in prostate cancer, where its levels are found to be increased (Giri *et al.*, 2001). PAR-activation by KLKs may facilitate activation of signaling that induces cell proliferation and other tumorigenic properties of the cells. Interestingly, Bindu-kumar *et al.* (2005) reported that PSA induces significant changes in expression of several cancer-related genes in prostate cancer cell lines, including down-regulation of vascular endothelial growth factor (VEGF). These studies collectively suggest that PSA, and other KLKs, may promote the growth of prostate cancer by stimulating cell proliferation.

4.3.2 Resisting cell death

In addition to evading growth-suppressing signals, cancer cells must be resistant to programmed cell death (apoptosis) and other forms of cell death, which usually sequester abnormal cells. PSA has been shown to decrease the rate of apoptosis (Niu *et al.*, 2008). PSA has also been found to promote survival of prostate cancer cells in which apoptosis was induced by an inhibitor of calcium-independent phospholi-pase A_2 (Nicotera *et al.*, 2009). Furthermore, knock-down of PSA expression in pros-tate cancer cells by RNA interference induced apoptosis and growth arrest (Niu *et al.*, 2008). However, due to the sequence similarity in the siRNA target site, the method is likely to also target KLK2, and when only one siRNA target site is used, off-target effects may well take place and explain some of the results.

4.3.3 Inducing angiogenesis

Like all tissues, the tumor tissue needs nutrients and oxygen, and ability to evacu-ate waste and carbon dioxide, in order to grow and survive (Hanahan and Weinberg, 2011). For this, tumors need new blood vessels to grow beyond microscopic size (Folkman, 2007). In several convincing studies, KLK1 has been shown to promote neovascularization via kininogen cleavage, which generates kinins (Chao *et al.*, 2010;

Emanueli *et al.*, 2001) (see also Chapter 10, Volume 1). KLK1 knockout mice have shown impaired muscle neovascularization in response to hindlimb ischemia, which was rescued by gene-transfer of wild-type KLK1 (Stone *et al.*, 2009). Moreover, KLK1 overexpression induced endothelial precursor cell migration and vascular remodeling in zebrafish (Stone *et al.*, 2009). The authors suggest that endogenous KLK1 plays a fundamental role in vascular repair. A functional kallikrein-kinin system may also be involved in angiogenesis in prostatic tumors (Wright *et al.*, 2008).

Other studies have addressed the antiangiogenic role of PSA, which has been demonstrated in cell culture models at concentrations more than 10 times lower than those found in the extracellular fluid of the prostate (Denmeade *et al.*, 2001; Heidtmann *et al.*, 1999; Mattsson *et al.*, 2008b). An initial finding, that showed PSA to be antiangiogenic was made by Fortier and coworkers (1999). They found that PSA inhibits endothelial cell tube formation, indicating a reduced angiogenic potential of the cells. Furthermore, PSA also inhibited endothelial cell growth, invasion, and migration, i.e., activities that are needed for blood vessel formation. Fortier and coworkers (2003) also showed that subcutaneous administration of PSA inhibits angiogenesis in an *in vivo* mouse model in which a basement membrane preparation with basic fibroblast growth factor (bFGF) was injected under the skin of mice, to form a plug inside in which blood vessel growth was monitored.

The mechanism by which PSA exerts its anti-angiogenic effect is unclear. Even the dependency on enzymatic activity is controversial. Two studies have shown that enzymatically inactive PSA exerted antiangiogenic activity similar to active PSA. One of these studies used an inactive mutant of PSA (Fortier *et al.*, 2003), while in the other (Chadha *et al.*, 2011) PSA activity was inhibited by zinc. Contrary to these studies, our studies strongly suggest that PSA activity is needed, or at least closely associated, with the antiangiogenic activity of PSA (Koistinen *et al.*, 2008; Mattsson *et al.*, 2008b). We first found that the enzymatic activity of different PSA forms that are present in seminal fluid correlates with the antiangiogenic activity (Mattsson *et al.*, 2008b). Furthermore, inhibition of PSA by small-molecule inhibitors of PSA abolished the antiangiogenic activity (Koistinen *et al.*, 2008), while stimulation of PSA-activity by peptides enhanced it (Mattsson *et al.*, 2012). PSA has been shown to cleave plasminogen into angiostatin-like fragments, which could mediate the anti-angiogenic effect of PSA (Heidtmann *et al.*, 1999). However, we have not been able to verify this (Hekim *et al.*, unpublished results). Several other KLKs, which may contaminate PSA preparations, have been found to cleave plasminogen as well (Lawrence *et al.*, 2010).

PSA has been shown to cleave several proteins present in ECM, e.g., laminin, fibronectin, and collagen type IV (Lawrence *et al.*, 2010), the fragments of which may generate cleavage products that act as angiogenesis inhibitors (Nyberg *et al.*, 2005). Another possible mechanism by which PSA modulates angiogenesis is via the regulation of gene expression. PSA has been found to inhibit expression of pro-angiogenic growth factors in human umbilical endothelial cells (HUVEC), including VEGF and bFGF (Chadha *et al.*, 2011). These gene expression changes, induced by PSA, were also found when PSA was inhibited by zinc. We have found only minor changes in gene expression (at the mRNA level) in HUVECs upon treatment with PSA (Mattsson *et al.*, 2008a). It is noteworthy that galectin-3, which is cleaved by PSA, is also a pro-angiogenic molecule, which has been found to mediate VEGF- and bFGF-mediated angiogenic response (Markowska *et al.*, 2010).

4.3.4 Activating invasion and metastasis

Several KLKs, including PSA and KLK2, may degrade ECM proteins and activate other ECM-degrading proteases, like the zymogen form of urokinase-type plasminogen activator (uPA), which is activated by KLK2 and some other KLKs (Frenette *et al.*, 1997; Lawrence *et al.*, 2010; Sotiropoulou *et al.*, 2009). Furthermore, KLK2 inactivates the plasminogen activator inhibitor type-1 (PAI-1) (Mikolajczyk *et al.*, 1999). KLK1 and PSA activate pro-MMP-2 and/or pro-MMP-9 (Lawrence *et al.*, 2010). These results suggest that KLKs are involved in proteolytic cascades that facilitate prostate cancer growth and metastasis (Sotiropoulou *et al.*, 2009).

PSA-treatment has been found to increase invasion of prostate cancer cells *in vitro* (Ishii *et al.*, 2004; Webber *et al.*, 1995). In contrast, subcutaneously administered PSA reduced metastatic propensity of intravenously injected melanoma cells in mice (Fortier *et al.*, 1999). It is possible that PSA stimulates the initial phase of the metastatic process, i.e. intravasation, which the *in vitro* invasion assay partially recapitulates. However, intravasation was bypassed in a mouse model, when the cells were directly injected into blood circulation. PSA may also inhibit the formation of metastatic tumors by inhibiting angiogenesis.

Several studies suggest that PSA may play a crucial role in the development of bone metastases (reviewed in Lawrence *et al.*, 2010; Williams *et al.*, 2007). These studies have shown that PSA cleaves and inactivates parathyroid hormone-related protein (PTHrP), promotes osteoblast proliferation and differentiation, induces apoptosis in osteoclast precursors, and promotes prostate cancer cell adhesion to bone marrow endothelial cells. KLK4 may also have a role in metastasis of prostate cancer cells in bone (Gao *et al.*, 2007).

Stable overexpression of PSA, KLK4, or KLK7 in prostate cancer cells changes their phenotype towards a mesenchymal phenotype (Klokk *et al.*, 2007; Veveris-Lowe *et al.*, 2005). In addition to morphological changes, the cells showed various other characteristics associated with epithelial-to-mesenchymal transition (EMT), such as reduced E-cadherin or increased vimentin expression, and increased migration and invasion. EMT is generally associated with increased aggressiveness of cancer cells, especially as concerns increased invasion and metastatic propensity (Hugo *et al.*, 2007). KLK1 has been found to promote prostate cancer cell migration and invasion via a PAR-1-dependent pathway (Gao *et al.*, 2010). In addition to KLK1, several other KLKs have been found to activate PARs (Lawrence *et al.*, 2010; Oikonomopoulou *et al.*, 2010).

4.3.5 Concerns about biological studies

PSA and KLK2 are not expressed in the mouse or rat. Therefore, several widely used transgenic prostate cancer models are either not relevant or knock-out studies cannot be performed. Instead, most biological studies aiming to solve functions of KLK2, PSA, and other KLKs have utilized cancer cell lines, although different cell lines may display very different responses. Also, the tumor environment is very complex and cells grown on/in an isolated environment may behave very differently from those grown within a tumor and in contact with the ECM and stromal cells (Bissell and Hines, 2011). The tumor microenvironment has a very strong effect on tumor development and metastasis, which are complex processes with several different steps, the balance of which eventually regulates the fate of the cancer cells (Bissell and Hines, 2011; Hanahan and Weinberg, 2011). Therefore, the *in vitro* growth characteristics of the cells may not necessarily predict tumorigenicity. In xenograft models using cells that have been transfected with KLKs, the KLK-producing cells may have differentiated already during growth on artificial plastic surfaces, before they are inoculated into mice to form tumors. This is feasible, as transfection of KLKs into prostate cancer cells has been shown to induce EMT-like changes (Klokk *et al.*, 2007; Veveris-Lowe *et al.*, 2005).

Therefore, transgenic models would be valuable. However, PSA-transgenic mice so far have not been reported to produce active PSA at levels comparable to those in prostate cancer, i.e. PSA is either inactive, or the concentrations are several orders of magnitutde lower than in the human prostate (Wei *et al.*, 1997; Williams *et al.*, 2010). PSA-transgenic mice or mice carrying both KLK2 and PSA transgenes have not shown any morphological changes in the prostate (Williams *et al.*, 2010).

4.4 Conclusions and outlook

KLKs, especially PSA, which is an established prostate cancer marker, have been used for prostate cancer diagnosis for several decades. PSA is considered a most useful cancer biomarker. Recently, it has been shown that PSA-based prostate cancer screening can reduce mortality. However, the use of prostate cancer screening is controversial, due to overdiagnosis. Several different molecular forms of PSA, like free PSA, have been found to improve diagnostic accuracy of PSA tests. Since the discovery of classical KLK1-3, several other KLKs have been found, and their expression in the prostate has been studied. When used together with PSA, some of these may improve prostate cancer diagnostics. However, it remains to be shown whether this approach can be used to improve detection of prostate tumors that need to be cured, and to reduce overdiagnosis.

KLKs expressed in the prostate may have functional role(s) in prostate cancer development, but conclusive evidence is still lacking. We propose that PSA promotes the growth of small tumors, but inhibits development of large tumors at the stage where new blood vessels are needed. PSA levels start to increase even decades before the prostate cancer is detected clinically (Lilja *et al.*, 2011). Thus, it is possible that PSA facilitates early development of prostate cancer. Several studies have suggested that PSA alters cells towards a more malignant direction. However, clinical evidence supports the hypothesis that the slow growth of prostate cancer is dependent on PSA expression. PSA levels are lower in malignant than in normal prostatic epithelium, and PSA expression is further reduced in poorly differentiated tumors (Abrahamsson *et al.*, 1988; Paju *et al.*, 2007). Furthermore, low tissue concentration of PSA is associated with poor prognosis (Stege *et al.*, 2000). PSA has antiangiogenic activity in *in vitro* and *in vivo* models, and high PSA expression in tumors is associated with low microvessel density (Papadopoulos *et al.*, 2001) and low angiogenesis activity (as determined by CD34 staining) (Ben Jemaa *et al.*, 2010). This may explain the generally slow growth of prostate cancer at the stage where new blood vessels are needed. Some mouse studies also support the notion that PSA inhibits tumor growth. However, the models used so far have several limitations and are not able to address the complexity of prostate cancer development. Perhaps transgenic mice, in which the expression of several KLKs is regulated, crossed with transgenic prostate cancer model mice, could be used in the future, in order to answer these questions.

Bibliography

Abrahamsson, P.A., Lilja, H., Falkmer, S., and Wadstrom, L.B. (1988). Immunohistochemical distribution of the three predominant secretory proteins in the parenchyma of hyperplastic and neoplastic prostate glands. Prostate 12, 39–46.

Ahn, J., Berndt, S.I., Wacholder, S., Kraft, P., Kibel, A.S., Yeager, M., Albanes, D., Giovannucci, E., Stampfer, M.J., Virtamo, J., Thun, M.J., Feigelson, H.S., Cancel-Tassin, G., Cussenot, O., Thomas, G., Hunter, D.J., Fraumeni, J.F. Jr., Hoover, R.N., Chanock, S.J., and Hayes, R.B. (2008). Variation in KLK genes, prostate-specific antigen and risk of prostate cancer. Nat. Genet. 40, 1032–1034.

Andriole, G.L., Crawford, E.D., Grubb, R.L.,3rd, Buys, S.S., Chia, D., Church, T.R., Fouad, M.N., Gelmann, E.P., Kvale, P.A., Reding, D.J., *et al.*, and Berg, C.D. (2009). Mortality results from a randomized prostate-cancer screening trial. N. Engl. J. Med. 360, 1310–1319.

Bangma, C.H., van Schaik, R.H., Blijenberg, B.G., Roobol, M.J., Lilja, H., and Stenman, U.H. (2010). On the use of prostate-specific antigen for screening of prostate cancer in European Randomised Study for Screening of Prostate Cancer. Eur. J. Cancer 46, 3109–3119.

Becker, C., Piironen, T., Pettersson, K., Hugosson, J., and Lilja, H. (2000). Clinical value of human glandular kallikrein 2 and free and total prostate-specific antigen in serum from a population of men with prostate-specific antigen levels 3.0 ng/mL or greater. Urology 55, 694–699.

Ben Jemaa, A., Bouraoui, Y., Sallami, S., Banasr, A., Ben Rais, N., Ouertani, L., Nouira, Y., Horchani, A., and Oueslati, R. (2010). Co-expression and impact of prostate specific membrane antigen and prostate specific antigen in prostatic pathologies. J. Exp. Clin. Cancer Res. 29, 171.

Benson, M.C., Whang, I.S., Pantuck, A., Ring, K., Kaplan, S.A., Olsson, C.A., and Cooner, W.H. (1992). Prostate specific antigen density: a means of distinguishing benign prostatic hypertrophy and prostate cancer. J. Urol. 147, 815–816.

Bill-Axelson, A., Holmberg, L., Ruutu, M., Garmo, H., Stark, J.R., Busch, C., Nordling, S., Haggman, M., Andersson, S.O., Bratell, S., Spångberg, A., Palmgren, J., Steineck, G., Adami, H.O., Johansson, J.E., and SPCG-4 Investigators (2011). Radical prostatectomy versus watchful waiting in early prostate cancer. N. Engl. J. Med. 364, 1708–1717.

Bindukumar, B., Schwartz, S.A., Nair, M.P., Aalinkeel, R., Kawinski, E., and Chadha, K.C. (2005). Prostate-specific antigen modulates the expression of genes involved in prostate tumor growth. Neoplasia 7, 241–252.

Bissell, M.J., and Hines, W.C. (2011). Why don't we get more cancer? A proposed role of the microenvironment in restraining cancer progression. Nat. Med. 17, 320–329.

Borgoño, C.A., and Diamandis, E.P. (2004). The emerging roles of human tissue kallikreins in cancer. Nat. Rev. Cancer. 4, 876–890.

Califice, S., Castronovo, V., Bracke, M., and van den Brule, F. (2004). Dual activities of galectin-3 in human prostate cancer: tumor suppression of nuclear galectin-3 vs tumor promotion of cytoplasmic galectin-3. Oncogene 23, 7527–7536.

Catalona, W.J., Smith, D.S., Ratliff, T.L., Dodds, K.M., Coplen, D.E., Yuan, J.J., Petros, J.A., and Andriole, G.L. (1991). Measurement of prostate-specific antigen in serum as a screening test for prostate cancer. N. Engl. J. Med. 324, 1156–1161.

Catalona, W.J., Hudson, M.A., Scardino, P.T., Richie, J.P., Ahmann, F.R., Flanigan, R.C., deKernion, J.B., Ratliff, T.L., Kavoussi, L.R., and Dalkin, B.L. (1994). Selection of optimal prostate specific antigen cutoffs for early detection of prostate cancer: receiver operating characteristic curves. J. Urol. 152, 2037–2042.

Chadha, K.C., Nair, B.B., Chakravarthi, S., Zhou, R., Godoy, A., Mohler, J.L., Aalinkeel, R., Schwartz, S.A., and Smith, G.J. (2011). Enzymatic activity of free-prostate-specific antigen (f-PSA) is not required for some of its physiological activities. Prostate 71, 1680–1690.

Chao, J., Shen, B., Gao, L., Xia, C.F., Bledsoe, G., and Chao, L. (2010). Tissue kallikrein in cardiovascular, cerebrovascular and renal diseases and skin wound healing. Biol. Chem. 391, 345–355.

Cohen, P., Graves, H.C., Peehl, D.M., Kamarei, M., Giudice, L.C., and Rosenfeld, R.G. (1992). Prostate-specific antigen (PSA) is an insulin-like growth factor binding protein-3 protease found in seminal plasma. J. Clin. Endocrinol. Metab. 75, 1046–1053.

Dallas, S.L., Zhao, S., Cramer, S.D., Chen, Z., Peehl, D.M., and Bonewald, L.F. (2005). Preferential production of latent transforming growth factor beta-2 by primary prostatic epithelial cells and its activation by prostate-specific antigen. J. Cell. Physiol. 202, 361–370.

Darson, M.F., Pacelli, A., Roche, P., Rittenhouse, H.G., Wolfert, R.L., Young, C.Y., Klee, G.G., Tindall, D.J., and Bostwick, D.G. (1997). Human glandular kallikrein 2 (hK2) expression in prostatic intraepithelial neoplasia and adenocarcinoma: a novel prostate cancer marker. Urology 49, 857–862.

De Angelis, G., Rittenhouse, H.G., Mikolajczyk, S.D., Blair Shamel, L., and Semjonow, A. (2007). Twenty years of PSA: from prostate antigen to tumor marker. Rev. Urol. 9, 113–123.

Denmeade, S.R., Sokoll, L.J., Chan, D.W., Khan, S.R., and Isaacs, J.T. (2001). Concentration of enzymatically active prostate-specific antigen (PSA) in the extracellular fluid of primary human prostate cancers and human prostate cancer xenograft models. Prostate 48, 1–6.

Diamandis, E.P., Okui, A., Mitsui, S., Luo, L.Y., Soosaipillai, A., Grass, L., Nakamura, T., Howarth, D.J., and Yamaguchi, N. (2002). Human kallikrein 11: a new biomarker of prostate and ovarian carcinoma. Cancer Res. 62, 295–300.

Draisma, G., Boer, R., Otto, S.J., van der Cruijsen, I.W., Damhuis, R.A., Schroder, F.H., and de Koning, H.J. (2003). Lead times and overdetection due to prostate-specific antigen screening: estimates from the European Randomized Study of Screening for Prostate Cancer. J. Natl. Cancer Inst. 95, 868–878.

Eeles, R.A., Kote-Jarai, Z., Giles, G.G., Olama, A.A., Guy, M., Jugurnauth, S.K., Mulholland, S., Leongamornlert, D.A., Edwards, S.M., Morrison, J., et al., Easton, D.F. (2008). Multiple newly identified loci associated with prostate cancer susceptibility. Nat. Genet. 40, 316–21.

Emanueli, C., Minasi, A., Zacheo, A., Chao, J., Chao, L., Salis, M.B., Straino, S., Tozzi, M.G., Smith, R., Gaspa, L., Bianchini, G., Stillo, F., Capogrossi, M.C., and Madeddu, P. (2001). Local delivery of human tissue kallikrein gene accelerates spontaneous angiogenesis in mouse model of hindlimb ischemia. Circulation 103, 125–132.

Finne, P., Finne, R., Bangma, C., Hugosson, J., Hakama, M., Auvinen, A., and Stenman, U.H. (2004). Algorithms based on prostate-specific antigen (PSA), free PSA, digital rectal examination and prostate volume reduce false-positive PSA results in prostate cancer screening. Int. J. Cancer 111, 310–315.

Finne, P., Auvinen, A., Maattanen, L., Tammela, T.L., Ruutu, M., Juusela, H., Martikainen, P., Hakama, M., and Stenman, U.H. (2008). Diagnostic value of free prostate-specific antigen among men with a prostate-specific antigen level of <3.0 μg per liter. Eur. Urol. 54, 362–370.

Folkman, J. (2007). Angiogenesis: an organizing principle for drug discovery? Nat. Rev. Drug. Discov. 6, 273–286.

Fortier, A.H., Nelson, B.J., Grella, D.K., and Holaday, J.W. (1999). Antiangiogenic activity of prostate-specific antigen. J. Natl. Cancer Inst. 91, 1635–1640.

Fortier, A.H., Holaday, J.W., Liang, H., Dey, C., Grella, D.K., Holland-Linn, J., Vu, H., Plum, S.M., and Nelson, B.J. (2003). Recombinant prostate specific antigen inhibits angiogenesis in vitro and in vivo. Prostate 56, 212–219.

Frenette, G., Tremblay, R.R., Lazure, C., and Dube, J.Y. (1997). Prostatic kallikrein hK2, but not prostate-specific antigen (hK3), activates single-chain urokinase-type plasminogen activator. Int. J. Cancer 71, 897–899.

Gao, J., Collard, R.L., Bui, L., Herington, A.C., Nicol, D.L., and Clements, J.A. (2007). Kallikrein 4 is a potential mediator of cellular interactions between cancer cells and osteoblasts in metastatic prostate cancer. Prostate 67, 348–360.

Gao, L., Smith, R.S., Chen, L.M., Chai, K.X., Chao, L., and Chao, J. (2010). Tissue kallikrein promotes prostate cancer cell migration and invasion via a protease-activated receptor-1-dependent signaling pathway. Biol. Chem. 391, 803–812.

Giri, D., Ozen, M., and Ittmann, M. (2001). Interleukin-6 is an autocrine growth factor in human prostate cancer. Am. J. Pathol. 159, 2159–2165.

Hanahan, D., and Weinberg, R.A. (2011). Hallmarks of cancer: the next generation. Cell 144, 646–674.

Heidtmann, H.H., Nettelbeck, D.M., Mingels, A., Jager, R., Welker, H.G., and Kontermann, R.E. (1999). Generation of angiostatin-like fragments from plasminogen by prostate-specific antigen. Br. J. Cancer 81, 1269–1273.

Hortin, G.L., Sviridov, D., and Anderson, N.L. (2008). High-abundance polypeptides of the human plasma proteome comprising the top 4 logs of polypeptide abundance. Clin. Chem. 54, 1608–1616.

Hugo, H., Ackland, M.L., Blick, T., Lawrence, M.G., Clements, J.A., Williams, E.D., and Thompson, E.W. (2007). Epithelial--mesenchymal and mesenchymal--epithelial transitions in carcinoma progression. J. Cell. Physiol. 213, 374–383.

Ingermann, A.R., Yang, Y.F., Han, J., Mikami, A., Garza, A.E., Mohanraj, L., Fan, L., Idowu, M., Ware, J.L., Kim, H.S., Lee, D.Y., and Oh, Y. (2010). Identification of a novel cell death receptor mediating IGFBP-3-induced anti-tumor effects in breast and prostate cancer. J. Biol. Chem. 285, 30233–30246.

Ishii, K., Otsuka, T., Iguchi, K., Usui, S., Yamamoto, H., Sugimura, Y., Yoshikawa, K., Hayward, S.W., and Hirano, K. (2004). Evidence that the prostate-specific antigen (PSA)/Zn^{2+} axis may play a role in human prostate cancer cell invasion. Cancer Lett. 207, 79–87.

Jamaspishvili, T., Scorilas, A., Kral, M., Khomeriki, I., Kurfurstova, D., Kolar, Z., and Bouchal, J. (2011). Immunohistochemical localization and analysis of kallikrein-related peptidase 7 and 11 expression in paired cancer and benign foci in prostate cancer patients. Neoplasma 58, 298–303.

Jansen, F.H., van Schaik, R.H., Kurstjens, J., Horninger, W., Klocker, H., Bektic, J., Wildhagen, M.F., Roobol, M.J., Bangma, C.H., and Bartsch, G. (2010). Prostate-specific antigen (PSA) isoform p2PSA in combination with total PSA and free PSA improves diagnostic accuracy in prostate cancer detection. Eur. Urol. 57, 921–927.

Jemal, A., Bray, F., Center, M.M., Ferlay, J., Ward, E., and Forman, D. (2011). Global cancer statistics. CA Cancer. J. Clin. 61, 69–90.

Jogie-Brahim, S., Feldman, D., and Oh, Y. (2009). Unraveling insulin-like growth factor binding protein-3 actions in human disease. Endocr. Rev. 30, 417–437.

Johansson, M., Holmstrom, B., Hinchliffe, S.R., Bergh, A., Stenman, U.H., Hallmans, G., Wiklund, F., and Stattin, P. (2012). Combining 33 genetic variants with prostate-specific antigen for prediction of prostate cancer: Longitudinal study. Int. J. Cancer 130, 129–137.

Klein, R.J., Hallden, C., Cronin, A.M., Ploner, A., Wiklund, F., Bjartell, A.S., Stattin, P., Xu, J., Scardino, P.T., Offit, K., Vickers, A.J., Grönberg, H., and Lilja, H. (2010). Blood biomarker levels to aid discovery of cancer-related single-nucleotide polymorphisms: kallikreins and prostate cancer. Cancer. Prev. Res. (Phila) 3, 611–619.

Klokk, T.I., Kilander, A., Xi, Z., Waehre, H., Risberg, B., Danielsen, H.E., and Saatcioglu, F. (2007). Kallikrein 4 is a proliferative factor that is overexpressed in prostate cancer. Cancer Res. 67, 5221–5230.

Klucky, B., Mueller, R., Vogt, I., Teurich, S., Hartenstein, B., Breuhahn, K., Flechtenmacher, C., Angel, P., and Hess, J. (2007). Kallikrein 6 induces E-cadherin shedding and promotes cell proliferation, migration, and invasion. Cancer Res. 67, 8198–8206.

Koistinen, H., Paju, A., Koistinen, R., Finne, P., Lovgren, J., Wu, P., Seppala, M., and Stenman, U.H. (2002). Prostate-specific antigen and other prostate-derived proteases cleave IGFBP-3, but prostate cancer is not associated with proteolytically cleaved circulating IGFBP-3. Prostate 50, 112–118.

Koistinen, H., Wohlfahrt, G., Mattsson, J.M., Wu, P., Lahdenpera, J., and Stenman, U.H. (2008). Novel small molecule inhibitors for prostate-specific antigen. Prostate 68, 1143–1151.

Kote-Jarai, Z., Amin Al Olama, A., Leongamornlert, D., Tymrakiewicz, M., Saunders, E., Guy, M., Giles, G.G., Severi, G., Southey, M., Hopper, J.L., *et al.*, and Eeles, R.A. (2011). Identification of a novel prostate cancer susceptibility variant in the KLK3 gene transcript. Hum. Genet. 129, 687–694.

Kurlender, L., Borgoño, C., Michael, I.P., Obiezu, C., Elliott, M.B., Yousef, G.M., and Diamandis, E.P. (2005). A survey of alternative transcripts of human tissue kallikrein genes. Biochim. Biophys. Acta 1755, 1–14.

Lawrence, M.G., Lai, J., and Clements, J.A. (2010). Kallikreins on steroids: structure, function, and hormonal regulation of prostate-specific antigen and the extended kallikrein locus. Endocr. Rev. 31, 407–446.

Lilja, H. (1985). A kallikrein-like serine protease in prostatic fluid cleaves the predominant seminal vesicle protein. J. Clin. Invest. 76, 1899–1903.

Lilja, H., Christensson, A., Dahlen, U., Matikainen, M.T., Nilsson, O., Pettersson, K., and Lovgren, T. (1991). Prostate-specific antigen in serum occurs predominantly in complex with alpha 1-antichymotrypsin. Clin. Chem. 37, 1618–1625.

Lilja, H., Ulmert, D., and Vickers, A.J. (2008). Prostate-specific antigen and prostate cancer: prediction, detection and monitoring. Nat. Rev. Cancer 8, 268–278.

Lilja, H., Cronin, A.M., Dahlin, A., Manjer, J., Nilsson, P.M., Eastham, J.A., Bjartell, A.S., Scardino, P.T., Ulmert, D., and Vickers, A.J. (2011). Prediction of significant prostate cancer diagnosed 20 to 30 years later with a single measure of prostate-specific antigen at or before age 50. Cancer 117, 1210–1219.

Lopez-Otin, C., and Matrisian, L.M. (2007). Emerging roles of proteases in tumour suppression. Nat. Rev. Cancer 7, 800–808.

Lose, F., Batra, J., O'Mara, T., Fahey, P., Marquart, L., Eeles, R.A., Easton, D.F., Al Olama, A.A., Kote-Jarai, Z., Guy, M., *et al.*, and Kedda, M.A. (2011). Common variation in kallikrein genes KLK5, KLK6, KLK12, and KLK13 and risk of prostate cancer and tumor aggressiveness. Urol. Oncol., [Epub ahead of print].

Markowska, A.I., Liu, F.T., and Panjwani, N. (2010). Galectin-3 is an important mediator of VEGF- and bFGF-mediated angiogenic response. J. Exp. Med. 207, 1981–1993.

Mattsson, J.M., Laakkonen, P., Kilpinen, S., Stenman, U.H., and Koistinen, H. (2008a). Gene expression changes associated with the anti-angiogenic activity of kallikrein-related peptidase 3 (KLK3) on human umbilical vein endothelial cells. Biol. Chem. 389, 765–771.

Mattsson, J.M., Valmu, L., Laakkonen, P., Stenman, U.H., and Koistinen, H. (2008b). Structural characterization and anti-angiogenic properties of prostate-specific antigen isoforms in seminal fluid. Prostate 68, 945–954.

Mattsson, J.M., Närvänen, A., Stenman, U.-H., and Koistinen H. (2012) Peptides Binding to Prostate-Specific Antigen Enhance Its Antiangiogenic Activity. Prostate 72, 1588–1594.

Memari, N., Diamandis, E.P., Earle, T., Campbell, A., van Dekken, H., and van der Kwast, T.H. (2007). Human kallikrein-related peptidase 12: antibody generation and immunohistochemical localization in prostatic tissues. Prostate 67, 1465–1474.

Merseburger, A.S., Kramer, M.W., Hennenlotter, J., Simon, P., Knapp, J., Hartmann, J.T., Stenzl, A., Serth, J., and Kuczyk, M.A. (2008). Involvement of decreased Galectin-3 expression in the pathogenesis and progression of prostate cancer. Prostate 68, 72–77.

Mikolajczyk, S.D., Millar, L.S., Kumar, A., and Saedi, M.S. (1999). Prostatic human kallikrein 2 inactivates and complexes with plasminogen activator inhibitor-1. Int. J. Cancer 81, 438–442.

Mikolajczyk, S.D., Millar, L.S., Wang, T.J., Rittenhouse, H.G., Wolfert, R.L., Marks, L.S., Song, W., Wheeler, T.M., and Slawin, K.M. (2000). „BPSA," a specific molecular form of free prostate-specific antigen, is found predominantly in the transition zone of patients with nodular benign prostatic hyperplasia. Urology 55, 41–45.

Misra, U.K., Payne, S., and Pizzo, S.V. (2011). Ligation of prostate cancer cell surface GRP78 activates a proproliferative and antiapoptotic feedback loop: a role for secreted prostate-specific antigen. J. Biol. Chem. 286, 1248–1259.

Mize, G.J., Wang, W., and Takayama, T.K. (2008). Prostate-specific kallikreins-2 and -4 enhance the proliferation of DU-145 prostate cancer cells through protease-activated receptors-1 and -2. Mol. Cancer Res. 6, 1043–1051.

Morote, J., Trilla, E., Esquena, S., Serrallach, F., Abascal, J.M., Muñoz, A., Id M'Hammed, Y., and de Torres, I.M. (2002). The percentage of free prostatic-specific antigen is also useful in men with normal digital rectal examination and serum prostatic-specific antigen between 10.1 and 20 ng/ml. Eur. Urol. 42, 333–337.

Nicotera, T.M., Schuster, D.P., Bourhim, M., Chadha, K., Klaich, G., and Corral, D.A. (2009). Regulation of PSA secretion and survival signaling by calcium-independent phopholipase A(2) beta in prostate cancer cells. Prostate 69, 1270–1280.

Niu, Y., Yeh, S., Miyamoto, H., Li, G., Altuwaijri, S., Yuan, J., Han, R., Ma, T., Kuo, H.C., and Chang, C. (2008). Tissue prostate-specific antigen facilitates refractory prostate tumor progression via enhancing ARA70-regulated androgen receptor transactivation. Cancer Res. 68, 7110–7119.

Nyberg, P., Xie, L., and Kalluri, R. (2005). Endogenous inhibitors of angiogenesis. Cancer Res. 65, 3967–3979.

Oesterling, J.E., Jacobsen, S.J., Chute, C.G., Guess, H.A., Girman, C.J., Panser, L.A., and Lieber, M.M. (1993). Serum prostate-specific antigen in a community-based population of healthy men. Establishment of age-specific reference ranges. JAMA 270, 860–864.

Oikonomopoulou, K., Diamandis, E.P., and Hollenberg, M.D. (2010). Kallikrein-related peptidases: proteolysis and signaling in cancer, the new frontier. Biol. Chem. 391, 299–310.

Paju, A., Bjartell, A., Zhang, W.M., Nordling, S., Borgström, A., Hansson, J., and Stenman, U.H. (2000). Expression and characterization of trypsinogen produced in the human male genital tract. Am. J. Pathol. 157, 2011–2021.

Paju, A., Hotakainen, K., Cao, Y., Laurila, T., Gadaleanu, V., Hemminki, A., Stenman, U.H., and Bjartell, A. (2007). Increased expression of tumor-associated trypsin inhibitor, TATI, in prostate cancer and in androgen-independent 22Rv1 Cells. Eur. Urol. 52, 1670–1679.

Paliouras, M., Borgoño, C., and Diamandis, E.P. (2007). Human tissue kallikreins: the cancer biomarker family. Cancer Lett. 249, 61–79.

Pampalakis, G., and Sotiropoulou, G. (2007). Tissue kallikrein proteolytic cascade pathways in normal physiology and cancer. Biochim. Biophys. Acta 1776, 22–31.

Pampalakis, G., Prosnikli, E., Agalioti, T., Vlahou, A., Zoumpourlis, V., and Sotiropoulou, G. (2009). A tumor-protective role for human kallikrein-related peptidase 6 in breast cancer mediated by inhibition of epithelial-to-mesenchymal transition. Cancer Res. 69, 3779–3787.

Papadopoulos, I., Sivridis, E., Giatromanolaki, A., and Koukourakis, M.I. (2001). Tumor angiogenesis is associated with MUC1 overexpression and loss of prostate-specific antigen expression in prostate cancer. Clin. Cancer Res. 7, 1533–1538.

Papsidero, L.D., Wang, M.C., Valenzuela, L.A., Murphy, G.P., and Chu, T.M. (1980). A prostate antigen in sera of prostatic cancer patients. Cancer Res. 40, 2428–2432.

Parikh, H., Deng, Z., Yeager, M., Boland, J., Matthews, C., Jia, J., Collins, I., White, A., Burdett, L., Hutchinson, A., et al., and Amundadottir, L. (2010). A comprehensive resequence analysis of the KLK15-KLK3-KLK2 locus on chromosome 19q13.33. Hum. Genet. 127, 91–99.

Partin, A.W., Catalona, W.J., Southwick, P.C., Subong, E.N., Gasior, G.H., and Chan, D.W. (1996). Analysis of percent free prostate-specific antigen (PSA) for prostate cancer detection: influence of total PSA, prostate volume, and age. Urology 48, 55–61.

Petraki, C.D., Gregorakis, A.K., Papanastasiou, P.A., Karavana, V.N., Luo, L.Y., and Diamandis, E.P. (2003). Immunohistochemical localization of human kallikreins 6, 10 and 13 in benign and malignant prostatic tissues. Prostate Cancer Prostatic Dis. 6, 223–227.

Rabien, A., Fritzsche, F.R., Jung, M., Tolle, A., Diamandis, E.P., Miller, K., Jung, K., Kristiansen, G., and Stephan, C. (2010). KLK15 is a prognostic marker for progression-free survival in patients with radical prostatectomy. Int. J. Cancer 127, 2386–2394.

Ramsay, A.J., Dong, Y., Hunt, M.L., Linn, M., Samaratunga, H., Clements, J.A., and Hooper, J.D. (2008). Kallikrein-related peptidase 4 (KLK4) initiates intracellular signaling via protease-activated receptors (PARs). KLK4 and PAR-2 are co-expressed during prostate cancer progression. J. Biol. Chem. 283, 12293–12304.

Rehault, S., Monget, P., Mazerbourg, S., Tremblay, R., Gutman, N., Gauthier, F., and Moreau, T. (2001). Insulin-like growth factor binding proteins (IGFBPs) as potential physiological substrates for human kallikreins hK2 and hK3. Eur. J. Biochem. 268, 2960–2968.

Roobol, M.J., Kerkhof, M., Schroder, F.H., Cuzick, J., Sasieni, P., Hakama, M., Stenman, U.H., Ciatto, S., Nelen, V., Kwiatkowski, M., et al., and Auvinen, A. (2009). Prostate cancer mortality reduction by prostate-specific antigen-based screening adjusted for nonattendance and contamination in the European Randomised Study of Screening for Prostate Cancer (ERSPC). Eur. Urol. 56, 584–591.

Samani, A.A., Yakar, S., LeRoith, D., and Brodt, P. (2007). The role of the IGF system in cancer growth and metastasis: overview and recent insights. Endocr. Rev. 28, 20–47.

Saraswati, S., Block, A.S., Davidson, M.K., Rank, R.G., Mahadevan, M., and Diekman, A.B. (2011). Galectin-3 is a substrate for prostate specific antigen (PSA) in human seminal plasma. Prostate 71, 197–208.

Schroder, F.H., Hugosson, J., Roobol, M.J., Tammela, T.L., Ciatto, S., Nelen, V., Kwiatkowski, M., Lujan, M., Lilja, H., Zappa, M., et al., and Auvinen, A. (2009). Screening and prostate-cancer mortality in a randomized European study. N. Engl. J. Med. 360, 1320–1328.

Shariat, S.F., Semjonow, A., Lilja, H., Savage, C., Vickers, A.J., and Bjartell, A. (2011). Tumor markers in prostate cancer I: blood-based markers. Acta Oncol. 50 (Suppl. 1), 61–75.

Shaw, J.L., and Diamandis, E.P. (2007). Distribution of 15 human kallikreins in tissues and biological fluids. Clin. Chem. 53, 1423–1432.

Sotiropoulou, G., Pampalakis, G., and Diamandis, E.P. (2009). Functional roles of human kallikrein-related peptidases. J. Biol. Chem. 284, 32989–32994.

Stamey, T.A., Yang, N., Hay, A.R., McNeal, J.E., Freiha, F.S., and Redwine, E. (1987). Prostate-specific antigen as a serum marker for adenocarcinoma of the prostate. N. Engl. J. Med. 317, 909–916.

Stege, R., Grande, M., Carlstrom, K., Tribukait, B., and Pousette, A. (2000). Prognostic significance of tissue prostate-specific antigen in endocrine-treated prostate carcinomas. Clin. Cancer Res. 6, 160–165.

Stenman, U.H., Leinonen, J., Alfthan, H., Rannikko, S., Tuhkanen, K., and Alfthan, O. (1991). A complex between prostate-specific antigen and alpha 1-antichymotrypsin is the major form of prostate-specific antigen in serum of patients with prostatic cancer: assay of the complex improves clinical sensitivity for cancer. Cancer Res. 51, 222–226.

Stenman, U.H., Hakama, M., Knekt, P., Aromaa, A., Teppo, L., and Leinonen, J. (1994). Serum concentrations of prostate specific antigen and its complex with alpha 1-antichymotrypsin before diagnosis of prostate cancer. Lancet 344, 1594–1598.

Stenman, U.H. (1997). Prostate-specific antigen, clinical use and staging: an overview. Br. J. Urol. 79 Suppl. 1, 53–60.

Stenman, U.H., Abrahamsson, P.A., Aus, G., Lilja, H., Bangma, C., Hamdy, F.C., Boccon-Gibod, L., and Ekman, P. (2005). Prognostic value of serum markers for prostate cancer. Scand. J. Urol. Nephrol. 216 Suppl. 64–81.

Stephan, C., Meyer, H.A., Cammann, H., Nakamura, T., Diamandis, E.P., and Jung, K. (2006). Improved prostate cancer detection with a human kallikrein 11 and percentage free PSA-based artificial neural network. Biol. Chem. 387, 801–805.

Stone, O.A., Richer, C., Emanueli, C., van Weel, V., Quax, P.H., Katare, R., Kraenkel, N., Campagnolo, P., Barcelos, L.S., Siragusa, M., Sala-Newby, G.B., Baldessari, D., Mione, M., Vincent, M.P., Benest, A.V., Al Haj Zen, A., Gonzalez, J., Bates, D.O., Alhenc-Gelas, F., and Madeddu, P. (2009). Critical role of tissue kallikrein in vessel formation and maturation: implications for therapeutic revascularization. Arterioscler. Thromb. Vasc. Biol. 29, 657–664.

Tabares, G., Radcliffe, C.M., Barrabes, S., Ramirez, M., Aleixandre, R.N., Hoesel, W., Dwek, R.A., Rudd, P.M., Peracaula, R., and de Llorens, R. (2006). Different glycan structures in prostate-specific antigen from prostate cancer sera in relation to seminal plasma PSA. Glycobiology 16, 132–145.

Turk, B. (2006). Targeting proteases: successes, failures and future prospects. Nat. Rev. Drug. Discov. 5, 785–799.

Ulmert, D., O'Brien, M.F., Bjartell, A.S., and Lilja, H. (2009). Prostate kallikrein markers in diagnosis, risk stratification and prognosis. Nat. Rev. Urol. 6, 384–391.

Venables, J.P., Klinck, R., Koh, C., Gervais-Bird, J., Bramard, A., Inkel, L., Durand, M., Couture, S., Froehlich, U., Lapointe, E., Lucier, J.F., Thibault, P., Rancourt, C., Tremblay, K., Prinos, P., Chabot, B., and Elela, S.A. (2009). Cancer-associated regulation of alternative splicing. Nat. Struct. Mol. Biol. 16, 670–676.

Veneziano, S., Pavlica, P., Querze, R., Nanni, G., Lalanne, M.G., and Vecchi, F. (1990). Correlation between prostate-specific antigen and prostate volume, evaluated by transrectal ultrasonography: usefulness in diagnosis of prostate cancer. Eur. Urol. 18, 112–116.

Veveris-Lowe, T.L., Lawrence, M.G., Collard, R.L., Bui, L., Herington, A.C., Nicol, D.L., and Clements, J.A. (2005). Kallikrein 4 (hK4) and prostate-specific antigen (PSA) are associated with the loss of E-cadherin and an epithelial-mesenchymal transition (EMT)-like effect in prostate cancer cells. Endocr. Relat. Cancer 12, 631–643.

Vickers, A.J., and Lilja, H. (2009). Prostate cancer: estimating the benefits of PSA screening. Nat. Rev. Urol. 6, 301–303.

Vickers, A.J., Cronin, A.M., Aus, G., Pihl, C.G., Becker, C., Pettersson, K., Scardino, P.T., Hugosson, J., and Lilja, H. (2010). Impact of recent screening on predicting the outcome of prostate cancer biopsy in men with elevated prostate-specific antigen: data from the European Randomized Study of Prostate Cancer Screening in Gothenburg, Sweden. Cancer 116, 2612–2620.

Vickers, A.J., Till, C., Tangen, C.M., Lilja, H., and Thompson, I.M. (2011a). An empirical evaluation of guidelines on prostate-specific antigen velocity in prostate cancer detection. J. Natl. Cancer Inst. 103, 462–469.

Vickers, A.J., Gupta, A., Savage, C.J., Pettersson, K., Dahlin, A., Bjartell, A., Manjer, J., Scardino, P.T., Ulmert, D., and Lilja, H. (2011b). A panel of kallikrein marker predicts prostate cancer in a large, population-based cohort followed for 15 years without screening. Cancer Epidemiol. Biomarkers Prev. 20, 255–261.

Wang, W., Mize, G.J., Zhang, X., and Takayama, T.K. (2010). Kallikrein-related peptidase-4 initiates tumor-stroma interactions in prostate cancer through protease-activated receptor-1. Int. J. Cancer 126, 599–610.

Wang, Y., Nangia-Makker, P., Tait, L., Balan, V., Hogan, V., Pienta, K.J., and Raz, A. (2009). Regulation of prostate cancer progression by galectin-3. Am. J. Pathol. 174, 1515–1523.

Webber, M.M., Waghray, A., and Bello, D. (1995). Prostate-specific antigen, a serine protease, facilitates human prostate cancer cell invasion. Clin. Cancer Res. 1, 1089–1094.

Wei, C., Willis, R.A., Tilton, B.R., Looney, R.J., Lord, E.M., Barth, R.K., and Frelinger, J.G. (1997). Tissue-specific expression of the human prostate-specific antigen gene in transgenic mice: implications for tolerance and immunotherapy. Proc. Natl. Acad. Sci. USA 94, 6369–6374.

Whitbread, A.K., Veveris-Lowe, T.L., Dong, Y., Tan, O.L., Gardiner, R., Samaratunga, H.M., Nicol, D.L., and Clements, J.A. (2010). Expression of PSA-RP2, an alternatively spliced variant from the PSA gene, is increased in prostate cancer tissues but the protein is not secreted from prostate cancer cells. Biol. Chem. 391, 461–466.

Williams, S.A., Singh, P., Isaacs, J.T., and Denmeade, S.R. (2007). Does PSA play a role as a promoting agent during the initiation and/or progression of prostate cancer? Prostate 67, 312–329.

Williams, S.A., Xu, Y., De Marzo, A.M., Isaacs, J.T., and Denmeade, S.R. (2010). Prostate-specific antigen (PSA) is activated by KLK2 in prostate cancer ex vivo models and in prostate-targeted PSA/KLK2 double transgenic mice. Prostate 70, 788–796.

Williams, S.A., Jelinek, C.A., Litvinov, I., Cotter, R.J., Isaacs, J.T., and Denmeade, S.R. (2011). Enzymatically active prostate-specific antigen promotes growth of human prostate cancers. Prostate 71, 1595–1607.

Wright, J.K., Botha, J.H., and Naidoo, S. (2008). Influence of the kallikrein-kinin system on prostate and breast tumour angiogenesis. Tumour Biol. 29, 130–136.

Xi, Z., Klokk, T.I., Korkmaz, K., Kurys, P., Elbi, C., Risberg, B., Danielsen, H., Loda, M., and Saatcioglu, F. (2004). Kallikrein 4 is a predominantly nuclear protein and is overexpressed in prostate cancer. Cancer Res. 64, 2365–2370.

Xuan, Q., Yang, X., Mo, L., Huang, F., Pang, Y., Qin, M., Chen, Z., He, M., Wang, Q., and Mo, Z.N. (2008). Expression of the serine protease kallikrein 7 and its inhibitor antileukoprotease is decreased in prostate cancer. Arch. Pathol. Lab. Med. 132, 1796–1801.

Young, C.Y., Seay, T., Hogen, K., Charlesworth, M.C., Roche, P.C., Klee, G.G., and Tindall, D.J. (1996). Prostate-specific human kallikrein (hK2) as a novel marker for prostate cancer. Prostate Suppl. 7, 17–24.

Yousef, G.M., Scorilas, A., Jung, K., Ashworth, L.K., and Diamandis, E.P. (2001). Molecular cloning of the human kallikrein 15 gene (KLK15). Up-regulation in prostate cancer. J. Biol. Chem. 276, 53–61.

Yousef, G.M., Scorilas, A., Chang, A., Rendl, L., Diamandis, M., Jung, K., and Diamandis, E.P. (2002). Down-regulation of the human kallikrein gene 5 (KLK5) in prostate cancer tissues. Prostate 51, 126–132.

Yousef, G.M., Stephan, C., Scorilas, A., Ellatif, M.A., Jung, K., Kristiansen, G., Jung, M., Polymeris, M.E., and Diamandis, E.P. (2003). Differential expression of the human kallikrein gene 14 (KLK14) in normal and cancerous prostatic tissues. Prostate 56, 287–292.

Zhang, W.M., Leinonen, J., Kalkkinen, N., and Stenman, U.H. (1997). Prostate-specific antigen forms a complex with and cleaves alpha 1-protease inhibitor in vitro. Prostate 33, 87–96.

Zhang, W.M., Finne, P., Leinonen, J., Salo, J., and Stenman, U.H. (2000). Determination of prostate-specific antigen complexed to alpha(2)-macroglobulin in serum increases the specificity of free to total PSA for prostate cancer. Urology 56, 267–272.

Zhu L. (2009). Thesis: Development of novel assays for measuring different molecular forms of prostate specific antigen (Yliopistopaino, Helsinki, Finland).

Ying Dong, Daniela Loessner, Shirly Sieh, Anna Taubenberger,
Ruth A. Fuhrman-Luck, Viktor Magdolen, Dietmar W. Hutmacher,
and Judith A. Clements

5 Cellular Model Systems to Study the Tumor Biological Role of Kallikrein-related Peptidases in Ovarian and Prostate Cancer

5.1 Introduction

Since the identification of the gene family of kallikrein-related peptidases (KLKs), their function has been robustly studied at the biochemical level. *In vitro* biochemical studies have shown that KLK proteases are involved in a number of extracellular processes that initiate intracellular signaling pathways by hydrolysis, as reviewed in Chapters 8, 9, and 15, Volume 1. These events have been associated with more invasive phenotypes of ovarian, prostate, and other cancers. Concomitantly, aberrant expression of KLKs has been associated with poor prognosis of patients with ovarian and prostate cancer (Borgoño and Diamandis, 2004; Clements *et al.*, 2004; Yousef and Diamandis, 2009), with prostate-specific antigen (PSA, KLK3) being a long standing, clinically employed biomarker for prostate cancer (Lilja *et al.*, 2008). Data generated from patient samples in clinical studies, along with biochemical activity, suggests that KLKs function in the development and progression of these diseases. To bridge the gap between their function at the molecular level and the clinical need for efficacious treatment and prognostic biomarkers, functional assessment at the *in vitro* cellular level, using various culture models, is increasing, particularly in a three-dimensional (3D) context (Abbott, 2003; Bissell and Radisky, 2001; Pampaloni *et al.*, 2007; Yamada and Cukierman, 2007).

5.2 Development of cellular model systems in cancer research

To date, the cellular model most widely used in cancer studies has been to culture cell lines established from primary and metastatic tumors on plastic surfaces of flasks or dishes as two-dimensional (2D) monolayers. 2D cell culture has the advantage of relatively low cost, renewable cell sources, retention of the original genomic features of the tumor, and it is amenable to genetic manipulation. Therefore, this methodology has been applied to maintain cell lines and perform functional assays of cell morphology, migration, invasion, and proliferation, with or without selected treatments. The 2D culture model, using a limited number of cell lines, has made invaluable contributions to our understanding of cancer biology, including cell signaling pathways, phenotypic plasticity and the effects of therapeutic agents. Using these 2D models, incremental understanding of factors in the host microenvironment, such as extracellular matrix (ECM) proteins, led to culturing cancer cells on top of coated type I and IV collagens, fibronectin, and vitronectin (Liotta *et al.*, 1980). In the last decade, growing awareness of the role of the tumor microenvironment, and the 3D aspects of tumors in their development and metastasis, has promoted our efforts to precisely model these features *in vitro* (Abbott, 2003; Bissell and Radisky, 2001; Cukierman *et al.*, 2002; Fischbach *et al.*, 2007). Incorporating microenvironmental factors into this dimension, 3D cellular model systems have included the use of natural matrices containing ECM proteins, in order to mimic growth of solid tumors (Ghajar and Bissell, 2010; Griffith and Swartz, 2006; Kim, 2005; Nelson and Bissell, 2006; Pampaloni *et al.*, 2007; Yamada and Cukierman, 2007). Additionally, cancer cells cultured in 3D-suspension as multicellular aggregates (MCAs) or spheroids have been applied to represent *in vivo* tumor growth where cancer cells are free-floating, for example, in the ascites (Friedrich *et al.*, 2009; Sutherland *et al.*, 1977). To mimic the stromal cellular component in the tumor and metastatic microenvironments, fibroblast, mesothelial, mesenchymal, endothelial, and inflammatory cells, as well as adipocytes, pre-osteoblasts, osteoblasts, and osteoclasts have been included, in order to explore their interaction with cancer cells (Barrett *et al.*, 2005; Hsiao *et al.*, 2009; Kaneko *et al.*, 2010; Schutyser *et al.*, 2002; Wong *et al.*, 2009). Furthermore, recent advances in research on biomimetic materials have led to the application of synthetic matrices, such as polyethylene glycol (PEG)-based hydrogels (Hutmacher, 2010; Prestwich, 2007) or poly(lactide-co-glycolide) (PLG) scaffolds (Fischbach *et al.*, 2007). These synthetic and biomimetic materials offer high flexibility for the design of new culture systems by adding known factors of interest to precisely mimic a physiologically relevant microenvironment (Hutmacher, 2010). In this chapter, we summarize the cell culture models used to mimic the *in vivo* microenvironment, and the application of these platforms to the study of ovarian and prostate cancer cell biology, with particular relevance to our understanding of the roles of KLK proteases in these cancers.

5.3 Traditional 3D cellular models commonly used in both ovarian and prostate cancers

5.3.1 Soft agar colony assay

More than 40 years ago, a soft agar assay was initiated as a cell culture method (Robb, 1970) for cultures of single cells in an agarose matrix, to assess anchorage-independent growth as an indicator of tumorigenicity. Single cells typically form colonies containing cells in a 3D arrangement. Since then, it has been widely used to evaluate tumor cell proliferation, tumorigenicity, and chemosensitivity in both ovarian and prostate cancer cells (Agre and Williams, 1983; Alberts *et al.*, 1980; Bertelsen *et al.*,

1984; Hamburger, 1987). Soft agar assays have been used to study the anchorage-independent growth of prostate cancer cells. For example, overexpression of KLK4 in prostate cancer PC-3 and DU145 cells significantly increased their colony formation when using this assay (Klokk *et al.*, 2007). Similarly, KLK7 overexpressing 22RV3 and DU145 cells formed significantly larger and more colonies than their vector/native control cells (Mo *et al.*, 2010), suggesting a functional role of these KLKs in enhancing prostate cancer growth. However, a lack of molecular components of the host tumor microenvironment in this model led to the application of matrices containing stromal ECM proteins.

5.3.2 3D-Matrigel™

A basement membrane-type matrix, derived from the Engelbreth-Holm-Swarm mouse tumor, has been used for nearly 3 decades, under the name Matrigel™ (BD Biosciences) and Cultrex® (R&D Systems) (Bissell *et al.*, 1982; Chu *et al.*, 1993). Early studies revealed that Matrigel™ is composed of ECM components found in the tumor stroma, namely laminin, collagens, fibronectin, and heparan sulfate proteoglycan (Hedman *et al.*, 1982; Kleinman *et al.*, 1986; Schubert and LaCorbiere, 1982). It has been used to mimic the *in vivo* tumor microenvironment, investigating cancer developmental and metastatic processes, such as angiogenesis, tumor growth, morphogenesis, chemo-response and, especially, invasion (Auerbach *et al.*, 2003; Harma *et al.*, 2010; Sweeney *et al.*, 1991; Yamamura *et al.*, 1993). Matrigel™ transwell assays have often been used to determine the capacity of tumor cell invasion, including the role of KLK2-4 in prostate cancer cells (Veveris-Lowe *et al.*, 2005). Simultaneous overexpression of KLK4-7 in ovarian cancer OV-MZ-6 cells increased invasion in the *in vitro* Matrigel™ assay, which was reflected in enhanced tumor growth in an *in vivo* mouse model (Prezas *et al.*, 2006). KLK7 transfection induced an epithelial-mesenchymal transition and an invasive phenotype in 22RV3 and DU145 prostate cancer cells using this *in vitro* Matrigel™ assay (Mo *et al.*, 2010). In addition, growing cells in Matrigel™ has also been used to study tumor architecture in association with cell function in cancers. For example, breast acini formed by benign mammalian epithelial cells in Matrigel™ can produce the milk component casein, but cancerous cells that grow as depolarized spheroids lack such a function, as reported by Bissell's group (Petersen *et al.*, 1992). Similarly, microarray gene expression analysis of a panel of 29 normal and cancerous prostate cancer cell lines, cultured in 3D Matrigel™ and 2D monolayers, also revealed biological and genetic features that correlate with changes to cell growth, morphogenesis, and invasiveness (Harma *et al.*, 2010). Particularly, PI3-Kinase, AKT, STAT/interferon, and integrin signaling pathways were activated in invasive cells, and specific small molecular inhibitors targeted against PI-3Kinase blocked invasive cell growth more effectively in 3D compared to 2D monolayer cultures (Harma *et al.*, 2010). To determine the specific functions of KLK proteases in different cancers, more defined *in vitro* models for the unique metastasis routes of ovarian and prostate cancer have now been generated.

5.4 Novel 3D cellular models in ovarian cancer biology

5.4.1 Ovarian cancer

Ovarian cancer is the leading cause of patient death in gynecologic tumors (AIHW, 2010; Cancer-Research-UK., 2011; Jemal *et al.*, 2011), with ~75% of patients diagnosed when tumors have already spread into the abdominal cavity (Goodman *et al.*, 2003). At this late stage, even advanced cytoreductive surgery is not curative and these patients are commonly given cytotoxic chemotherapy. Unfortunately, most patients initially respond to chemotherapy, but the tumor eventually becomes resistant. As a result, the prognosis is poor, with less than 30% survival for 5 years after diagnosis (Berrino *et al.*, 2007). A clinical feature of ovarian cancer is that more than 70% of patients show presence of a pool of fluid in their abdominal cavity, namely ascitic fluid, or effusions, which harbors a population of tumor cells (Puls *et al.*, 1996). Thus, ovarian cancer metastasis occurs predominantly within the peritoneal cavity. Due to the lack of a capsule around the ovary and its exposure to the peritoneal space, ovarian cancer cells are shed from the primary site into the abdominal cavity and form MCAs or spheroids. The floating MCAs adhere to mesothelial cells on the surface of the peritoneum, invade into the underlying ECM, and grow as secondary tumors (**Fig. 5.1a**). One key step in this tumor progression is that ovarian cancer cells survive in the 3D-suspension ascites microenvironment, followed by the crucial events of adhering to and invading into the peritoneal membrane, which is a 3D solid matrix. Therefore, a better understanding of the biology of ovarian cancer and development of novel approaches to help identify effective treatments is needed, in order to improve patient outcomes (Bast *et al.*, 2009; Cannistra, 2004). Thus, the use of *in vitro* cellular model systems that faithfully mimic the *in vivo* microenvironment may lead to more efficacious strategies towards prolonging the survival time of these patients. Aberrant expression of the KLKs, suggesting their roles in this cancer, has been reported previously. In this section, we summarize 3D cellular model systems, used most recently in ovarian cancer biology, with an emphasis on those used to assess the cellular function of KLKs (**Fig. 5.1, Tab. 5.1**).

5.4.2 3D-suspension model to mimic ascites suspension

3D-suspension models have been applied in various malignancies, in particular for ovarian cancer (Casey *et al.*, 2001; Frankel *et al.*, 2000; Moss *et al.*, 2009a; Shield *et al.*, 2007), to study tumorigenesis and cellular responses to different treatments *in vitro*.

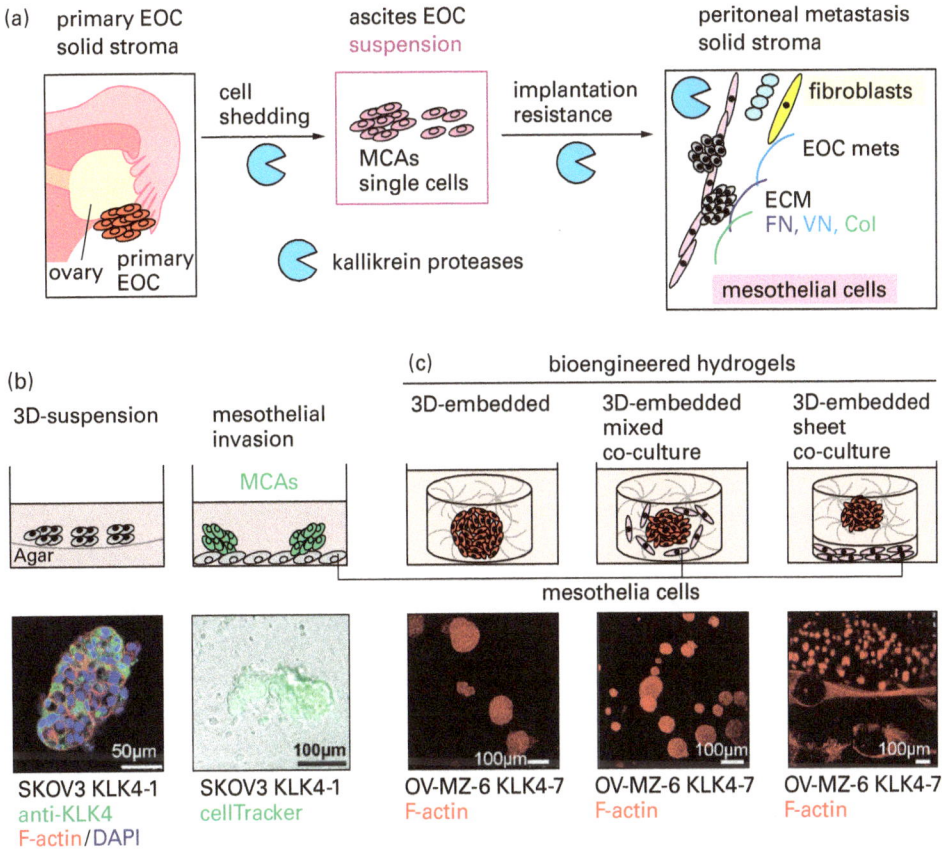

Fig. 5.1 Cellular models to mimic ovarian cancer metastasis. (a) Schematic diagram of ovarian cancer peritoneal dissemination and potential roles of KLKs. **(b)** Top panel, 3D-suspension cultures to mimic ovarian cancer cells growing in the ascitic fluid microenvironment, as well as invasion of mesothelial cell monolayers to mimic peritoneal dissemination. Bottom panel, left, MCAs formed by KLK4-expressing SKOV3 clone 1 (KLK4-1) stained with anti-V5 and AlexaFluor488 (KLK4, green), F-actin stained with Alexa Fluor 568 phalloidin (red) and 4′-6-diamidino-2-phenylindole (DAPI) stained nuclei (blue); right, invasion into mesothelial cell monolayer by KLK4-1 MCA labeled with CellTracker[492] (green). **(c)** Top panel, schematic diagram showing ovarian cancer cells embedded and cultured in 3D-bioengineered hydrogels, alone (left) and direct (middle), and sheet (right) co-culture with mesothelial cells. Bottom panel: KLK4-7 simultaneous overexpressing OV-MZ-6 cells grown in the above models. F-actin filaments stained red using rhodamine-415 conjugated phalloidin and imaged by confocal laser scanning microscopy. Scale bars as indicated.

These methodologies include cell cultures in non-adhesive plastic culture dishes. Cell-culture plasticware, treated with poly-2-hydroxyethyl methacrylate (polyHEMA) (Friedrich *et al.*, 2007), and stirred spinner flasks, such as rotating wall vessel bioreactors, in order to inhibit adhesion to the plastic and to maintain the cells in suspension (Vamvakidou *et al.*, 2007). However, historically, the approach most often used is an overlay or hanging-drop method, in which cell suspensions are seeded in culture

Tab. 5.1 Established cellular models to study ovarian and prostate cancer

Model	Method	Application in ovarian cancer	Application in prostate cancer
Agarose	Colony formation by single cells cultured in agarose (Robb, 1970)	Anchorage-independent growth, for chemosensitivity testing (Alberts *et al.*, 1980)	Tumor growth and survival, to assess tumorigenicity, KLK4 (Klokk *et al.*, 2007); KLK7 (Mo *et al.*, 2010)
Matrigel™	Invasion: single cells on top of Matrigel™ in the upper chamber of a transwell with attractant in its lower chamber	Significantly increased invasion of simultaneously overexpressing KLK4-7 OV-MZ-6 cells (Prezas *et al.*, 2006)	KLK7 induced invasion in 22RV3 and DU145 cells (Mo *et al.*, 2010)
	Proliferation, growth: culturing cells in or on top of Matrigel		Morphogenesis, interaction of prostate cancer cells and ECM, gene regulation and response to treatment, angiogenesis (Auerbach *et al.*, 2003; Harma *et al.*, 2010)
Collagen	Culturing cells in or on top of collagen matrices	*In vitro* tumor growth, invasive and expansive (Moss *et al.*, 2009b), gene expression (Barbolina *et al.*, 2009a)	Molecular functions of cancer cells (Podgorski *et al.*, 2005)
Liquid-overlay suspension	Culturing cells on a non-adhesive surface to enforce cell growth in suspension	Study of the functions of integrins (Casey *et al.*, 2001; Shield *et al.*, 2007), proteases MMP-14 (Barbolina *et al.*, 2007) and KLKs (Dong *et al.*, 2010); cellular response to chemotherapy (Frankel *et al.*, 2000)	Tumorigenesis, cell cycle analysis, cellular response to treatment (Frankel *et al.*, 2000)
Organotypic models	Culturing fibroblasts in collagen, mesothelial cells on top to constitute the peritoneum; culturing ovarian cancer cells on top of mesothelial cells	Mimicking the peritoneal invasion of ovarian cancer, and the role of MMP-2, ECM proteins, and integrins (Kenny *et al.*, 2008; Kenny *et al.*, 2007)	Not yet established
Synthetic matrices	Culturing cells within synthetic matrices, such as PEG (Hutmacher, 2010) and PLG scaffold (Fischbach *et al.*, 2007)	Mimicking the growth of ovarian cancer cells in the stroma, roles of KLKs and the interaction of ovarian cancer and stromal cells (Loessner *et al.*, 2010)	Mimicking the interaction of prostate cancer and stromal cells: in progress

wells coated with a thin layer of agarose (Casey *et al.*, 2001; Friedrich *et al.*, 2009; Shield *et al.*, 2007; Sutherland *et al.*, 1977) (**Fig. 5.1b**). Under this enforced 3D-suspension culture condition, with carefully adjusted amounts of media, single cells do not adhere to the surface of the culture wells or flasks. Instead, they can adhere to each other and form unpolarized MCAs that mimic ovarian cancer spheroids floating in the 3D-suspension ascites microenvironment.

It is believed that the formation of homotypic MCAs aids cell survival in the 3D-suspension microenvironment (Pickl and Ries, 2009; Sodek *et al.*, 2009). Consequently, heterogeneous cellular phenotypes in MCAs have an impact on their behavior and response to treatment (Durand and Olive, 2001; Frankel *et al.*, 1997). Recent studies indicated that compacted ovarian cancer MCAs are more invasive, because the contracting force drives peritoneal mesothelial cells apart, thus aiding invasion and dissemination *via* myosin (Iwanicki *et al.*, 2011) and integrins (Sodek *et al.*, 2009). Early studies revealed that α5β1 (Casey *et al.*, 2001) and α2β1 (Shield *et al.*, 2007) integrins mediate spheroid formation by ovarian cancer cells in the 3D-suspension microenvironment. Consistent with this observation, we recently found that in 3D-suspension (**Fig. 5.2a**) KLK4-overexpressing ovarian cancer SKOV3 cells (KLK4-1) formed compact MCAs, compared to the vector control cells (**Fig. 5.2b**). We also found increased α5/β1 integrins in both KLK4-overexpressing and vector/native control SKOV-3 cells cultured in 3D-suspension, independent of KLK4 expression, compared to those in 2D monolayers (**Fig. 5.2c**), which suggests that KLK4 was not involved in the activation of the integrin pathway in this model. Interestingly, KLK7 likely is associated with integrin pathway activation in 3D-suspension, as KLK7 transfection of SKOV3 cells increased α5β1 integrin (Dong *et al.*, 2010).

We also showed that ovarian cancer cells derived from the ascites expressed higher levels of KLK7 and α5β1 integrin than their matched primary tumor cells, in a sub-group of patients. Lipocalin 2 (LCN2), which is an early marker of epithelial phenotypic induction and which is associated with chemoresistance (Li *et al.*, 2009; Lim *et al.*, 2007), also in ovarian cancer (Stewart *et al.*, 2006), is induced in 3D-suspension in KLK4 transfected SKOV3 cells (**Fig. 5.2c**). Notably, this induction is more pronounced in the KLK4 transfected cells, especially in 2D monolayers, and suggests that KLK4 induces mesenchymal-epithelial cell plasticity in the mesenchymal SKOV3 cell line. Additionally, Stack's group showed that matrix metalloprotease 14 (MMP-14) or membrane type 1 metalloprotease (MT1-MMP) regulate both MCA formation in 3D-suspension and disaggregation of spheroids within the ECM, indicating a role of these proteases in ovarian cancer metastasis (Moss *et al.*, 2009a). Our group has shown that KLK4- and KLK7-overexpressing SKOV3 cells formed compact MCAs that are more resistant to the first-line chemotherapeutic, paclitaxel, than those formed by vector/native control cells (Dong *et al.*, 2010; Dong *et al.*, unpublished observation). Of clinical relevance is the finding that those patients with high *KLK4* and *KLK7* levels in their ovarian tumors were not chemo-responsive (Dong *et al.*, 2010; Xi *et al.*, 2004). Together, this data indicates that this 3D-suspension model can,

Fig. 5.2 Differential expression of α5/β1 integrins, lipocalin 2 (LCN2), MMP-14, and caspase 8 in ▶
KLK-expressing ovarian cancer cells in 2D and 3D. (a) Schematic diagram showing 3D-suspension
cultures of ovarian cancer cells. **(b)** Phase contrast image showing the KLK4-expressing (KLK4-1) and
vector-1 SKOV3 cells cultured in 3D-suspension. **(c)** Western blot analysis showing expression of α5
and β1 integrin subunits, and LCN2 in KLK4-1, KLK4-2, vector-1, and native SKOV3 control cells cultured as 2D-monolayers and in 3D-suspension on day 1, 4, 7, and 14, as indicated. *LCN2, consistent
increase in LCN2 in KLK4-1 and KLK4-2 clones, compared to vector and native control cells. **(d)** Confocal laser scanning microscopy imaging of 2D monolayer (top panel: green = α-tubulin), showing
increased tubulin disruption in OV-KLK4-7 cells and increased caspase 8 expression in spheroids, in
3D-embedded cultures (bottom panel: green = caspase 8) on paclitaxel treatment. F-actin filaments
stained red using rhodamine 415-conjugated phalloidin and nuclei in blue with DAPI. Scale bars:
25 µm. **(e)** Expression of α5/β1 integrins in 2D and 3D cultures of OV-KLK4-7 cells compared to
OV-vector cells measured by quantitative real time PCR, with or without paclitaxel treatment
(10 nmol/L). Relative expression shown as ratio of respective integrin to 18S (n = 3; mean ± SEM;
* – $P < 0.05$; ** – $P < 0.01$; *** – $P < 0.001$). **(f)** Western blot analysis showing expression of MMP-14
on OV-vector and OV-KLK4-7 cells in 3D hydrogel-based cultures, but not in 2D-monolayers with or
without paclitaxel administration.

at least partially, recapitulate the *in vivo* ascites microenvironment seen in ovarian
cancer patients.

5.4.3 In vitro models for ovarian cancer invasion into the peritoneal membrane

Inside the abdominal and pelvic cavity, internal organs are covered by a layer of mesothelial cells and an underlying thin layer of connective ECM-containing tissues, called
the peritoneum. As the peritoneum is the major site of metastasis, effort has been made
to mimic this invasion by using mesothelial cells that line the peritoneal membrane
(**Fig. 5.1b**). A recent study revealed that myosin is the key factor for ovarian cancer cell
MCAs to invade into the peritoneum (Iwanicki *et al.*, 2011). It is known that collagens,
fibronectin, vitronectin, and laminin are the major ECM components of the peritoneal membrane (Davies *et al.*, 2004; Witz *et al.*, 2001). Skubitz's group reported that
blocking α5β1 integrin function reduced adhesion of ovarian cancer cells to fibronectin (Casey *et al.*, 2001). It was also observed that ascites-derived ovarian cancer spheroids prefer adhesion to fibronectin, type I and IV collagens, and laminin (Burleson
et al., 2004a). Ovarian cancer cells and those from patient's ascites were cultured on
top of peritoneal mesothelial cells, showing their invasion into the mesothelial monolayer, which was mediated by integrins (Burleson *et al.*, 2004b). Shield *et al.* (2007)
showed that α2β1 integrin plays a role in the attachment of ovarian cancer cells to type
I collagen and the disaggregation of spheroids. Using a similar model (as shown in
Fig. 5.1b), our group reported that spheroids formed by KLK7-overexpressing SKOV3
cells with high α5β1 integrin expression invade into peritoneal mesothelial cell monolayers and form cancer cell foci (Dong *et al.*, 2010). These studies demonstrated the
ability of ovarian cancer cells and of those from patient ascites, mediated by integ-

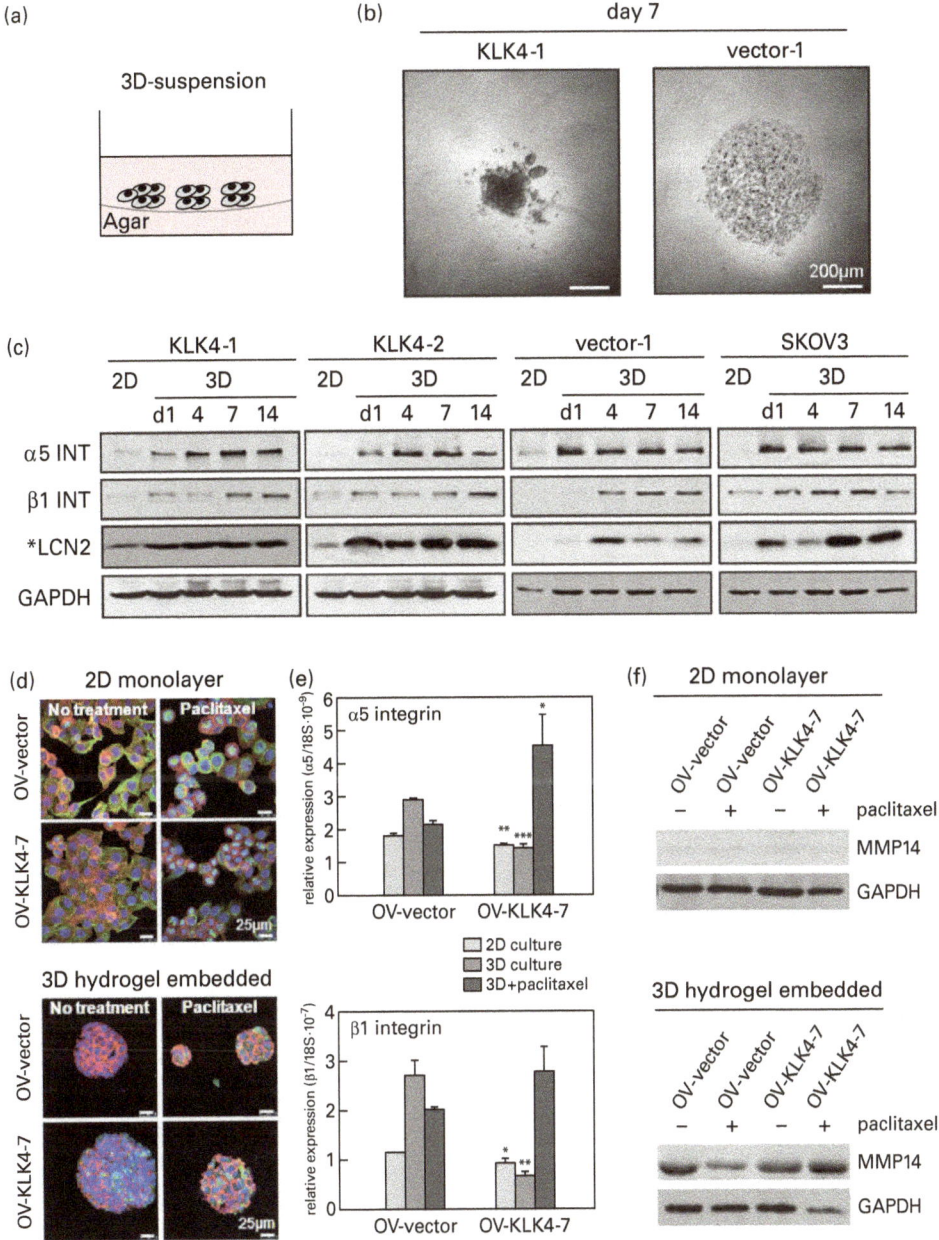

(a)

3D-suspension

Agar

(b) day 7

KLK4-1 vector-1

200μm

(c)

| | KLK4-1 | KLK4-2 | vector-1 | SKOV3 |

α5 INT

β1 INT

*LCN2

GAPDH

(d) 2D monolayer

OV-vector

OV-KLK4-7

No treatment   Paclitaxel

25μm

3D hydrogel embedded

OV-vector

OV-KLK4-7

No treatment   Paclitaxel

25μm

(e)

α5 integrin

relative expression (α5/18S·10⁻⁹)

OV-vector   OV-KLK4-7

☐ 2D culture
◾ 3D culture
◾ 3D+paclitaxel

β1 integrin

relative expression (β1/18S·10⁻⁷)

OV-vector   OV-KLK4-7

(f) 2D monolayer

OV-vector OV-vector OV-KLK4-7 OV-KLK4-7

−   +   −   +   paclitaxel

MMP14

GAPDH

3D hydrogel embedded

OV-vector OV-vector OV-KLK4-7 OV-KLK4-7

−   +   −   +   paclitaxel

MMP14

GAPDH

rins and KLK proteases, to adhere to and invade into the mesothelial monolayer, thus mimicking crucial steps in peritoneal metastasis.

5.4.4 The role of 3D-collagen I matrix in ovarian cancer cell behavior

The ECM protein components in Matrigel™ were found in the submesothelial matrix of the peritoneal membrane (Witz *et al.*, 2001) and in the ovarian tumor stroma (Davies *et al.*, 2004). A differential gene expression pattern was induced by Matrigel™ and its matrix component collagen (Bissell and Barcellos-Hoff, 1987; Schuetz *et al.*, 1988). Subsequently, individual components of these matrices have been examined with respect to their roles in different aspects of cancer progression. For example, 3D collagen I provided a stromal barrier for MMP-mediated events in peritoneal dissemination related to ovarian cancer (Sodek *et al.*, 2008). Stack's group reported that the 3D collagen I matrix, which is a major component of the ovarian tumor stroma, has an impact on both invasive and expansive growth of ovarian cancer cells (Moss *et al.*, 2009b). Ovarian cancer cells grown in a 3D collagen I microenvironment showed high expression of the transcriptional factor, early growth response protein (EGR1), and subsequently MMP-14, compared to those cells cultured on a collagen-I-coated surface (Barbolina *et al.*, 2007). Microarray analysis of ovarian cancer cells cultured within the 3D collagen I matrix revealed that the connective tissue growth factor (CTGF) was downregulated, while overexpression of this factor retarded 3D collagen invasion (Barbolina *et al.*, 2009a). These studies provided insights into molecular mechanisms underlying ovarian cancer cell invasion that are influenced by this major stromal component.

5.4.5 A 3D-organotypic model to mimic ovarian cancer metastasis

The omentum, which comprises both the mesothelial layer and underlying matrix, is the most frequent metastatic site for ovarian cancer. To study the mechanisms of peritoneal dissemination, Lengyel's group first established an *in vitro* model in order to mimic the microenvironment of the omentum (Kenny *et al.*, 2007). To establish the major cellular component of the omentum, they collected primary peritoneal mesothelial and fibroblast cells from patient omentum. To mimic the stromal microenvironment of the omentum, isolated fibroblasts were embedded in the major protein component of the ECM type I collagen, with the peritoneal mesothelial cells seeded on top. Using this 3D-organotypic model, differential roles of mesothelial cells, omental fibroblasts, and ECM in the attachment and invasion of ovarian cancer cells were reported. Subsequently, MMP-2 enhanced adhesion of ovarian cancer cells to the peritoneal membrane, by cleavage of the ECM proteins vitronectin and fibronectin into small fragments, which increased binding to ovarian cancer cells *via* their receptors, α5β1 and αvβ3 integrins. In fact, enhanced adhesion was abrogated by blocking antibodies to these integrins and siRNA knock-down approaches (Kenny *et al.*, 2008; Kenny and Lengyel, 2009). This would be an excellent model for further exploration of KLK involvement in peritoneal dissemination.

5.4.6 Bioengineered 3D culture systems for ovarian cancer

Advances in material sciences and bioengineering, combined with a growing body of knowledge in cancer cell biology and progress in understanding crucial mechanisms that govern interactions between cells and their extracellular microenvironments, have revolutionized the development of biomaterials. Consequently, criteria for the design of biomaterials have expanded from specifically structural and mechanical towards bio-functional aspects, also favoring biologically active and bi-directional cell-material interactions. Novel families of synthetic hydrogels have been developed that mimic key features of the natural ECM. In particular, cell-mediated proteolytic degradability and display of cell-binding sites found exclusively in naturally-derived materials were translated into synthetic materials. We have been using a bioengi-neered hydrogel platform and have shown its potential as an *in vitro* 3D cell culture model, culturing cancer cells alone or with peritoneal mesothelial cells (**Fig. 5.1c**), in order to study ovarian cancer biology (Hutmacher, 2010; Loessner *et al.*, 2010). Within this platform, synthetic protease-sensitive hydrogels are formed from peptide-functionalized, multi-arm polyethylene glycol (PEG) macromolecules *via* a factor XIII-catalyzed reaction, similar to fibrinogen cross-linking during natural fibrin coagula-tion. Simultaneously, by means of the same reaction, bio-functional peptides, *e.g.* the Arg-Gly-Asp (RGD) integrin cell attachment motif, and other proteins, can be stably incorporated into the hydrogel network (Rizzi *et al.*, 2006; Rizzi and Hubbell, 2005).

KLKs likely initiate peritoneal invasion, which has been evidenced by our pre-vious report, which demonstrates that simultaneous overexpression of KLK4-7 in ovarian cancer OV-MZ-6 cells increased invasion in the Matrigel™ assay and in an *in vivo* mouse xenograft model (Prezas *et al.*, 2006). In accordance with this published data, we showed dramatic differences in cell behavior when these cells are embed-ded in 3D matrices (**Fig. 5.1c**; **Fig. 5.2d**), compared to 2D monolayers. Spheroid forma-tion of ovarian cancer cells was observed exclusively in 3D within biomimetic PEG hydrogels (**Fig. 5.2d**). It is of interest, that KLK4-7-overexpressing OV-MZ-6 cells grew into larger spheroids, compared to vector control cells in 3D, particularly within RGD-functionalized hydrogels (**Fig. 5.2d**). We observed reduced α5/β1 integrins in both 2D monolayers and 3D hydrogel-cultured, KLK-overexpressing OV-MZ-6 cells compared to vector control cells (**Fig. 5.2e**). Interestingly, an increase in α5β1 integrin expression was detected in 3D hydrogel-cultured, KLK-transfected OV-MZ-6 cells following pacli-taxel treatment but not in vector control cells, although increased caspase 8 expres-sion showed that both cell types are responsive to paclitaxel (**Fig. 2.2d**). At the protein level, western blot analysis showed the presence of MMP-14 in cell lysates from 3D hydrogel cultures, but not in 2D monolayers, independent of paclitaxel administration (**Fig. 5.2f**). This is in line with a previous report showing MMP-14 expression exclusively in ovarian cancer cells cultured in a 3D collagen I matrix rather than as 2D monolayers (Barbolina *et al.*, 2009b).

To further expand the utility of these bioengineered models, we have established an integrated 3D co-culture model of epithelial (ovarian cancer) and mesothelial (stromal) cells, replicating tumor-stroma interactions (**Fig. 5.1a, c**, right hand panels). Cancer spheroids were grown in the PEG-based hydrogels noted above, which comprise ECM features due to the incorporation of MMP cleavage sites and RGD integrin-binding motifs (Rizzi *et al.*, 2006; Rizzi and Hubbell, 2005). In the direct 3D co-culture of ovarian cancer spheroids in PEG-based hydrogels with mesothelial cells, an increase in spheroid size and proliferation was noted for both KLK4-7-overexpressing OV-MZ-6 cells and vector control cells, compared to 3D mono-cultures of these cells (**Fig. 5.3a, b**). A further advance currently under development is to embed the cancer cells in the hydrogel above a sheet of mesothelial cells, either embedded in hydrogel or supported on a mesh (**Fig. 5.1c**, far right panel). This integrated model allows the crosstalk between these cells to be more easily defined, given that cells reside in distinct matrices, which can be separated for analysis. Our findings provide additional evidence that biomimetic engineered matrices can be used as *in vitro* 3D platforms in cancer research. The flexibility of the physical and chemical properties of these

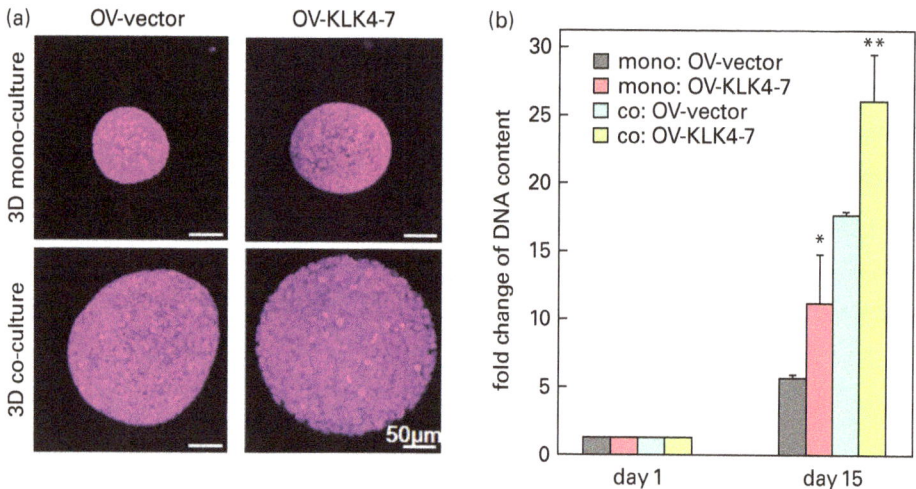

Fig. 5.3 **Enhanced spheroid formation upon KLK expression and co-culture of ovarian cancer with mesothelial cells in bioengineered hydrogels. (a)** Larger spheroids were formed upon direct co-culture of epithelial ovarian cancer (OV-vector and OV-KLK4-7) with mesothelial cells, as depicted by DAPI staining. Scale bars: 50 μm. **(b)** Enhanced proliferation, measured by CyQuant assays, was detected upon KLK4-7 expression (OV-KLK4-7), compared to vector controls (OV-vector) in 3D mono-cultures after 15 days, and 3D co-cultures of ovarian cancer (OV-vector and OV-KLK4-7) with mesothelial cells. Indicated as fold change of DNA content (n = 3; mean ± SEM; * – $P<0.05$; ** – $P<0.01$).

bioengineered platforms reveals their unique potential for satisfying the increasing demand for more versatile and physiologically relevant 3D cell culture models, used to assess specific aspects of cancer cell biology.

In summary, the data generated by using these cellular models, which more faithfully mimic the host microenvironment, has laid the foundation for a greater understanding of the biology of ovarian cancer and the role played by the KLKs. The role of other stromal components, endothelial cells, inflammatory cells, and ECM proteins present in the ovary and the peritoneal membrane can be determined by incorporating them into these models. In this way, our increasing knowledge will help design efficacious treatment to help prolong the survival time for women afflicted with this type of cancer.

5.5 Cellular models in prostate cancer

5.5.1 Prostate cancer

Prostate cancer is the second most commonly diagnosed cancer in men after that of the lung (Jemal *et al.*, 2011). With current procedures, the 5-year survival rate is nearly 100% for patients with localized prostate cancer, but there is no effective treatment for those with advanced disease (Walsh *et al.*, 2007). Most prostate cancer patients with advanced disease experience painful bone metastasis, for which the effect of palliative treatment is limited to relieving symptoms and the prognosis is very poor. Prostate cancer is present in a 3D-context surrounded by local solid stroma at the primary site and in the microenvironment of bone metastatic lesions (**Fig. 5.4a**). Survival of tumor cells and their evolution, linked to the host microenvironment, is crucial for the success of the metastatic process (Josson *et al.*, 2010; Sung *et al.*, 2008). Reprogramming cell behavior and the different signaling pathways in the organ-specific microenvironment is key to these cancer cells' survival. In this respect, interactions between the cancer and neighboring prostate or bone stromal cells alter the malignant potential of tumor cells. Pro-tumorigenic factors contributed by prostate fibroblasts alone include growth factors (Kaminski *et al.*, 2006; Zhao and Peehl, 2009), angiogenic factors (Tuxhorn *et al.*, 2002; Zhao and Peehl, 2009), kinases (Wang *et al.*, 2010), cytokines (Paland *et al.*, 2009; Sung *et al.*, 2008; Zhao and Peehl, 2009), interferon-inducible molecules (Shou *et al.*, 2002), protease inhibitors (Madar *et al.*, 2009), cell surface glycoproteins (Vanpoucke *et al.*, 2007), reactive oxygen species, and structural components (Sung *et al.*, 2008). Therefore, cellular models that mimic these microenvironments are valuable tools in the context of these specific processes, particularly those related to metastasis. Since the introduction of the PSA test in the 1980s, we have seen a 25% reduction in mortality due to prostate cancer. Stage migration also means 80% of men now present with localized disease, which can be either small volume and clinically insignificant or high grade, clinically significant disease.

(a)

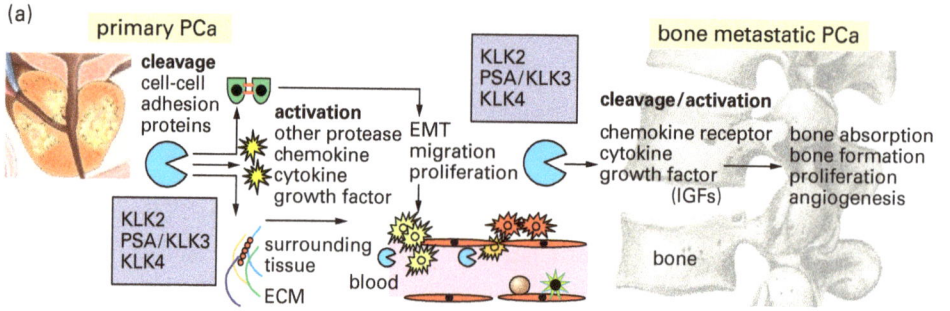

(b)

bioengineered hydrogels

3D-embedded   3D-embedded mixed co-culture   3D-embedded sheet co-culture

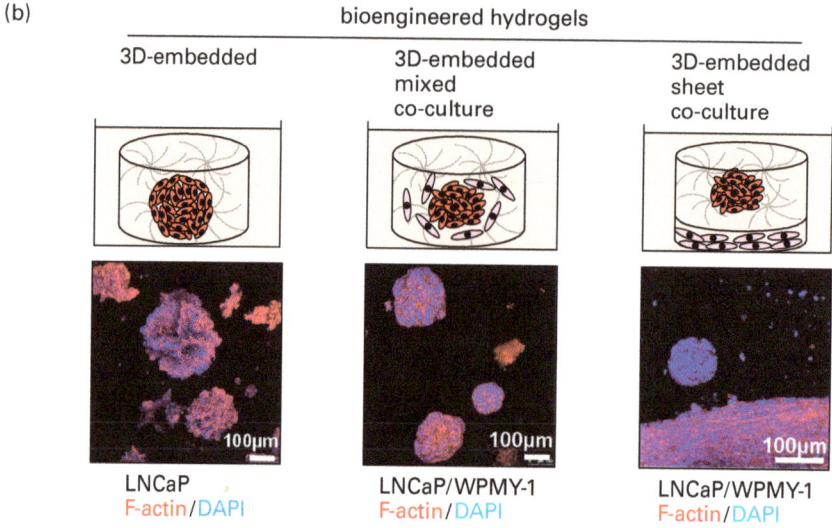

LNCaP
F-actin/DAPI   LNCaP/WPMY-1
F-actin/DAPI   LNCaP/WPMY-1
F-actin/DAPI

(c)

bone microenvironment mimicking models

PCa cells on OBM   PCa cells on TEB   PCa cells/hydrogel on TEB

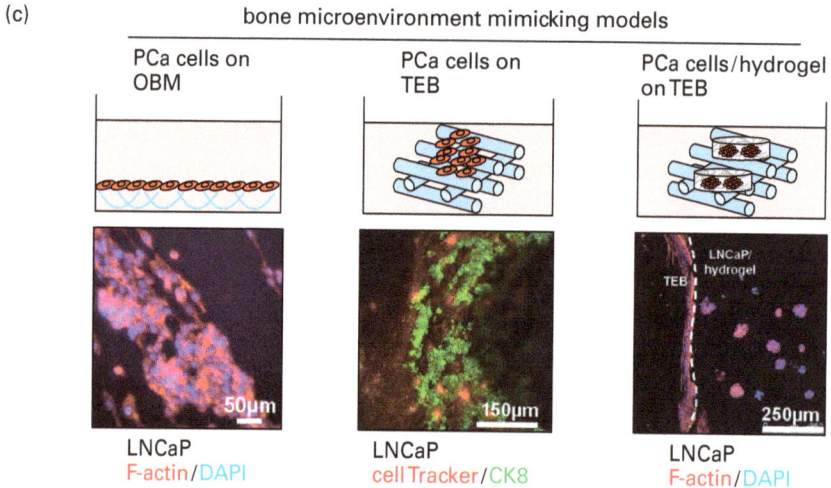

LNCaP
F-actin/DAPI   LNCaP
cell Tracker/CK8   LNCaP
F-actin/DAPI

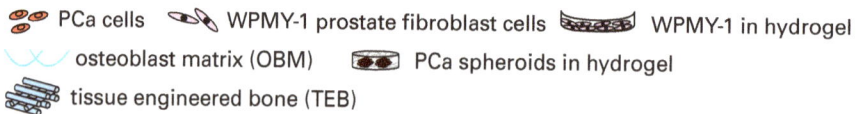

PCa cells   WPMY-1 prostate fibroblast cells   WPMY-1 in hydrogel
osteoblast matrix (OBM)   PCa spheroids in hydrogel
tissue engineered bone (TEB)

◄ Fig. 5.4 **Cellular models used by our group to mimic prostate cancer (PCa) development and metastasis (a)** Schematic diagram, showing the potential functions of KLKs in PCa metastasis. **(b)** Top panel, schematic diagram showing PCa cells cultured in 3D-bioengineered hydrogels, alone (left), direct (middle), and sheet (right) co-cultures with WPMY-1 prostate myofibroblast cells. Bottom panel, KLK2-4 expressing LNCaP cells cultured in the models in top panel stained with phalloidin for F-actin (red) and DAPI for nuclei. **(c)** Top panel, schematic diagram showing PCa cells on osteoblast matrix (OBM), tissue bioengineered bone (TEB), and PCa spheroids in hydrogel on top of TEB. Bottom panel, endogenous KLK2-4 expressing LNCaP cells cultured on OBM (left hand panel) and LNCaP spheroids in hydrogel on top of TEB (right hand panel) stained with phalloidin for F-actin (red) and DAPI for nuclei, and LNCaP cells growing on TEB stained with cytokeratin (CK8) (middle panel). Scale bars as indicated.

(Lilja *et al.*, 2008; Walsh *et al.*, 2007). Recent studies, using different cellular models (**Fig. 5.4b, c, Tab. 5.1**), have revealed the function of other prostate-related KLKs in the progression of this disease, as reviewed in Chapter 4 and elsewhere (Lawrence *et al.*, 2010; Sotiropoulou *et al.*, 2009).

5.5.2 3D-suspension models for prostate cancer growth and metastasis

As with ovarian cancer, a number of 3D prostate cancer models have utilized 3D-suspension cultures. Although prostate cancer cells are surrounded by solid matrices at both the primary and secondary tumor sites, tumor cells in the blood, lymph, and possibly in the bone marrow, are in suspension. Furthermore, suspension cultures are a relatively simple method of forcing tumor cells and stromal cells of various origins to form MCAs. Not unlike ovarian cancer, prostate cancer 3D suspension culture models include the liquid overlay method (Hedlund *et al.*, 1999), cells cultured in spinner flasks (Essand *et al.*, 1995), and rotating-wall vessels (RWVs) (Ingram *et al.*, 1997). To provide a more meaningful molecular analysis of prostate cancer progression and metastasis, Chung's lab established a 3D co-culture model using RWVs, and characterized the interaction between prostate cancer cells, prostate stromal cells, and bone stromal cells (Zhau *et al.*, 1997). This RWV 3D co-culture model allowed reciprocal genetic and behavioral changes to be observed in the prostate cancer and bone stromal or prostate stromal cells. Although KLK function was not evaluated in this model, other, more conventional 2D approaches have been used to determine the involvement of KLKs in interactions of prostate cancer cells with bone cells. We have performed co-culture of endogenously KLK4-expressing LNCaP cells and KLK4-overexpressing PC3 cells, with conditioned medium from SaOs2 (an osteosarcoma line often used to replicate the bone metastatic condition) cells (Rodan *et al.*, 1987), which increased KLK4 levels in both cell lines, as well as alkaline phosphatase activity in SaOs2 cells (Gao *et al.*, 2007). KLK4-transfected PC3 cells also showed increased migration towards the SaOs2-conditioned medium and greater attachment to the bone-matrix proteins type I and IV collagens (Gao *et al.*, 2007). These studies, albeit in 2D culture, demonstrate that the KLKs are affected by a bone microenvironment and warrant further study in this respect.

5.5.3 *In vitro* models for prostate cancer angiogenesis

Taking other stromal components into account, the establishing of vasculature networks to mimic the important angiogenic process can be determined by culturing endothelial cells with the use of Matrigel™ assays (Bissell *et al.*, 1987). Here, a reduced vasculature formation by endothelial cells from patients with systemic sclerosis was related to a decrease in KLK12, but not in KLK9 or KLK11, using Matrigel™ assays (Giusti *et al.*, 2005). Increased angiogenesis and related factors are associated with poor prognosis for prostate cancer patients (Mucci *et al.*, 2009). Intriguingly, although elevated PSA (KLK3) is the biomarker for prostate cancer detection and progression, this enzyme displayed anti-angiogenic activity on human umbilical vein endothelial cells (HUVEC) in *in vitro* Matrigel™ assays (Chadha *et al.*, 2011; Fortier *et al.*, 1999). Further studies on the functions of KLKs should be carried out in different contexts, such as co-cultures of endothelial cells with site organ-specific stromal cells, as well as for MT-MMPs in endothelial cell and lung fibroblast 3D co-cultures (Ghajar *et al.*, 2010).

5.5.4 Bioengineered 3D culture systems for prostate cancer growth and metastasis

3D cell culture in solid matrices has been a long-standing *in vitro* cell culture technique in prostate cancer research. Prostate cancer cells form MCAs or glandular-like acini in protein matrices, including collagen type-1, fibrin, and laminin-rich gels, such as the commercially available Matrigel™ (Doillon *et al.*, 2004; Harma *et al.*, 2010). Modified, dye-quenched collagens have also been used to image prostate cancer cell-derived MMP-mediated invasion into the ECM (Jedeszko *et al.*, 2008; Sameni *et al.*, 2008). For a better-defined, controllable, and reproducible 3D matrix for prostate cancer cell cultures, the application of synthetic hydrogels allows for definition of biochemical and physical properties that are important to cellular function, such as porosity, mechanical rigidity, and fixed diffusion rates for incorporated factors, such as small molecule drugs (Lin and Anseth, 2009; Schmidt *et al.*, 2008; Thanos and Emerich, 2008). Using a similar biomimetic PEG-based hydrogel culture (Hutmacher, 2010) (**Fig. 5.4b**, left hand panel) as described above for ovarian cancer, we observed that LNCaP cells grew as irregular shaped spheroids over 28 days in 3D culture (**Fig. 5.5a**).

Interestingly, mRNA levels of *KLK2*, *KLK3*, and *KLK4* were still expressed at this time and were significantly upregulated upon R1881 (synthetic androgen) treatment for 48 hours, at day 26, compared to ethanol controls in 3D (**Fig. 5.5b**), indicating that LNCaP cells were still androgen-responsive, which is a measure of their functional viability. We also have successfully established 3D co-cultures of LNCaP cells

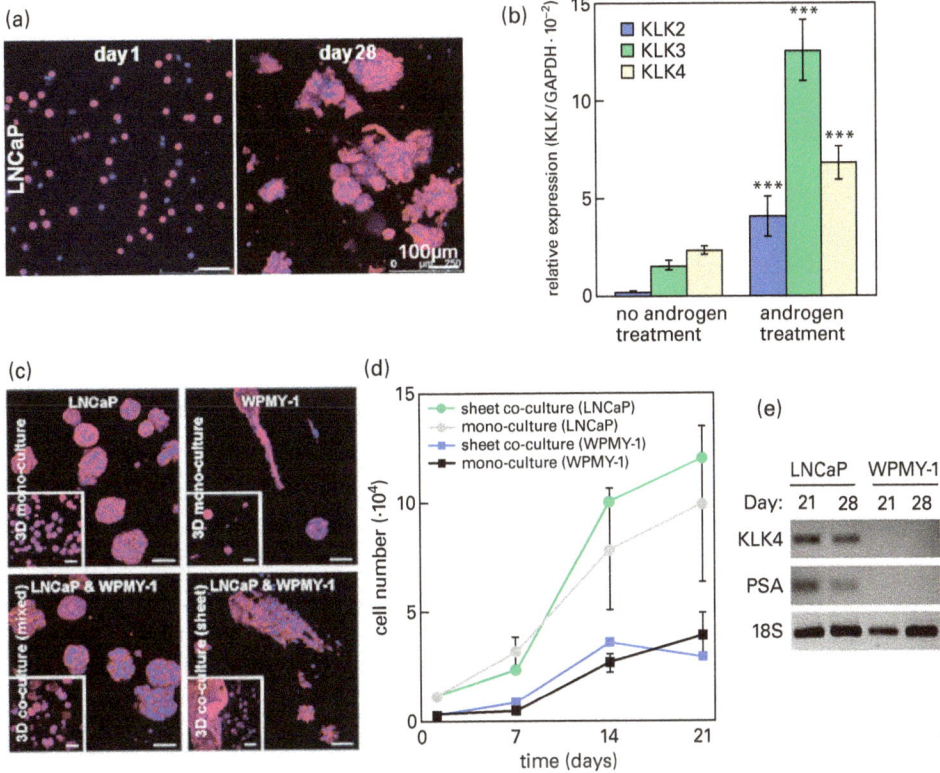

Fig. 5.5 **Spheroid formation of prostate cancer LNCaP cells, grown within hydrogels. (a)** Confocal laser scanning microscopy depicts the growth of LNCaP spheroids within hydrogels over 28 days. F-actin stained with rhodamine 415-conjugated phalloidin (red), nuclei stained with DAPI (blue). Scale bars: 100 μm. **(b)** Expression of *KLK2*, *KLK3*, and *KLK4* mRNA is significantly upregulated upon androgen treatment (1 nM R1881 for 48 hr beginning on day 26; 0.008% v/v ethanol treatment as control) in 3D hydrogel cultures, compared to non-treated controls, as measured by quantitative real-time PCR. Relative expression, shown as the ratio of respective *KLK* to *GAPDH* (n = 3; mean ± SEM; *** – *P* < 0.005). **(c)** Confocal laser scanning microscopy depicts the growth of LNCaP and WPMY-1 cells in hydrogels over 21 days, both alone and co-cultured, as indicated. F-actin filaments stained with rhodamine 415-conjugated phalloidin (red), nuclei stained with DAPI (blue). Scale bars: 100 μm (insets, 250 μm). **(d)** Cell number of LNCaP and WPMY-1 hydrogel mono-cultures or sheet co-cultures, measured over 21 days using AlamarBlue assays (n = 3, monoculture; n = 2, co-culture; mean ± SE). **(e)** RT-PCR showing expression of *KLK3/PSA* and *KLK4* in LNCaP and WPMY-1 cells cultured in hydrogels with *18S* as a loading control.

seeded directly with prostate-derived WPMY-1 myofibroblast cells in these hydrogels (embedded direct co-culture; **Fig. 5.4b**, middle panel), or with LNCaP and WPMY-1 cells seeded in separate hydrogels, which were subsequently joined, in order to facilitate indirect cellular interactions (called 'sheet co-culture', due to the morphology of WPMY-1 cells in these cultures) (**Fig. 5.4b**, right hand panel). Mono-cultures of each cell type in these 3D hydrogels served as controls (**Fig. 5.5c**). LNCaP cells appeared to exhibit increased proliferation upon sheet co-culture with WPMY-1 cells, compared to the LNCaP mono-cultures, while there was no obvious difference in cell number between WPMY-1 cells in co-culture versus mono-culture (**Fig. 5.5d**). *PSA/KLK3* and *KLK4* expression was retained after 28 days in the LNCaP cell mono-cultures. As expected, WPMY-1 cells did not express *PSA* or *KLK4* (**Fig. 5.5e**).

Since type I collagen is the major component of bone matrices, it has been used in many studies, including those of cathepsin B function, in the bone metastatic breast cancer environment, by Sloan's laboratory (Podgorski *et al.*, 2005). To better explore the role of the bone matrix on prostate cancer cells *ex vivo*, we established a bone matrix derived from human primary osteoblast cultures (OBM) (Reichert *et al.*, 2010) (**Fig. 5.4c**, left hand panel). These matrices closely resemble native bone matrix with regard to their structural features and their composition, as confirmed by scanning electron microscopy (SEM), mass spectroscopy, histochemistry, and immunofluorescence (Reichert *et al.*, 2010). As shown in **Fig. 5.6a, b,** this OBM has a dense network of fibrils, and is highly mineralized (Hutmacher *et al.*, 2009). Previously, we have studied interactions of PC3 and LNCaP prostate cancer cells within these matrices, and found that OBM has profound effects on cell morphology and gene expression (Reichert *et al.*, 2010). Herein, PC3 cells adopted an elongated spreading morphology along the matrix fibrils (**Fig. 5.6c**). Cell adhesion to the bone microenvironment is considered an important step in bone colonization of metastatic cancer cells (Schneider *et al.*, 2011). To characterize the adhesion strength of prostate cancer cells to OBM, we have employed a technique called Atomic Force Microscopy-based Single Cell Forces Spectroscopy (AFM-SCFS) (Helenius *et al.*, 2008; Zhang *et al.*, 2002) (**Fig. 5.6d**). AFM-SCFS enables us to detect forces ranging from 10 piconewtons (pN) up to several hundreds of nanonewtons (nN). This large force-range, together with precise control of the time during which the cell interacts with the adhesive substrate (from millisec-

Fig. 5.6 **Using an *in vitro* model to study prostate cancer cell interactions with the bone microenvi-** ►
ronment. (a) Scanning electron microscopy (SEM) image of the human primary osteoblast-derived bone matrix (OBM), revealing its fibrillar composition. **(b)** Alizarin red staining confirms the high extent of matrix mineralization. The inset shows a stained control matrix grown under non-osteogenic conditions. **(c)** PC3 cells, grown for five days on the matrix. F-actin and nuclei are stained using rhodamine-phalloidin/DAPI (red/blue). Fibronectin fibrils within the matrix are visualized by immunofluorescence labelling (green). The PC3 cells adopt an elongated, spreading morphology on the matrix. **(d)** Quantifying cell adhesion using atomic force microscopy-based single cell force spectroscopy (AFM-SCFS). A single living PC3 cell is attached to a concanavalin A-functionalized AFM cantilever. **(e)** Schematic illustration of the adhesion experiment. The PC3 cell attached to the AFM cantilever is brought into contact with OBM. During the approach, the force acting on the cantilever is recorded (grey force-distance (F–D) curve). After a preset contact period the cantilever is withdrawn and the cell is thereby separated from the bone matrix (red F–D curve). From the retraction curve (red), the adhesive interactions, established between the cell and OBM, can be quantified. Typically, the maximal force required to detach the cell (= detachment force, arrow) is used for the quantification of overall cell adhesion. **(f)** Cell detachment forces, represented as boxplots, showing the time-dependent build-up of PC3-WT and PC3-K3 (KLK3) cell adhesion on OBM over 120 sec. PC3-K3 cells establish stronger adhesion to OBM compared to PC3-WT cells.

onds to minutes), enables us to monitor cell adhesion from single molecule interactions to the formation of higher-order adhesion structures at the force level (Puech *et al.*, 2006; Taubenberger *et al.*, 2007). To quantify adhesion between PC3 cells and OBM, a single cell is gently attached to the force probe, which is the AFM cantilever. By precise movement of piezo elements, the cell is then brought into contact with OBM for a predefined contact time. Thereafter, the cantilever is retracted, which results in cell detachment. During the process described, the force acting on the AFM cantilever is measured, and a force-distance curve is plotted (**Fig. 5.6d–f**). From the

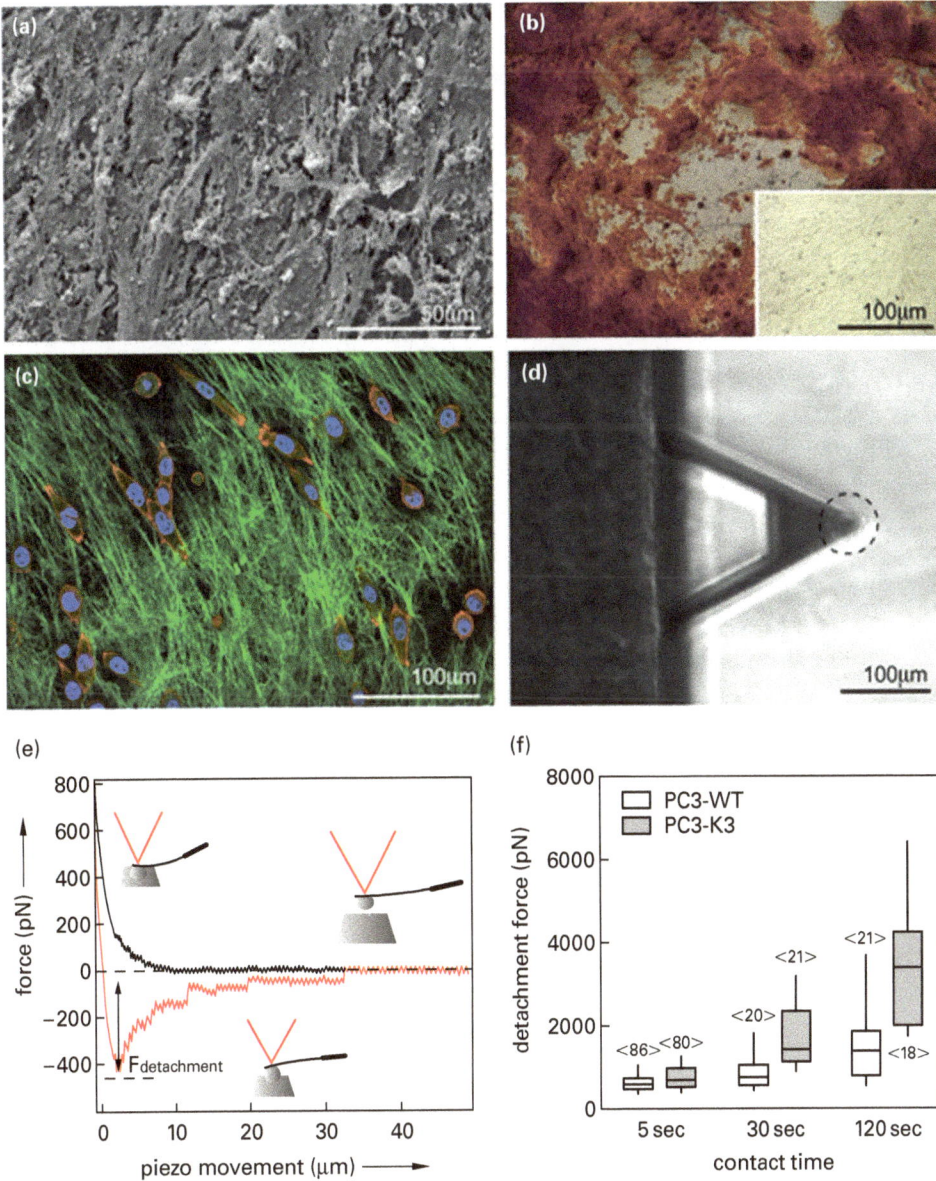

retraction part of the curve, the so-called detachment force is measured, which can be used to describe overall cell adhesion (**Fig. 5.6e**). We have adapted this technique to quantitatively compare cell adhesion of PC3 cells transfected with KLK3 (PC3-K3) and untransfected, wildtype PC3 cells (PC3-WT) to OBM (**Fig. 5.6f**). Herein, we detected significantly stronger cell adhesion for PC3-K3 cells than for PC3-WT cells, suggesting an increased adhesion to OBM by PSA/KLK3.

A further development of this model is to use a tissue-engineered bone (TEB) construct, in which the osteoblast cells are seeded onto a polycaprolone (PCL) scaffold, in order to provide a more rigid backbone, akin to the bone microenvironment. Prostate cancer cells are then seeded on top of this TEB construct, either directly or within a hydrogel support so as to facilitate direct or indirect cell-to-cell communication (**Fig. 5.4c**, middle and right hand panels).

In summary, both natural and biomimetic-engineered matrices, used as *in vitro* cellular platforms, are beginning to provide new knowledge on specific aspects of the KLKs in the progression of prostate cancer. Our further efforts will be directed towards improving these tumor microenvironment models, in order to fully understand the definitive roles played by KLKs both in localized and metastatic disease.

5.6 Challenges and future direction

For the past three decades, PSA/KLK3 has been the biomarker of choice for prostate cancer, while more recently other KLKs have shown potential as biomarkers for ovarian cancer, suggesting their regulatory functions in these cancers. Using various cellular model systems, in conjunction with analysis of clinical samples, we are now beginning to demonstrate the key role of KLKs in the development, progression, and treatment resistance of ovarian and prostate cancer. We still face the challenge of providing the most appropriate models for studying the dramatic variation in cell composition, overall architecture, and specific function in the individual tissues and organs involved in these processes, in particular in metastasis. Specific inhibitors of individual KLKs have been successfully generated and some of them have shown a functional response at a cellular level suggesting their potential as pharmaceutical targets (Chapters 6, 7, and 13, Volume 1). Our further efforts are directed to discovering effective approaches that target KLK action in both the tumor and its microenvironment, in order to improve the outcome and quality of life of patients afflicted with these cancers.

Acknowledgment

We like to acknowledge Dr. Helen Irving-Rodgers for her critical review of the manuscript.

Bibliography

Abbott, A. (2003). Cell culture: biology's new dimension. Nature 424, 870–872.

Agre, P., and Williams, T.E. (1983). The human tumor cloning assay in cancer drug development. A review. Invest. New Drugs 1, 33–45.

AIHW (2010). Ovarian cancer in Australia: an overview. In Cancer Series Number 52 (Canberra, Australia, Australian Institute of Health and Welfare).

Alberts, D.S., Samon, S.E., Chen, H.S., Surwit, E.A., Soehnlen, B., Young, L., and Moon, T.E. (1980). In-vitro clonogenic assay for predicting response of ovarian cancer to chemotherapy. Lancet 2, 340–342.

Auerbach, R., Lewis, R., Shinners, B., Kubai, L., and Akhtar, N. (2003). Angiogenesis assays: a critical overview. Clin. Chem. 49, 32–40.

Barbolina, M.V., Adley, B.P., Ariztia, E.V., Liu, Y., and Stack, M.S. (2007). Microenvironmental regulation of membrane type 1 matrix metalloproteinase activity in ovarian carcinoma cells via collagen-induced EGR1 expression. J. Biol. Chem. 282, 4924–4931.

Barbolina, M.V., Adley, B.P., Kelly, D.L., Shepard, J., Fought, A.J., Scholtens, D., Penzes, P., Shea, L.D., and Stack, M.S. (2009a). Downregulation of connective tissue growth factor by three-dimensional matrix enhances ovarian carcinoma cell invasion. Int. J. Cancer 125, 816–825.

Barbolina, M.V., Moss, N.M., Westfall, S.D., Lui, Y., Burkhalter, R.J., Marga, F., Forgacs, G., Hudson, L.G., and Stack, M.S. (2009b). Microenvironmental regulation of ovarian cancer metastasis. In: Ovarian Cancer, Stack, M.S., and Fishman, D.A., eds. (New York, Springer), pp. 319–334.

Barrett, J.M., Mangold, K.A., Jilling, T., and Kaul, K.L. (2005). Bi-directional interactions of prostate cancer cells and bone marrow endothelial cells in three-dimensional culture. Prostate 64, 75–82.

Bast, R.C. Jr., Hennessy, B., and Mills, G.B. (2009). The biology of ovarian cancer: new opportunities for translation. Nat. Rev. Cancer 9, 415–428.

Berrino, F., De Angelis, R., Sant, M., Rosso, S., Bielska-Lasota, M., Coebergh, J.W., and Santaquilani, M. (2007). Survival for eight major cancers and all cancers combined for European adults diagnosed in 1995–99: results of the EUROCARE-4 study. Lancet Oncol. 8, 773–783.

Bertelsen, C.A., Sondak, V.K., Mann, B.D., Korn, E.L., and Kern, D.H. (1984). Chemosensitivity testing of human solid tumors. A review of 1582 assays with 258 clinical correlations. Cancer 53, 1240–1245.

Bissell, D.M., Arenson, D.M., Maher, J.J., and Roll, F.J. (1987). Support of cultured hepatocytes by a laminin-rich gel. Evidence for a functionally significant subendothelial matrix in normal rat liver. J. Clin. Invest. 79, 801–812.

Bissell, M.J., and Barcellos-Hoff, M.H. (1987). The influence of extracellular matrix on gene expression: is structure the message? J. Cell Sci. Suppl. 8, 327–343.

Bissell, M.J., Hall, H.G., and Parry, G. (1982). How does the extracellular matrix direct gene expression? J. Theor. Biol. 99, 31–68.

Bissell, M.J., and Radisky, D. (2001). Putting tumours in context. Nat. Rev. Cancer 1, 46–54.

Borgoño, C.A., and Diamandis, E.P. (2004). The emerging roles of human tissue kallikreins in cancer. Nat. Rev. Cancer 4, 876–890.

Burleson, K.M., Casey, R.C., Skubitz, K.M., Pambuccian, S.E., Oegema, T.R. Jr., and Skubitz, A.P. (2004a). Ovarian carcinoma ascites spheroids adhere to extracellular matrix components and mesothelial cell monolayers. Gynecol. Oncol. 93, 170–181.

Burleson, K.M., Hansen, L.K., and Skubitz, A.P. (2004b). Ovarian carcinoma spheroids disaggregate on type I collagen and invade live human mesothelial cell monolayers. Clin. Exp. Metastasis 21, 685–697.

Cancer-Research-UK. (2011). Ovarian cancer survival statistics.

Cannistra, S.A. (2004). Cancer of the ovary. N. Engl. J. Med. 351, 2519–2529.

Casey, R.C., Burleson, K.M., Skubitz, K.M., Pambuccian, S.E., Oegema, T.R. Jr., Ruff, L.E., and Skubitz, A.P. (2001). ß1-integrins regulate the formation and adhesion of ovarian carcinoma multicellular spheroids. Am. J. Pathol. 159, 2071–2080.

Chadha, K.C., Nair, B.B., Chakravarthi, S., Zhou, R., Godoy, A., Mohler, J.L., Aalinkeel, R., Schwartz, S.A., and Smith, G.J. (2011). Enzymatic activity of free-prostate-specific antigen (f-PSA) is not required for some of its physiological activities. Prostate 71, 1680–1690.

Chu, Y.W., Runyan, R.B., Oshima, R.G., and Hendrix, M.J. (1993). Expression of complete keratin filaments in mouse L cells augments cell migration and invasion. Proc. Natl. Acad. Sci. USA 90, 4261–4265.

Clements, J.A., Willemsen, N.M., Myers, M., and Dong, Y. (2004). The tissue kallikrein family of serine proteases: functional roles in human disease and potential as clinical biomarkers. Crit. Rev. Clin. Lab. Sci. 41, 265–312.

Cukierman, E., Pankov, R., and Yamada, K.M. (2002). Cell interactions with three-dimensional matrices. Curr. Opin. Cell Biol. 14, 633–639.

Davies, E.J., Blackhall, F.H., Shanks, J.H., David, G., McGown, A.T., Swindell, R., Slade, R.J., Martin-Hirsch, P., Gallagher, J.T., and Jayson, G.C. (2004). Distribution and clinical significance of heparan sulfate proteoglycans in ovarian cancer. Clin. Cancer Res. 10, 5178–5186.

Doillon, C.J., Gagnon, E., Paradis, R., and Koutsilieris, M. (2004). Three-dimensional culture system as a model for studying cancer cell invasion capacity and anticancer drug sensitivity. Anticancer Res. 24, 2169–2177.

Dong, Y., Tan, O.L., Loessner, D., Stephens, C., Walpole, C., Boyle, G.M., Parsons, P.G., and Clements, J.A. (2010). Kallikrein-related peptidase 7 promotes multicellular aggregation via the alpha(5)beta(1) integrin pathway and paclitaxel chemoresistance in serous epithelial ovarian carcinoma. Cancer Res. 70, 2624–2633.

Durand, R.E., and Olive, P.L. (2001). Resistance of tumor cells to chemo- and radiotherapy modulated by the three-dimensional architecture of solid tumors and spheroids. Methods Cell Biol. 64, 211–233.

Essand, M., Gronvik, C., Hartman, T., and Carlsson, J. (1995). Radioimmunotherapy of prostatic adenocarcinomas: effects of 131I-labelled E4 antibodies on cells at different depth in DU 145 spheroids. Int. J. Cancer 63, 387–394.

Fischbach, C., Chen, R., Matsumoto, T., Schmelzle, T., Brugge, J.S., Polverini, P.J., and Mooney, D.J. (2007). Engineering tumors with 3D scaffolds. Nat. Methods 4, 855–860.

Fortier, A.H., Nelson, B.J., Grella, D.K., and Holaday, J.W. (1999). Antiangiogenic activity of prostate-specific antigen. J. Natl. Cancer Inst. 91, 1635–1640.

Frankel, A., Buckman, R., and Kerbel, R.S. (1997). Abrogation of taxol-induced G2-M arrest and apoptosis in human ovarian cancer cells grown as multicellular tumor spheroids. Cancer Res. 57, 2388–2393.

Frankel, A., Man, S., Elliott, P., Adams, J., and Kerbel, R.S. (2000). Lack of multicellular drug resistance observed in human ovarian and prostate carcinoma treated with the proteasome inhibitor PS-341. Clin. Cancer Res. 6, 3719–3728.

Friedrich, J., Ebner, R., and Kunz-Schughart, L.A. (2007). Experimental anti-tumor therapy in 3-D: spheroids – old hat or new challenge? Int. J. Radiat. Biol. 83, 849–871.

Friedrich, J., Seidel, C., Ebner, R., and Kunz-Schughart, L.A. (2009). Spheroid-based drug screen: considerations and practical approach. Nat. Protoc. 4, 309–324.

Gao, J., Collard, R.L., Bui, L., Herington, A.C., Nicol, D.L., and Clements, J.A. (2007). Kallikrein 4 is a potential mediator of cellular interactions between cancer cells and osteoblasts in metastatic prostate cancer. Prostate 67, 348–360.

Ghajar, C.M., and Bissell, M.J. (2010). Tumor engineering: the other face of tissue engineering. Tissue Eng. Part A 16, 2153–2156.

Ghajar, C.M., Kachgal, S., Kniazeva, E., Mori, H., Costes, S.V., George, S.C., and Putnam, A.J. (2010). Mesenchymal cells stimulate capillary morphogenesis via distinct proteolytic mechanisms. Exp. Cell Res. 316, 813–825.

Giusti, B., Serrati, S., Margheri, F., Papucci, L., Rossi, L., Poggi, F., Magi, A., Del Rosso, A., Cinelli, M., Guiducci, S., Kahaleh, B., Matucci-Cerinic, M., Abbate, R., Fibbi, G., and Del Rosso, M. (2005). The antiangiogenic tissue kallikrein pattern of endothelial cells in systemic sclerosis. Arthritis Rheum. 52, 3618–3628.

Goodman, M.T., Correa, C.N., Tung, K.H., Roffers, S.D., Cheng Wu, X., Young, J.L. Jr., Wilkens, L.R., Carney, M.E., and Howe, H.L. (2003). Stage at diagnosis of ovarian cancer in the United States, 1992–1997. Cancer 97, 2648–2659.

Griffith, L.G., and Swartz, M.A. (2006). Capturing complex 3D tissue physiology in vitro. Nat. Rev. Mol. Cell. Biol. 7, 211–224.

Hamburger, A.W. (1987). The human tumor clonogenic assay as a model system in cell biology. Int. J. Cell Cloning 5, 89–107.

Harma, V., Virtanen, J., Makela, R., Happonen, A., Mpindi, J.P., Knuuttila, M., Kohonen, P., Lotjonen, J., Kallioniemi, O., and Nees, M. (2010). A comprehensive panel of three-dimensional models for studies of prostate cancer growth, invasion and drug responses. PLoS One 5, e10431.

Hedlund, T.E., Duke, R.C., and Miller, G.J. (1999). Three-dimensional spheroid cultures of human prostate cancer cell lines. Prostate 41, 154–165.

Hedman, K., Johansson, S., Vartio, T., Kjellen, L., Vaheri, A., and Hook, M. (1982). Structure of the pericellular matrix: association of heparan and chondroitin sulfates with fibronectin-procollagen fibers. Cell 28, 663–671.

Helenius, J., Heisenberg, C.P., Gaub, H.E., and Muller, D.J. (2008). Single-cell force spectroscopy. J. Cell Sci. 121, 1785–1791.

Hsiao, A.Y., Torisawa, Y.S., Tung, Y.C., Sud, S., Taichman, R.S., Pienta, K.J., and Takayama, S. (2009). Microfluidic system for formation of PC-3 prostate cancer co-culture spheroids. Biomaterials 30, 3020–3027.

Hutmacher, D.W., Horch, R.E., Loessner, D., Rizzi, S., Sieh, S., Reichert, J.C., Clements, J.A., Beier, J.P., Arkudas, A., Bleiziffer, O., and Kneser, U. (2009). Translating tissue engineering technology platforms into cancer research. J. Cell. Mol. Med. 13, 1417–1427.

Hutmacher, D.W. (2010). Biomaterials offer cancer research the third dimension. Nat. Mater. 9, 90–93.

Ingram, M., Techy, G.B., Saroufeem, R., Yazan, O., Narayan, K.S., Goodwin, T.J., and Spaulding, G.F. (1997). Three-dimensional growth patterns of various human tumor cell lines in simulated microgravity of a NASA bioreactor. In Vitro Cell Dev. Biol. Anim. 33, 459–466.

Iwanicki, M.P., Davidowitz, R.A., Ng, M.R., Besser, A., Muranen, T., Merritt, M., Danuser, G., Ince, T., and Brugge, J.S. (2011). Ovarian cancer spheroids use myosin-generated force to clear the mesothelium. Cancer Discovery 1, 144–157.

Jedeszko, C., Sameni, M., Olive, M.B., Moin, K., and Sloane, B.F. (2008). Visualizing protease activity in living cells: from two dimensions to four dimensions. Curr. Protoc. Cell Biol., Chapter 4, Unit 4, 20.

Jemal, A., Bray, F., Center, M.M., Ferlay, J., Ward, E., and Forman, D. (2011). Global cancer statistics. CA Cancer J. Clin. 61, 69–90.

Josson, S., Matsuoka, Y., Chung, L.W., Zhau, H.E., and Wang, R. (2010). Tumor-stroma co-evolution in prostate cancer progression and metastasis. Semin. Cell. Dev. Biol. 21, 26–32.

Kaminski, A., Hahne, J.C., Haddouti el, M., Florin, A., Wellmann, A., and Wernert, N. (2006). Tumour-stroma interactions between metastatic prostate cancer cells and fibroblasts. Int. J. Mol. Med. 18, 941–950.

Kaneko, A., Satoh, Y., Tokuda, Y., Fujiyama, C., Udo, K., and Uozumi, J. (2010). Effects of adipocytes on the proliferation and differentiation of prostate cancer cells in a 3-D culture model. Int. J. Urol. 17, 369–376.

Kenny, H.A., Kaur, S., Coussens, L.M., and Lengyel, E. (2008). The initial steps of ovarian cancer cell metastasis are mediated by MMP-2 cleavage of vitronectin and fibronectin. J. Clin. Invest. 118, 1367–1379.

Kenny, H.A., Krausz, T., Yamada, S.D., and Lengyel, E. (2007). Use of a novel 3D culture model to elucidate the role of mesothelial cells, fibroblasts and extra-cellular matrices on adhesion and invasion of ovarian cancer cells to the omentum. Int. J. Cancer 121, 1463–1472.

Kenny, H.A., and Lengyel, E. (2009). MMP-2 functions as an early response protein in ovarian cancer metastasis. Cell Cycle 8, 683–688.

Kim, J.B. (2005). Three-dimensional tissue culture models in cancer biology. Semin. Cancer Biol. 15, 365–377.

Kleinman, H.K., McGarvey, M.L., Hassell, J.R., Star, V.L., Cannon, F.B., Laurie, G.W., and Martin, G.R. (1986). Basement membrane complexes with biological activity. Biochemistry 25, 312–318.

Klokk, T.I., Kilander, A., Xi, Z., Waehre, H., Risberg, B., Danielsen, H.E., and Saatcioglu, F. (2007). Kallikrein 4 is a proliferative factor that is overexpressed in prostate cancer. Cancer Res. 67, 5221–5230.

Lawrence, M.G., Lai, J., and Clements, J.A. (2010). Kallikreins on steroids: structure, function, and hormonal regulation of prostate-specific antigen and the extended kallikrein locus. Endocr. Rev. 31, 407–446.

Li, S.H., Hawthorne, V.S., Neal, C.L., Sanghera, S., Xu, J., Yang, J., Guo, H., Steeg, P.S., and Yu, D. (2009). Upregulation of neutrophil gelatinase-associated lipocalin by ErbB2 through nuclear factor-kappaB activation. Cancer Res. 69, 9163–9168.

Lilja, H., Ulmert, D., and Vickers, A.J. (2008). Prostate-specific antigen and prostate cancer: prediction, detection and monitoring. Nat. Rev. Cancer 8, 268–278.

Lim, R., Ahmed, N., Borregaard, N., Riley, C., Wafai, R., Thompson, E.W., Quinn, M.A., and Rice, G.E. (2007). Neutrophil gelatinase-associated lipocalin (NGAL) an early-screening biomarker for ovarian cancer: NGAL is associated with epidermal growth factor-induced epithelio-mesenchymal transition. Int. J. Cancer 120, 2426–2434.

Lin, C.C., and Anseth, K.S. (2009). PEG hydrogels for the controlled release of biomolecules in regenerative medicine. Pharm. Res. 26, 631–643.

Liotta, L.A., Tryggvason, K., Garbisa, S., Hart, I., Foltz, C.M., and Shafie, S. (1980). Metastatic potential correlates with enzymatic degradation of basement membrane collagen. Nature 284, 67–68.

Loessner, D., Stok, K.S., Lutolf, M.P., Hutmacher, D.W., Clements, J.A., and Rizzi, S.C. (2010). Bioengineered 3D platform to explore cell-ECM interactions and drug resistance of epithelial ovarian cancer cells. Biomaterials 31, 8494–8506.

Madar, S., Brosh, R., Buganim, Y., Ezra, O., Goldstein, I., Solomon, H., Kogan, I., Goldfinger, N., Klocker, H., and Rotter, V. (2009). Modulated expression of WFDC1 during carcinogenesis and cellular senescence. Carcinogenesis 30, 20–27.

Mo, L., Zhang, J., Shi, J., Xuan, Q., Yang, X., Qin, M., Lee, C., Klocker, H., Li, Q.Q., and Mo, Z. (2010). Human kallikrein 7 induces epithelial-mesenchymal transition-like changes in prostate carcinoma cells: a role in prostate cancer invasion and progression. Anticancer Res. 30, 3413–3420.

Moss, N.M., Barbolina, M.V., Liu, Y., Sun, L., Munshi, H.G., and Stack, M.S. (2009a). Ovarian cancer cell detachment and multicellular aggregate formation are regulated by membrane type 1 matrix metalloproteinase: a potential role in I.p. metastatic dissemination. Cancer Res. 69, 7121–7129.

Moss, N.M., Liu, Y., Johnson, J.J., Debiase, P., Jones, J., Hudson, L.G., Munshi, H.G., and Stack, M.S. (2009b). Epidermal growth factor receptor-mediated membrane type 1 matrix metalloproteinase endocytosis regulates the transition between invasive versus expansive growth of ovarian carcinoma cells in three-dimensional collagen. Mol. Cancer Res. 7, 809–820.

Mucci, L.A., Powolny, A., Giovannucci, E., Liao, Z., Kenfield, S.A., Shen, R., Stampfer, M.J., and Clinton, S.K. (2009). Prospective study of prostate tumor angiogenesis and cancer-specific mortality in the health professionals follow-up study. J. Clin. Oncol. 27, 5627–5633.

Nelson, C.M., and Bissell, M.J. (2006). Of extracellular matrix, scaffolds, and signaling: tissue architecture regulates development, homeostasis, and cancer. Annu. Rev. Cell. Dev. Biol. 22, 287–309.

Paland, N., Kamer, I., Kogan-Sakin, I., Madar, S., Goldfinger, N., and Rotter, V. (2009). Differential influence of normal and cancer-associated fibroblasts on the growth of human epithelial cells in an *in vitro* cocultivation model of prostate cancer. Mol. Cancer Res. 7, 1212–1223.

Pampaloni, F., Reynaud, E.G., and Stelzer, E.H. (2007). The third dimension bridges the gap between cell culture and live tissue. Nat. Rev. Mol. Cell. Biol. 8, 839–845.

Petersen, O.W., Ronnov-Jessen, L., Howlett, A.R., and Bissell, M.J. (1992). Interaction with basement membrane serves to rapidly distinguish growth and differentiation pattern of normal and malignant human breast epithelial cells. Proc. Natl. Acad. Sci. USA 89, 9064–9068.

Pickl, M., and Ries, C.H. (2009). Comparison of 3D and 2D tumor models reveals enhanced HER2 activation in 3D associated with an increased response to trastuzumab. Oncogene 28, 461–468.

Podgorski, I., Linebaugh, B.E., Sameni, M., Jedeszko, C., Bhagat, S., Cher, M.L., and Sloane, B.F. (2005). Bone microenvironment modulates expression and activity of cathepsin B in prostate cancer. Neoplasia 7, 207–223.

Prestwich, G.D. (2007). Simplifying the extracellular matrix for 3-D cell culture and tissue engineering: a pragmatic approach. J. Cell. Biochem. 101, 1370–1383.

Prezas, P., Arlt, M.J., Viktorov, P., Soosaipillai, A., Holzscheiter, L., Schmitt, M., Talieri, M., Diamandis, E.P., Krüger, A., and Magdolen, V. (2006). Overexpression of the human tissue kallikrein genes KLK4, 5, 6, and 7 increases the malignant phenotype of ovarian cancer cells. Biol. Chem. 387, 807–811.

Puech, P.H., Poole, K., Knebel, D., and Muller, D.J. (2006). A new technical approach to quantify cell-cell adhesion forces by AFM. Ultramicroscopy 106, 637–644.

Puls, L.E., Duniho, T., Hunter, J.E., Kryscio, R., Blackhurst, D., and Gallion, H. (1996). The prognostic implication of ascites in advanced-stage ovarian cancer. Gynecol. Oncol. 61, 109–112.

Reichert, J.C., Quent, V.M., Burke, L.J., Stansfield, S.H., Clements, J.A., and Hutmacher, D.W. (2010). Mineralized human primary osteoblast matrices as a model system to analyse interactions of prostate cancer cells with the bone microenvironment. Biomaterials 31, 7928–7936.

Rizzi, S.C., and Hubbell, J.A. (2005). Recombinant protein-co-PEG networks as cell-adhesive and proteolytically degradable hydrogel matrixes. Part I: Development and physicochemical characteristics. Biomacromolecules 6, 1226–1238.

Rizzi, S.C., Ehrbar, M., Halstenberg, S., Raeber, G.P., Schmoekel, H.G., Hagenmuller, H., Muller, R., Weber, F.E., and Hubbell, J.A. (2006). Recombinant protein-co-PEG networks as cell-adhesive and proteolytically degradable hydrogel matrixes. Part II: biofunctional characteristics. Biomacromolecules 7, 3019–3029.

Robb, J.A. (1970). Microcloning and replica plating of mammalian cells. Science 170, 857–858.

Rodan, S.B., Imai, Y., Thiede, M.A., Wesolowski, G., Thompson, D., Bar-Shavit, Z., Shull, S., Mann, K., and Rodan, G.A. (1987). Characterization of a human osteosarcoma cell line (Saos-2) with osteoblastic properties. Cancer Res. 47, 4961–4966.

Sameni, M., Dosescu, J., Yamada, K.M., Sloane, B.F., and Cavallo-Medved, D. (2008). Functional live-cell imaging demonstrates that beta1-integrin promotes type IV collagen degradation by breast and prostate cancer cells. Mol. Imaging 7, 199–213.

Schmidt, J.J., Rowley, J., and Kong, H.J. (2008). Hydrogels used for cell-based drug delivery. J. Biomed. Mater. Res. A 87, 1113–1122.

Schneider, J.G., Amend, S.R., and Weilbaecher, K.N. (2011). Integrins and bone metastasis: integrating tumor cell and stromal cell interactions. Bone 48, 54–65.

Schubert, D., and LaCorbiere, M. (1982). Properties of extracellular adhesion-mediating particles in myoblast clone and its adhesion-deficient variant. J. Cell. Biol. 94, 108–114.

Schuetz, E.G., Li, D., Omiecinski, C.J., Muller-Eberhard, U., Kleinman, H.K., Elswick, B., and Guzelian, P.S. (1988). Regulation of gene expression in adult rat hepatocytes cultured on a basement membrane matrix. J. Cell. Physiol. 134, 309–323.

Schutyser, E., Struyf, S., Proost, P., Opdenakker, G., Laureys, G., Verhasselt, B., Peperstraete, L., Van de Putte, I., Saccani, A., Allavena, P., Mantovani, A., and van Damme, J. (2002). Identification of biologically active chemokine isoforms from ascitic fluid and elevated levels of CCL18/pulmonary and activation-regulated chemokine in ovarian carcinoma. J. Biol. Chem. 277, 24584–24593.

Shield, K., Riley, C., Quinn, M.A., Rice, G.E., Ackland, M.L., and Ahmed, N. (2007). Alpha2beta1 integrin affects metastatic potential of ovarian carcinoma spheroids by supporting disaggregation and proteolysis. J. Carcinog. 6, 11.

Shou, J., Soriano, R., Hayward, S.W., Cunha, G.R., Williams, P.M., and Gao, W.Q. (2002). Expression profiling of a human cell line model of prostatic cancer reveals a direct involvement of interferon signaling in prostate tumor progression. Proc. Natl. Acad. Sci. USA 99, 2830–2835.

Sodek, K.L., Brown, T.J., and Ringuette, M.J. (2008). Collagen I but not Matrigel matrices provide an MMP-dependent barrier to ovarian cancer cell penetration. BMC Cancer 8, 223.

Sodek, K.L., Ringuette, M.J., and Brown, T.J. (2009). Compact spheroid formation by ovarian cancer cells is associated with contractile behavior and an invasive phenotype. Int. J. Cancer 124, 2060–2070.

Sotiropoulou, G., Pampalakis, G., and Diamandis, E.P. (2009). Functional roles of human kallikrein-related peptidases. J. Biol. Chem. 284, 32989–32994.

Stewart, J.J., White, J.T., Yan, X., Collins, S., Drescher, C.W., Urban, N.D., Hood, L., and Lin, B. (2006). Proteins associated with cisplatin resistance in ovarian cancer cells identified by quantitative proteomic technology and integrated with mRNA expression levels. Mol. Cell. Proteomics 5, 433–443.

Sung, S.Y., Hsieh, C.L., Law, A., Zhau, H.E., Pathak, S., Multani, A.S., Lim, S., Coleman, I.M., Wu, L.C., Figg, W.D., Dahut, W.L., Nelson, P., Lee, J.K., Amin, M.B., Lyles, R., Johnstone, P.A., Marshall, F.F., and Chung, L.W. (2008). Coevolution of prostate cancer and bone stroma in three-dimensional coculture: implications for cancer growth and metastasis. Cancer Res. 68, 9996–10003.

Sutherland, R.M., MacDonald, H.R., and Howell, R.L. (1977). Multicellular spheroids: a new model target for *in vitro* studies of immunity to solid tumor allografts. J. Natl. Cancer Inst. 58, 1849–1853.

Sweeney, T.M., Kibbey, M.C., Zain, M., Fridman, R., and Kleinman, H.K. (1991). Basement membrane and the SIKVAV laminin-derived peptide promote tumor growth and metastases. Cancer Metastasis Rev. 10, 245–254.

Taubenberger, A., Cisneros, D.A., Friedrichs, J., Puech, P.H., Muller, D.J., and Franz, C.M. (2007). Revealing early steps of alpha2beta1 integrin-mediated adhesion to collagen type I by using single-cell force spectroscopy. Mol. Biol. Cell 18, 1634–1644.

Thanos, C.G., and Emerich, D.F. (2008). On the use of hydrogels in cell encapsulation and tissue engineering system. Recent Pat. Drug Deliv. Formul. 2, 19–24.

Tuxhorn, J.A., McAlhany, S.J., Dang, T.D., Ayala, G.E., and Rowley, D.R. (2002). Stromal cells promote angiogenesis and growth of human prostate tumors in a differential reactive stroma (DRS) xenograft model. Cancer Res. 62, 3298–3307.

Vamvakidou, A.P., Mondrinos, M.J., Petushi, S.P., Garcia, F.U., Lelkes, P.I., and Tozeren, A. (2007). Heterogeneous breast tumoroids: An *in vitro* assay for investigating cellular heterogeneity and drug delivery. J. Biomol. Screen. 12, 13–20.

Vanpoucke, G., Orr, B., Grace, O.C., Chan, R., Ashley, G.R., Williams, K., Franco, O.E., Hayward, S.W., and Thomson, A.A. (2007). Transcriptional profiling of inductive mesenchyme to identify molecules involved in prostate development and disease. Genome Biol. 8, R213.

Veveris-Lowe, T.L., Lawrence, M.G., Collard, R.L., Bui, L., Herington, A.C., Nicol, D.L., and Clements, J.A. (2005). Kallikrein 4 (hK4) and prostate-specific antigen (PSA) are associated with the loss of E-cadherin and an epithelial-mesenchymal transition (EMT)-like effect in prostate cancer cells. Endocr. Relat. Cancer 12, 631–643.

Walsh, P.C., DeWeese, T.L., and Eisenberger, M.A. (2007). Clinical practice. Localized prostate cancer. N. Engl. J. Med. 357, 2696–2705.

Wang, J., Ying, G., Jung, Y., Lu, J., Zhu, J., Pienta, K.J., and Taichman, R.S. (2010). Characterization of phosphoglycerate kinase-1 expression of stromal cells derived from tumor microenvironment in prostate cancer progression. Cancer Res. 70, 471–480.

Witz, C.A., Montoya-Rodriguez, I.A., Cho, S., Centonze, V.E., Bonewald, L.F., and Schenken, R.S. (2001). Composition of the extracellular matrix of the peritoneum. J. Soc. Gynecol. Investig. 8, 299–304.

Wong, C.P., Bray, T.M., and Ho, E. (2009). Induction of proinflammatory response in prostate cancer epithelial cells by activated macrophages. Cancer Lett. 276, 38–46.

Xi, Z., Kaern, J., Davidson, B., Klokk, T.I., Risberg, B., Trope, C., and Saatcioglu, F. (2004). Kallikrein 4 is associated with paclitaxel resistance in ovarian cancer. Gynecol. Oncol. 94, 80–85.

Yamada, K.M., and Cukierman, E. (2007). Modeling tissue morphogenesis and cancer in 3D. Cell 130, 601–610.

Yamamura, K., Kibbey, M.C., Jun, S.H., and Kleinman, H.K. (1993). Effect of matrigel and laminin peptide YIGSR on tumor growth and metastasis. Semin. Cancer Biol. 4, 259–265.

Yousef, G.M., and Diamandis, E.P. (2009). The human kallikrein gene family: new biomarkers for ovarian cancer. In: Ovarian Cancer, Stack, M.S., and Fishman, D.A., eds. (New York, Springer), pp. 165–188.

Zhang, X., Wojcikiewicz, E., and Moy, V.T. (2002). Force spectroscopy of the leukocyte function-associated antigen-1/intercellular adhesion molecule-1 interaction. Biophys. J. 83, 2270–2279.

Zhao, H., and Peehl, D.M. (2009). Tumor-promoting phenotype of CD90hi prostate cancer-associated fibroblasts. Prostate 69, 991–1000.

Zhau, H.E., Goodwin, T.J., Chang, S.M., Baker, T.L., and Chung, L.W. (1997). Establishment of a three-dimensional human prostate organoid coculture under microgravity-simulated conditions: evaluation of androgen-induced growth and PSA expression. In Vitro Cell. Dev. Biol. Anim. 33, 375–380.

Manfred Schmitt, Julia Dorn, Marion Kiechle,
Eleftherios P. Diamandis, and Liu-Ying Luo

6 Clinical Relevance of Kallikrein-related Peptidases in Breast Cancer

6.1 Introduction

Breast cancer is the most commonly diagnosed cancer among women around the world. Although many treatment options are currently available, such as surgery, radiotherapy, chemotherapy, immunotherapy, and endocrine therapy, it remains the second leading cause of cancer-related deaths among women, after lung cancer (CDC National Program of Cancer Registries, USA; Stuckey, 2011). It is estimated that worldwide more than one million women are affected annually. The development of breast cancer represents a multistep malignant transformation of the epithelial cells as a result of multiple genetic changes and environmental insults. A number of factors have long been recognized to contribute to such a malignant transformation process, such as oncogenes, tumor suppressor genes, hormones, growth factors, and proteases.

Currently, mammography screening is the mainstay for early detection of breast cancer. According to randomized control trials, it can reduce breast cancer mortality for women aged 39 to 69 by 15% (Nelson *et al.*, 2009). Nevertheless, with this screening method, false-positive results are common and often lead to additional imaging and biopsies, especially in younger women. Consequently, there remains a need for a minimally invasive, cost-effective procedure that could be used alone or alongside mammography to improve screening specificity.

Potentially, serum/plasma-based biomarkers may be helpful for an early diagnosis of the disease, for assessment of the course of the disease, prediction of response or resistance to cancer therapeutics, or monitoring of efficacy of therapy in breast cancer patients. Yet, several serum-based biomarkers have been described in the literature and are in clinical application, such as CA 15-3, BR 27.29 (CA 27.29), carcinoembryonic antigen (CEA), tissue polypeptide antigen, tissue polypeptide specific antigen, or p105HER2 (the shed extracellular domain of HER2) (Marić *et al.*, 2011). However, none of these markers is specific or sensitive enough to allow early diagnosis of the malignant or the non-cancerous breast cases (Duffy, 2006).

During the past decade, with the discovery of all the new members of the kallikrein-related peptidase (KLK) family, more and more evidence has indicated that KLKs play pivotal roles in breast cancer progression and metastasis. KLKs might potently influence the bioavailability or functionality of important growth factors, thus disrupting the balance between growth and antigrowth signals in the tumor microenvironment, supporting unregulated cell proliferation, also in breast cancer (Krenzer *et al.*, 2011; Naidoo *et al.*, 2009; Sher *et al.*, 2006; Tham *et al.*, 2011).

Identification of biomarkers that can assist in early detection and refine disease management is one of the areas on which current clinical research efforts are focused (Misek and Kim, 2011). Currently, breast cancer biomarkers in regular clinical practice mainly encompass histomorphological markers (TNM status: tumor size, nodal status, incidence of metastasis, nuclear grade, histological subtype, lymphovascular invasion), plus receptors for the steroid hormones estrogen and progesterone, as well as newer cancer biomarkers such as the multigene panel Oncotype DX® (Genomic Health, Redwood City, CA, USA), MammaPrint® (Agendia, Amsterdam, The Netherlands), and Endopredict® (Sividon Diagnostics, Cologne, Germany) and tumor invasion factors uPA/PAI-1 (Harris *et al.*, 2007).

Recent gene-expression profiling studies, however, have brought up the concept that breast cancer is not a single disease, but a group of molecularly distinct neoplastic disorders (Sotiriou and Pusztai, 2009). Thus, the concept of "biomarker-based, individualized cancer patient treatment and targeted therapies" has come to public attention. In the past years, serum- and tumor-tissue-associated biomarkers have also come into the focus of clinical application in breast cancer and have been given an essential role in selecting breast cancer patients for treatment and follow-up of the cancer disease.

Tumor metastasis is a multistep process, to which cell-cell and/or cell-ECM (extracellular matrix) adhesion, ECM degradation, cell migration, and angiogenesis are thought to contribute significantly. Mounting evidence supports the view that extracellular proteases mediate many of the changes in the tumor microenvironment during tumor progression. A major component of tumor invasion is the action of tumor-associated proteases such as matrix metalloproteases (MMP), plasmin, and cathepsins, but also the KLKs, in disrupting the surrounding basement membrane and the adjacent ECM. Regarding this topic, the potential diagnostic value of tumor tissue-associated KLKs and/or circulating KLKs present in the peripheral blood of breast cancer patients has already been explored (Avgeris *et al.*, 2011; Black *et al.*, 2000b; Mangé *et al.*, 2008; Papachristopoulou *et al.*, 2009). Interestingly, when assessed at the mRNA level, most KLKs are expressed at lower levels in breast cancer tissue, compared to normal breast tissue (Mangé *et al.*, 2008).

Another tumor metastasis-related mechanism involving proteases points to the interaction of KLKs with proteinase-activated receptors (PAR) (see Chapter 15, Volume 1). PAR, including the four variants PAR-1, -2, -3, and -4, are G-protein-coupled receptors. They not only have distinct functions in thrombosis and inflammation but they also promote tumor invasion, by inducing cancer cell migration upon proteolytic cleavage of the receptor moiety. PAR-1, -3, and -4 are mainly activated by thrombin, whereas PAR-2 can be activated by trypsin and other trypsin-like serine proteases. Several of the KLKs can also activate PAR-1 and PAR-2, thereby triggering intracellular signaling in the stromal-tumor microenvironment and altering the behavior of cancer cells in promoting tumor cell migration and invasion. Expression of PAR-1 is both required and sufficient to promote growth and invasion of breast carcinoma cells

(Boire *et al.*, 2005). High levels of PAR-1 expression are directly correlated with epithelial tumor progression in both clinically obtained breast cancer biopsy specimens and in a wide spectrum of differentially metastatic cell lines (Booden *et al.*, 2004; Even-Ram *et al.*, 1998). Similarly, PAR-2 is overexpressed in breast cancer tumor tissue specimens. It can induce proangiogenic growth factors and chemokines (Collier *et al.*, 2008; Su *et al.*, 2009; Versteeg *et al.*, 2008).

6.2 Expression of KLKs in normal breast tissue

According to data collected by Northern blot analysis, RT-PCR, or ELISA, several of the KLKs display broad tissue expression patterns both at the mRNA and protein levels (Shaw and Diamandis, 2007). The only exception is KLK3, also known as PSA (prostate specific antigen), which is predominantly expressed in the prostate and is released into the blood mainly due to local prostatic tissue destruction, which leads to increased leakage of KLK3 into the blood circulation. Thus, KLK3 is utilized as a serum-based biomarker for prostate cancer screening, monitoring, and clinical management.

About two decades ago, however, with the aid of ultrasensitive immunoassays, it was discovered that KLK3 is not prostate-specific but is also expressed in a wide variety of other tissues, including the breast (Monne *et al.*, 1994; Shaw and Diamandis, 2007). In fact, KLK3 was the first member of the KLK family that has been extensively investigated in the healthy breast and in breast cancer. In normal breast tissue, nearly all KLKs have been identified, with the highest levels observed for KLK6-8, 10, 11, and 14 (Mangé *et al.*, 2008; Shaw and Diamandis, 2007). It has been shown by immunohistochemical staining that these KLKs are mainly expressed in the breast's glandular epithelium (Petraki *et al.*, 2006b). Since KLKs are secreted proteases, a number of them are detectable in breast secretions, e.g. milk of lactating women, breast cyst fluid, and nipple aspirate fluid (Shaw and Diamandis, 2007). In nipple aspirate fluid, KLK6 has the highest level, KLK2 the lowest (Sauter *et al.*, 2004a).

6.3 Clinical relevance of KLKs in breast cancer

In clinical practice, it is often seen that breast cancer patients with similar clinical and histomorphological tumor characteristics have markedly different outcomes and response to a given therapy. Therapeutic options for breast cancer range from primary surgery to adjuvant chemotherapy, radiotherapy, endocrine therapy, and immunotherapy. Fortunately, cancer biomarkers can give support to breast cancer patient stratification and risk assessment and to identifying those patients who are expected to respond to certain anticancer drugs (Saijo, 2012; Schmitt *et al.*, 2007). Of the tissue-based markers, assessment of estrogen and progesterone receptor status is mandatory in the selection of patients for treatment with endocrine therapy (such as tamoxifen or aromatase inhibitors), while assessment of HER2 expression status is essential in selecting patients for immunotherapy with the humanized antibody trastuzumab (Herceptin®) or the tyrosine kinase inhibitor lapatinib (Tyverb®).

KLKs exhibit differential expression in the breast tissue during various stages of breast cancer development and progression, suggesting that they might be implicated in regulating tumor growth and metastasis. For this malignant disease, most of the KLKs, except KLK4 and KLK15, show reduced mRNA and/or protein expression levels, compared to expression of the KLKs in normal breast tissue (**Tab. 6.1**) (Avgeris *et al.*, 2010, 2012; Shaw and Diamandis, 2007; Yousef *et al.*, 2004a).

Beyond that, nine of the fifteen members of the KLK family have shown promise as potential prognostic and/or predictive cancer biomarkers (**Tab. 6.1**). Five KLKs predict favorable prognosis (KLK3, 9, 12, 13, and 15), four are markers of unfavorable, poor prognosis (KLK5, 7, 10, and 14) (Borgoño and Diamandis, 2004; Emami and Diamandis, 2008; Obiezu *et al.*, 2005). KLK3 and KLK10 also are predictive markers of response to endocrine therapy (Diamandis *et al.*, 1999; Foekens *et al.*, 1999; Luo *et al.*, 2002). Furthermore, breast cancer risk is associated with presence of single nucleotide polymorphisms (SNP) of KLK2 (Ex5 þ 118C > T) or KLK4 (4207C > G) and of plasma kallikrein (KLKB1: 3293C > T) (Lee *et al.*; 2009). No studies were published regarding any possible prognostic/predictive value of KLK1, 2, 4, 6, 8, and 11 in breast cancer.

The question is as to how KLKs can promote breast cancer with reduced amounts of KLK protein. One possible explanation is that *KLK* mRNA expression levels do not necessarily correlate with KLK protein expression, so that even though their mRNA expression is reduced, KLK protein expression may be high. Another possibility relates to often uncontrolled preanalytical situations, inasmuch that several of the *KLK* mRNAs may be unstable as soon as the biopsies have been taken and hence, reduced mRNA values are recorded. Otherwise, it has been proposed that some of the KLKs may actively participate in proteolytic cascades (Yoon *et al.*, 2007). In such a case, the amounts of KLK protein in a given tumor might be influenced by the levels of their activators/inhibitors, or by autocatalytic activation/degradation (Yoon *et al.*, 2007).

KLK1 (tissue kallikrein; pancreatic /renal kallikrein). *KLK1* mRNA is expressed in breast cancer tissue, but less than in healthy breast tissue (Mangé *et al.*, 2008). The primary function of KLK1 (hK1) is cleavage of kininogen substrates to release Lys-bradykinin (kallidin). KLK1 has been implicated in many important biological processes, such as the activation of growth and peptide hormones, the homeostatic control of blood pressure, pathogenesis of acute pancreatitis, arthritis, asthma, neutrophil diapedesis, and cancer (see Chapter 10, Volume 1). Regarding breast cancer, Hermann *et al.* (1995) and Rehbock *et al.* (1995) localized KLK1, by immunohistochemistry, in the apical portion of ductal breast carcinoma cells. Lobular carcinomas were negative. KLK1 protein expression appeared to be associated with nuclear grade. Dedifferentiated tumors were negative. Surprisingly, no data has been published yet regard-

Tab. 6.1 Clinical utility of KLKs present in tumor tissues of breast cancer patients.

	Normal		Cancerous		Clinical relevance of KLK in breast cancer (technique applied)	Reference
	mRNA	Protein	mRNA	Protein		
KLK1	High	Low	Decreased	Present	In breast cancer, KLK1 protein is present in ductal but not lobular carcinomas. KLK1 protein is absent in dedifferentiated tumors.	Mangé *et al.*, 2008 Hermann *et al.*, 1995 Rehbock *et al.*, 1995
KLK2	Moderate	Moderate	Decreased	Not determined	SNP and gene-based analyses associated with breast cancer risk for *KLK2* gene. Higher concentrations of KLK2 were found in tumor tissues expressing steroid hormone receptors.	Black *et al.*, 2000a Lee *et al.*, 2009 Mangé *et al.*, 2008
KLK3	Moderate	Absent	Not changed	Decreased or absent	Favorable prognosis (ELISA). Expression of KLK3 correlates with poor response to tamoxifen therapy in recurrent breast cancer. Most KLK3-positive cases are lymph node-negative. Inverse correlation between HER2 and KLK3. Higher concentrations of KLK3 were found in tumor tissues expressing steroid hormone receptors.	Black *et al.*, 2000 Foekens *et al.*, 1999 Lai *et al.*, 1996 Mangé *et al.*, 2008 Narita *et al.*, 2006a Yu *et al.*, 1995b; 1998
KLK4	Low	Low	Increased	Increased in tumor cells surrounding stromal cells	SNP and gene-based analyses associated with breast cancer risk for *KLK4* gene. Low *KLK4* expression levels in well-differentiated tumors and stage I patients. Inverse correlation between *KLK4* mRNA expression and progesterone receptor expression.	Lee *et al.*, 2009 Mangé *et al.*, 2008 Papachristopoulou *et al.*, 2009
KLK5	Moderate	Moderate	Decreased	Present	Unfavorable prognosis (qRT-PCR). Relative high expression levels of variant *KLK5-SV2* in normal breast tissues but no expression in breast cancer tissues.	Avgeris *et al.*, 2011 Li *et al.*, 2009 Talieri *et al.*, 2011 Yousef *et al.*, 2002b, 2003b, 2004a

Tab. 6.1 (continued)

	Normal		Cancerous		Clinical relevance of KLK in breast cancer (technique applied)	Reference
	mRNA	Protein	mRNA	Protein		
KLK6	High	Moderate	Decreased (Mangé et al., 2008). Increased, but decreased in cases with lymph-node metastasis (Wang et al., 2008)	Increased, but decreased in cases with lymph-node metastasis	KLK6 protein expressed less frequently in metasta-sized breast cancer tumor tissues	Diamandis et al., 2000 Mangé et al., 2008 Petraki et al., 2001 Yousef et al., 2004b Wang et al., 2008
KLK7	High	Low	Decreased	Not determined	Unfavorable prognosis (RT-PCR, Talieri et al., 2004) Favorable prognosis (qRT-PCR, Holzscheiter et al., 2006)	Holzscheiter et al., 2006 Li et al., 2009 Kishi et al., 2004 Mangé et al., 2008 Talieri et al., 2004 Talieri et al., 2011
KLK8	High	Moderate	Decreased (Yousef et al., 2004a). Not changed (Mangé et al., 2008).	Not determined	No clinically relevant data for KLK8 reported	Kishi et al., 2003 Mangé et al., 2008 Yousef et al., 2004a
KLK9	Low	High	Decreased	Not determined	Favorable prognosis (qRT-PCR)	Mangé et al., 2008 Yousef et al., 2003c

KLK10	High	Absent	Decreased	Present	DNA-methylation of KLK10 associated with unfavorable prognosis. Lack of KLK10 in DCIS predicts for invasive cancer. Predictive marker for response to tamoxifen therapy	Dhar et al., 2001 Kioulafa et al., 2009 Luo et al., 2002 Mangé et al., 2008 Yousef et al., 2004a Yunes et al., 2003 Zhang et al., 2006 Mangé et al., 2008 Sano et al., 2007
KLK11	High	High	Not changed (Sano et al., 2007). Decreased (Mangé et al., 2008).	Not determined	Significantly higher KLK11 expression was found in grade I/II patients	
KLK12	Low	Absent	Decreased	Not determined	KLK12sv3 variant expression is associated with favorable prognosis (qRT-PCR)	Talieri et al., 2012 Mangé et al., 2008 Yousef et al., 2000, 2004a
KLK13	Moderate	Low	Expressed	Not determined	Favorable prognosis (qRT-PCR)	Chang et al., 2002 Borgoño et al., 2003b Fritzsche et al., 2006 Mangé et al., 2008 Papachristopoulou et al., 2011 Yousef et al., 2001b, 2002c
KLK14	High	Moderate	Decreased (Yousef et al., 2002c). Increased (Fritzsche et al., 2006).	Decreased (Yousef et al., 2002c). Increased (Fritzsche et al., 2006)	Unfavorable prognosis (qRT-PCR)	
KLK15	Absent	Moderate	Increased	Not determined	Favorable prognosis (qRT-PCR)	Yousef et al., 2002a

ing a possible clinical impact of KLK1 for prognosis or therapy response in breast cancer.

Recent reports, however, implicate a crucial role of KLK1 in other human cancer diseases (Chee *et al.*, 2008; Planque *et al.*, 2008). KLK1 is expressed by tumor tissues in esophageal and gastric cancer, clear cell renal carcinoma, astrocytoma, pituitary adenoma, and chronic myeloid leukemia (Bhoola and Fink, 2006). Support for the presence of KLK1 in lung cancer derives from the detection of elevated levels of KLK1 in the serum of lung cancer patients (Planque *et al.*, 2008).

KLK2 (human glandular kallikrein-1; tissue kallikrein-2). Like most of the KLKs, *KLK2* mRNA is expressed in breast cancer tissue, but less than in healthy breast tissue (Mangé *et al.*, 2008). KLK2 protein was also detected in breast cancer tumor tissues by immunoassays (Black *et al.*, 2000a; Magklara *et al.*, 1999). Similar to *KLK3*, *KLK2* displays strong androgen responsiveness and restricted prostatic expression. Functional androgen response elements (ARE) have also been discovered in similar regions in the *KLK2* gene (Mitchell *et al.*, 2000; Murtha *et al.*, 1993; Young *et al.*, 1992). In breast tissues, the expression of *KLK2* and *KLK3* are mainly androgen-dependent (Hsieh *et al.*, 1997; Magklara *et al.*, 1999). Magklara *et al.* (2000) investigated the steroid hormone-dependent regulation of *KLK2*. The data discloses that *KLK2* gene expression in breast cancer cell lines is under the control of androgens and progestins, similar to KLK3. Additionally, the expression levels in the breast of most of the KLKs do not differ by menopausal status, except for KLK2 and KLK3 (Sauter *et al.*, 2004a). Amazingly, no data has been published regarding a possible clinical impact of KLK2 for prognosis or therapy response in breast cancer or any other malignancy.

KLK3 (prostate specific antigen; semenogelase). KLK3 can be produced by breast tumors, but only low levels of KLK3 are actually released into the female blood circulation (Black *et al.*, 2000b). Although blood KLK3 levels are low, they are not different from those found in the blood of healthy women (Giai *et al.*, 1995; Poh *et al.*, 2008). In contrast, *KLK3* mRNA is downregulated in tumor tissues of breast cancer patients, compared to healthy women (Mangé *et al.*, 2008). Yu *et al.* (1995a) provided evidence of KLK3 protein being present in at least every fourth patient, by assessment of KLK3 protein in tumor tissue extracts by ELISA. However, levels are reduced, compared to normal breast tissues (Poh *et al.*, 2008; Kraus *et al.*, 2010; Yu *et al.*, 1995a).

KLK3 has chymotrypsin-like enzymatic activity and physiologically, in the prostate, it plays a major role in liquefying semen after ejaculation, by cleaving semenogelins and fibronectin (Emami *et al.*, 2008). Although KLK3 is highly elevated in the blood of prostate cancer patients compared to healthy men, due to the destruction, KLK3 protein expression is actually reduced in prostate cancer tissue, compared to its healthy counterparts. The major molecular form of serum KLK3 (PSA) in breast cancer patients is its uncomplexed, free form (Dash *et al.*, 2011; Giai *et al.*, 1995; Borchert *et al.*, 1997). Free serum KLK3, though not showing high diagnostic sensitivity for breast cancer, has a high diagnostic specificity (Black *et al.*, 2000c). Zarghami *et al.*

(1997) demonstrated that in premenopausal women serum KLK3 fluctuates during the menstrual cycle, in response to progesterone surge. One may conclude from this data that KLK3 is not an ideal serum biomarker for distinguishing between healthy women and women afflicted with breast cancer.

In contrast to serum KLK3, nipple aspirate fluid (NAF), due to its easy accessibility, seems to be an alternative source for breast cancer diagnosis (Sauter *et al.*, 1996; 2004a and b). It could be demonstrated that NAF KLK3 levels decrease with larger tumor size, nodal involvement, and more advanced disease stage. Also, NAF KLK3 levels were lower in breast cancer patients with metastases. At the time, the authors speculated that NAF may therefore be useful as a breast cancer screening tool for young women who are not recommended to undergo mammography, and as an adjunct to screening women who have mammograms performed. Yet, other studies have shown that NAF KLK3 is higher in normal subjects, compared to women with hyperplastic breast lesions, women with precancerous mastopathy or invasive cancer have low NAF KLK3 levels. Also, nipple fluid KLK3 does not differ by tumor status. Thus, Zhao *et al.* (2001) concluded that NAF KLK3 detection is not a useful biomarker for breast cancer detection. There was also no correlation between serum and NAF KLK3 (Mitchell *et al.*, 2002).

KLK3-positive breast cancer patients may have a better prognosis than KLK3-negative ones (Black *et al.*, 2000b). Yu *et al.* (1998) determined KLK3 protein in tumor tissue extracts of breast cancer patients by ELISA. KLK3 expression was correlated with smaller tumor size and tumors with low S-phase fraction. Survival analyses indicate that the relative risks for both disease recurrence and death are significantly lower in the KLK3-positive patient group than in the KLK3-negative one (Yu *et al.*, 1998). Contrasting data were published by Heyl *et al.* (1999), who semi-quantified KLK3 protein expression in breast cancer tumor tissues by immunohistochemistry. Using this technique, they did not find a correlation between KLK protein expression and survival. According to these authors, for breast cancer, immunohistochemical assessment of KLK3 protein expression is not to be recommended.

At the same time, but independently, the prevalence and prognostic value of histologically confirmed KLK3 immunoreactivity in breast carcinoma was investigated by Alanen *et al.* (1999). KLK3 protein was expressed in about one third of breast cancer specimens. Staining results were compared with mitotic activity, tumor size, nuclear grade, steroid hormone receptor status, and clinical follow-up data. Only among the group of premenopausal breast cancer patients, concomitant estrogen receptor negativity and KLK3-negativity proved to be associated with a high risk of breast cancer death, also after adjustment for tumor size, nuclear grade, and axillary lymph node status.

Thus, parallel absence of KLK3 and estrogen receptor expression is supposed to be an indicator of unfavorable prognosis among premenopausal patients. Narita *et al.* (2006a and b) have also assessed the clinical value of breast cancer KLK3 immunoreactivity, and correlated KLK3 protein expression with estrogen/progesterone receptor and HER2 status. KLK3 was expressed in about two-third of the cases. Statistical analysis demonstrated a significant correlation between KLK3 and the histological type of the tumor, and showed an inverse correlation between KLK3-positive cases and HER2 overexpression. Survival data were not presented.

Already in 1996, Lai *et al.* reported that KLK3 may act as a negative growth regulator in hormone-dependent breast cancers. Zarghami *et al.* (1997) argued that the expression of KLK3 in the female breast is under the control of androgens and progestins. Shortly after, it was reported that after treatment of advanced stage breast cancer patients with megestrol acetate (a progesterone derivative with antineoplastic properties), who did not demonstrate any changes in their plasma KLK3 level, that these patients had a significantly better prognosis. Those patients who did demonstrate an increase in their plasma KLK3 represented a subset of breast cancer patients who most probably will benefit more from megestrol acetate withdrawal (Diamandis *et al.*, 1999).

At the same time, Foekens *et al.* (1999) also reported that high expression of KLK3 correlates with poor response to endocrine therapy. The study revealed that increased KLK3 protein levels in breast cancer tumor tissues (assessed by ELISA in tumor cytosols) is related to poor response to tamoxifen therapy, and shorter progression-free and overall survival after start of treatment for the recurrent disease. In their study, KLK3 was not related to disease-free survival for patients with primary breast cancer. No newer or updated reports regarding KLK3 expression in breast cancer patients treated with endocrine therapy have been published so far.

Another interesting study was published by Bharaj *et al.* (2000). They reported the prognostic significance of a single nucleotide polymorphism in the proximal androgen response element (ARE) of the breast cancer *KLK3* gene promoter. A G → A base change at position -158 evolved as a polymorphism that may lead to a significant reduction in risk of death.

KLK4 (KLK-L1; enamel matrix serine protease 1; prostase serine protease 17). Different from the other fourteen human KLKs, significant elevation of *KLK4* tissue mRNA levels has been observed in tumor tissues of breast cancer patients, compared to benign breast tumor patients or to normal breast tissue (Papachristopoulou *et al.*, 2009; Mangé *et al.*, 2008). KLK4 cleaves ECM proteins (fibrin and collagens I and IV). Fibrin degradation products have been associated with cancer of the breast, bladder, prostate, and ovary (Obiezu *et al.*, 2006).

To explore the roles of KLK4 in cancer development and progression, investigations that used prostate cancer as a model have been undertaken. For instance, in PC-3 prostate cancer cells that overexpress KLK4, cell migration is increased (Veveris-Lowe *et al.*, 2005; Whitbread *et al.*, 2006). The associated loss of E-cadherin from tumor cells paralleled by elevation of the ECM component vimentin are indicative of EMT (epithelial-mesenchymal transition), which provides compelling evidence that KLK4 plays a functional role in cancer progression through its promotion of tumor cell migration (Veveris-Lowe *et al.*, 2005). During the initial phase of tumor cell invasion, EMT occurs at the primary tumor site. EMT is a biological process that allows a polarized epithelial cell, which normally interacts with the basement membrane via its basal surface, to undergo multiple biochemical changes, which enable it to assume a mesenchymal cell phenotype, which includes enhanced migratory capacity, invasiveness, elevated resistance to apoptosis, and greatly increased production of ECM components. EMT allows tumor cells to degrade the underlying basement membrane and to migrate away from the epithelial layer in which it is originated to invade distant sites. In order to initiate an EMT, a number of distinct molecular processes are engaged, such as expression of specific cell-surface proteins, expression of cytoskeletal proteins, and production of proteolytic enzymes (Kalluri, 2009; Thiery, 2002). For example, decreased E-cadherin function signals ECM and is an early event of tumor metastasis.

In breast cancer, up-regulation of KLK4 protein occurs only in the tumor-cell-surrounding stromal cells and not in the tumor cells, as was demonstrated by immunohistochemical staining of breast cancer tissues (Mangé *et al.*, 2008). *KLK4* is only expressed in breast cancer cells that express the progesterone receptor, a finding which was extended by Lai *et al.* (2009), employing a cultured breast cancer cell line, by demonstrating progesterone receptor interaction with the *KLK4* gene promoter. Low *KLK4* mRNA expression occurs more frequently in well-differentiated breast tumors and in stage I breast cancer patients. There is a significant negative correlation between *KLK4* mRNA expression and progesterone receptor staining (Papachristopoulou *et al.*, 2009).

All this data demonstrates that *KLK4* gene expression can be considered a candidate cancer biomarker predicting unfavorable prognosis in breast cancer patients (Mangé *et al.*, 2008; Papachristopoulou *et al.*, 2009). Yet, high-level expression of KLK4 is not only observed in normal and diseased breast tissue, but also in prostate and ovarian cancer tissues (Klokk *et al.*, 2007; Mangé *et al.*, 2008; Papachristopoulou *et al.*, 2009; Prezas *et al.*, 2006; Xi *et al.*, 2004a) as well as squamous oral carcinoma (Zhao *et al.*, 2011). In ovarian cancer, overexpression of KLK4 predicts resistance to the chemotherapeutic drug paclitaxel (Xi *et al.*, 2004b). These clinical correlations suggest that high levels of KLK4 are associated with more aggressive tumors. Otherwise, small-interfering-RNA-mediated knockdown of endogenous *KLK4* in LNCaP prostate cancer cells resulted in impeded cell growth, which qualifies KLK4 as a proliferative factor effecting cancer development and progression (Klokk *et al.*, 2007).

Besides its role in cancer, KLK4 is one of the major proteases secreted into the enamel matrix of the developing teeth (see Chapter 11, Volume 1). It degrades the retained organic matrix, following the termination of enamel protein secretion, thus facilitating the orderly replacement of organic matrix with mineral, generating an enamel layer that is harder, less porous, and unstained by retained enamel proteins. Mutations in the *KLK4* gene can cause autosomal recessive amelogenesis imperfecta, a condition featuring soft, porous enamel that contains residual protein (Lu *et al.*, 2008).

Several of the KLKs, including KLK4, can activate PAR-1 and PAR-2, thereby triggering intracellular signaling via the PARs. In this respect it is worthwhile to mention that KLK4 and PAR-2 have been found to be coexpressed during prostate cancer progression, where KLK4 initiates the loss of PAR-2 from the tumor cell surface with receptor internalization (Ramsay *et al.*, 2008). KLK4 can also stimulate prostate cancer cell proliferation and phosphorylation of ERK1/2 through interaction with PAR-1 and PAR-2 (Mize *et al.*, 2008).

KLK4 does modulate the tumor-associated plasminogen activation system by either activating the pro-enzyme form of uPA or by cleaving the tumor cell surface-associated uPA receptor, CD87 (Beaufort *et al.*, 2006) (see Chapter 9, Volume 1). Several growth factors and hormones have also been identified as potential substrates of KLKs, such as insulin-like growth factor binding proteins (IGFBP) and tumor necrosis factor β (TNF-β). Insulin-like growth factors (IGF) are well-known mitogens and are notorious for promoting breast cancer development. IGFBPs bind to IGF-I and -II and alter the interaction with their receptors (Pollack, 2008).

KLK5 (KLK-L2; stratum corneum tryptic enzyme). Although almost undetectable in serum of normal individuals, elevated serum concentrations of KLK5 are present in about 40–50% of patients with breast cancer (Yousef *et al.*, 2003b). In contrast, *KLK5* mRNA is higher in normal breast tissues than in breast cancer tumor tissues (Yousef *et al.*, 2004a). Although lower than in the healthy breast (Avgeris *et al.*, 2011; Li *et al.*, 2009), relatively elevated *KLK5* mRNA levels are more common in pre-/perimenopausal, node-positive, and estrogen receptor-negative patients. *KLK5* mRNA is downregulated in metastases, compared to primary cancers (Li *et al.*, 2009).

The prognostic relevance of *KLK5* mRNA for breast cancer was first demonstrated by Yousef *et al.* (2002b). In this study, *KLK5* mRNA overexpression emerged as a significant predictor of poor disease-free and overall survival. Talieri *et al.* (2011) also demonstrated that a relative increase in tumor tissue *KLK5* mRNA expression is a prognostic factor for disease-free and overall survival in breast cancer patients. Both Talieri *et al.* (2011) and Li *et al.* (2009) reported parallel downregulation of *KLK5* and *KLK7* in breast cancer. KLK5 expression was also reported to be of prognostic relevance in other cancer diseases, e.g. cancers of the colon (Talieri *et al.*, 2009a), lung (Planque *et al.*, 2010), ovary (Dorn *et al.*, 2011), and the kidney (Petraki *et al.*, 2006a).

KLK5 is detectable in several tissues, with the highest expression levels seen in the skin, vagina, breast, salivary gland, cervix uteri, thyroid, and esophagus (Shaw and Diamandis, 2007). Besides that, KLK5 is present, with a relatively high concentration, in the milk of lactating women. Utilizing Expressed Sequence Tag database analysis and RT-PCR reaction, Yousef *et al.* (2004b) identified an alternatively spliced form of KLK5 (KLK5-splice variant 2, KLK5-SV2) which is expressed by a variety of different tissues (mammary gland, cervix, salivary gland, and trachea). KLK5 is involved in stratum corneum turnover of the skin and desquamation of its epidermis (Brattsand *et al.*, 1999) (see Chapter 13, Volume 1). Uncontrolled KLK5 proteolytic activity in the epidermis can trigger atopic dermatitis-like lesions (Briot *et al.*, 2009).

KLK5 can efficiently digest ECM components, such as collagens type-I,-II,-III, and -IV, fibronectin, and laminin. KLK5 may activate plasminogen and low-molecular-weight kininogen and regulate binding of the plasminogen activator inhibitor type-1 (PAI-1) to the ECM compound vitronectin (Michael *et al.*, 2005). Through cleavage of desmoglein 1 it may promote loss of junctional integrity (Jiang *et al.*, 2011). Collectively, these findings suggest that through proteolytic cleavage of growth factors and ECM proteins, KLK5 may be involved in tumor progression, particularly in tumor cell invasion and angiogenesis. Michael *et al.* (2006) reported that KLK5 also cleaves IGFBP, endorsing its role in cancer progression through growth factor regulation. In the study by Prassas *et al.* (2008), three small molecule libraries were screened for their potential to decrease KLK5 protein expression in the breast cancer cell line MDA-MB 468. They found that cardiac glycosides did effectively suppress the transcription of *KLK5*, a process linked to proto-oncogene (c-myc/fos) expression.

KLK6 (protease M; zyme; neurosin; serine protease 9; serine protease 18). KLK6 was originally identified due to its differential expression in metastatic breast cancer cells (Anisowicz *et al.*, 1996). In breast cancer tissue, *KLK6* mRNA is expressed at a lower rate than in healthy breast tissue (Mangé *et al.*, 2008; Yousef *et al.*, 2004c), but KLK6 protein expression is elevated in breast cancer tissues, compared to normal breast tissues (Petraki *et al.*, 2001). Along that line, Wang *et al.* (2008) investigated the impact of KLK6 expression on the rate of metastasis in breast cancer patients. They reported that KLK6 protein was expressed less frequently in breast carcinoma patients displaying lymph node metastases and positive estrogen receptor status. Thus, KLK6 expression in cancerous tissues may play an important role in the invasion and metastasis of primary breast carcinomas. No further clinically relevant data regarding KLK6-depending prognosis or therapy response prediction have been published for breast cancer. For other malignancies (ovary, colorectum, lung, kidney, pancreas), however, the prognostic relevance of KLK6 has been demonstrated (Kim *et al.*, 2011; Nagahara *et al.*, 2005; Nathalie *et al.*, 2009; Petraki *et al.*, 2006a; 2012; Rückert *et al.*, 2008; Seiz *et al.*, 2012; White *et al.*, 2009).

KLK7 (stratum corneum chymotryptic enzyme, serine protease 6). Similar to other KLKs (Mangé *et al.*, 2008), KLK7 mRNA expression is lower in breast cancer than in healthy breast tissue. No data was published demonstrating a direct role of KLK7 in tumor invasion and metastasis. However, KLK7 is known to generate active matrix metalloprotease 9 (MMP-9) by proteolytic action (Ramani *et al.*, 2011). MMP-9 is involved in tumor invasion and metastasis in breast cancer (Cupić *et al.*, 2011). Indirect evidence of a critical role for KLK7 expression in breast cancer comes from two independent studies, in which the clinical relevance of KLK7 in breast cancer was documented. Talieri *et al.* (2004), using semi-quantitative RT-PCR, reported that *KLK7* gene expression was significantly lower in breast cancer patients of low stage (I/II) and patients with positive progesterone receptors, and that breast cancer patients with *KLK7*-positive tumors had relatively shorter disease-free and overall survival than patients with *KLK7*-negative tumors.

In contrast, Holzscheiter *et al.* (2006), employing quantitative RT-PCR, showed that elevated *KLK7* mRNA expression was significantly associated with a better outcome in untreated breast cancer patients. When comparing the patients participating in these investigations, Holzscheiter *et al.* (2006) pointed out that in the first study the majority of patients was subjected to adjuvant therapy, which might explain the observed discrepancy. Patients in the second study were the least influenced by postoperative treatments, and thus might best disclose the impact of *KLK7* mRNA expression on the natural course of breast cancer. Talieri *et al.* (2012) and Li *et al.* (2009) reported parallel downregulation of *KLK7* with *KLK5* in breast cancer. KLK7 mRNA or protein expression was also reported to be of prognostic relevance to ovarian cancer (Kyriakopoulou *et al.*, 2003; Oikonomopoulou *et al.*, 2008; Shan *et al.*, 2006), colorectal cancer (Inoue *et al.*, 2010; Talieri *et al.*, 2009b), pancreatic cancer (Iakovlev *et al.*, 2012; Johnson *et al.*, 2007), prostate cancer (Mo *et al.*, 2010), and oral squamous carcinoma (Zhao *et al.*, 2011).

KLK8 (neuropsin; ovasin; serine protease 19; tumor-associated differentially expressed gene 14 protein). *KLK8* mRNA is also less expressed in breast cancer tissues than in healthy breast tissue (Mangé *et al.*, 2008; Yousef *et al.*, 2004a). No data has been reported concerning the possible clinical relevance of KLK8 in breast cancer with regard to prognosis or therapy response prediction. Its prognostic value was demonstrated, however, for lung cancer (Planque *et al.*, 2010; Sher *et al.*, 2006) and ovarian cancer (Borgoño *et al.*, 2006; Kountourakis *et al.*, 2009; Magklara *et al.*, 2001; Shigemasa *et al.*, 2004).

KLK9 (KLK-L3). *KLK9* mRNA is also expressed less in breast cancer tissues than in healthy breast tissue (Mangé *et al.*, 2008). *KLK9* mRNA expression is significantly higher in patients with early-stage cancers and with small tumor size (Yousef *et al.*, 2003c). *KLK9*-positive patients have longer disease-free and overall survival, especially in the estrogen receptor- and progesterone receptor-negative subgroups of patients. Thus, *KLK9* mRNA is a marker of favorable prognosis in breast cancer. Except for ovarian cancer (Yousef *et al.*, 2001a), KLK9 was not reported as a marker to predict prognosis or therapy response in other malignant diseases.

KLK10 (normal epithelial cell-specific-1; protease serine-like 1). Like several other KLKs, *KLK10* mRNA is expressed less in breast cancer tissues than in healthy breast tissue (Mangé *et al.*, 2008; Yousef *et al.*, 2004a). The gene that encodes for *KLK10* was initially isolated by subtractive hybridization, by virtue of its downregulation in radiation-transformed breast epithelial cells, in comparison to its normal counterpart and, hence, it was termed normal epithelial cell specific 1 (NES1) (Goyal, 1998; Liu *et al.*, 1996). Further investigations showed that almost all of the normal breast specimens expressed KLK10 mRNA/protein (Dhar *et al.*, 2001; Petraki *et al.*, 2002; Zhang *et al.*, 2006) with loss of KLK10 expression during tumor progression. More specifically, about half of ductal carcinoma *in situ* (DCIS), and the majority of infiltrating ductal carcinoma specimens lacked *KLK10* mRNA. Importantly, *KLK10*-negative DCIS cases assessed at the time of biopsy were subsequently diagnosed as infiltrating ductal carcinomas at the time of definitive surgery (Zhang *et al.*, 2006).

KLK10 DNA-methylation does not occur in normal breast tissues and benign diseases, but may do so in breast cancer. CpG island hypermethylation of the *KLK10* gene seems to be responsible for the tumor-specific loss of *KLK10* gene expression in breast cancer (Kioulafa, *et al.*, 2009; Sidiropoulos *et al.*, 2005). KLK10 protein concentration, determined by ELISA in blood of breast cancer patients, was found to be doubled, compared to blood of healthy women (Ewan King *et al.*, 2007).

Although statistical analysis has shown that breast cancer tumor tissue KLK10 protein levels are associated with younger age, pre-menopausal status, and tumors that are negative for estrogen and progesterone receptors, KLK10 expression is not related to disease-free or overall survival (Luo *et al.*, 2002). However, higher levels of KLK10 are significantly associated with a poor response rate to endocrine therapy with tamoxifen. Furthermore, higher KLK10 levels are significantly related to shorter progression-free and overall survival after the start of tamoxifen therapy in the metastatic setting. Taken together, these results suggest that KLK10 is a predictive marker for response to tamoxifen therapy.

Since silencing of tumor suppressor genes through hypermethylation is a frequent event in carcinogenesis, these alterations have the potential to be used as biomarkers. The prognostic significance of *KLK10* exon 3 methylation in patients has thus been explored (Li *et al.*, 2001; Sidiropoulos *et al.*, 2005). *KLK10* is not methylated in normal breast tissues and fibroadenomas, whereas it is methylated in about 50% of early-stage breast cancer tumors. *KLK10* gene methylation seems to be present more frequently in those patients who recur and die. Thus, *KLK10* exon 3 methylation can provide important prognostic information in early breast cancer patients (Kioulafa *et al.*, 2009). The prognostic value of KLK10 was demonstrated for colon, renal, and pancreatic cancer (Petraki *et al.*, 2006a; Rückert *et al.*, 2008; Talieri *et al.*, 2009a).

KLK11 (trypsin-like serine protease; hippostasin; serine protease 20). *KLK11* mRNA expression is similar (Sano *et al.*, 2007) or lower (Mangé *et al.*, 2008) in breast cancer tumor tissues, compared to healthy breast tissue. Remarkably, estrogen receptor-positive breast cancer cells show relatively high expression of KLK11. Since KLK11 may degrade IGFBP-3, these results indicate that KLK11, expressed in estrogen receptor-positive breast cancer cells, can be crucial for breast cancer progression, by providing growth factors, e.g. by degradation of IGFBP-3 (Sano *et al.*, 2007).

Data concerning a prognostic or predictive role for KLK11 in breast cancer was not reported, but KLK11 expression was documented to be a cancer biomarker for predicting the prognosis of several other cancer diseases (larynx, stomach, colorectum, lung, kidney, ovary, prostate) (Borgoño *et al.*, 2003a; Diamandis *et al.*, 2004; Patsis *et al.*, 2012; Petraki *et al.*, 2006a; Planque *et al.*, 2010; Sasaki *et al.*, 2006; Shigemasa *et al.*, 2004; Stavropoulou *et al.*, 2005; Talieri *et al.*, 2009a; Wen *et al.*, 2011; Yu *et al.*, 2010).

KLK12 (KLK-L5). Like several other KLKs, *KLK12* mRNA is expressed less in breast cancer tissues than in healthy breast tissue (Mangé *et al.*, 2008; Yousef *et al.*,

2004a). Data concerning a prognostic or predictive role of KLK12 was not reported for breast cancer or any other malignancy. However, the *KLK* splice variant *KLK12sv3* was reported to be associated with improved disease-free survival (Talieri *et al.*, 2012). Alternative splicing of *KLKs* emerges as a process that might play a significant role in tissue development and physiology, but also in cancer. The *KLK* family encompasses at least 82 alternative transcripts.

Guillon-Munos *et al.* (2011) demonstrated that KLK12 can target all six members of the cell-function-regulating CCN family at different proteolytic sites. e.g. fragmentation of CCN1 or CCN5 by KLK12 prevents VEGF (vascular endothelial growth factor) binding. In addition, it triggers the release of intact VEGF and BMP2 (bone morphogenetic protein-2) from the CCN complexes. These findings suggest that KLK12 may indirectly regulate the bioavailability and activity of several growth factors through the processing of their CCN binding partners.

KLK13 (KLK-L4). Chang *et al.* (2002) studied the expression of KLK13 in patients with epithelial breast carcinoma. They discovered that a relatively higher level of *KLK13* mRNA expression is present more frequently in older, estrogen receptor-positive patients. *KLK13* expression was a significant predictor of improved disease-free and overall survival in patients with grade I-II tumors, or in patients who were estrogen receptor- and progesterone receptor-positive, as well as lymph node-positive. KLK13 may cleave major ECM components (Kapadia *et al.*, 2004).

Higher *KLK13* mRNA expression was associated with an up to 80% reduction of the risk for disease recurrence or death. Thus, in breast cancer, *KLK13* mRNA is a favorable prognostic marker. In good agreement with these findings, tumor tissue KLK13 protein level for ovarian and gastric cancer is also a marker of a favorable prognosis (Konstantoudakis *et al.*, 2010; Scorilas *et al.*, 2004). In ovarian cancer, however, elevated *KLK13* mRNA expression is a marker of poor prognosis (White *et al.*, 2009). The same is true for colon cancer (Talieri *et al.*, 2009a).

KLK14 (KLK-L6). *KLK14* mRNA is also expressed less in breast cancer tissues than in healthy breast tissue (Mangé *et al.*, 2008; Yousef *et al.*, 2002c). Yousef *et al.* (2001c) cloned *KLK14*, which is also expressed at the gene and protein level in a variety of tissues, including the breast (Borgoño *et al.*, 2003b). Yousef *et al.* (2002c) reported that elevated levels of *KLK14* mRNA are more common in patients with advanced stage cancer, and that overexpression of *KLK14* is associated with decreased disease-free and overall survival, thus qualifying *KLK14* as a marker of unfavorable prognosis in breast cancer.

Fritzsche *et al.* (2006) published the association of KLK14 protein expression with higher tumor grade and positive nodal status. These clinical findings were confirmed recently by Papachristopoulou *et al.* (2011), who found that *KLK14* mRNA expression is correlated with estrogen receptor status. Although almost undetectable in serum of normal individuals, elevated concentrations of KLK14 are present in about 40–50% of patients with breast cancer (Borgoño *et al.*, 2003b; Yousef *et al.*, 2003b).

To further characterize the value of KLK14 as a breast cancer biomarker, Fritzsche *et al.* (2006) analyzed KLK14 expression in normal breast tissue and in breast cancer tumor tissue specimens both at the mRNA and the protein level. Different from the initial study by Yousef *et al.* (2001b), they reported that *KLK14* mRNA expression was elevated in breast cancer tumor tissues, compared to normal breast tissue, data which was later confirmed by Papachristopoulou *et al.* (2011). Likewise, KLK14 protein also was overexpressed in diseased breast tissues (Fritzsche *et al.*, 2006).

The substrate repertoire of trypsin-like KLK14 includes collagens I and IV, fibronectin, laminin, and high-molecular-weight kininogen, fibrinogen, plasminogen, vitronectin, and IGFBP-3, indicating that KLK14 may be involved in several facets of tumor progression and metastasis (Borgoño *et al.*, 2007; Rajapakse and Takahashi, 2007). KLK14 is expressed in a variety of tissues. The highest levels of KLK14 are present in the central nervous system (brain, cerebellum, spinal cord). *KLK14* mRNA also is down-regulated in testicular, prostatic, and ovarian cancer (Yousef *et al.*, 2001b). Little data is available concerning a possible prognostic or predictive relevance to other malignancies, except for colon cancer. Talieri *et al.* (2009a) reported that, for this malignant disease, *KLK14* mRNA is also a marker of poor prognosis.

KLK15 (ACO; HSRNASPH; prostinogen). Higher concentrations of *KLK15* mRNA have been found more frequently in node-negative patients (Yousef *et al.*, 2002c). Although there were no statistically significant associations between KLK15 status and age, menopausal status, tumor size, stage, nuclear grade, histological type, estrogen receptor or progesterone receptor status, and adjuvant chemotherapy, an increase in *KLK15* predicted longer progression-free and overall survival, thus qualifying KLK15 as a favorable prognostic factor (Yousef *et al.*, 2002c). Clinical investigations regarding the clinical impact of KLK15 on other malignancies are scarce, except for ovarian cancer. Here, it was shown by Batra *et al.* (2011) that *KLK15* single nucleotide polymorphism located close to a novel exon shows evidence of an association with poor survival of ovarian cancer .

6.4 Hormonal regulation of KLKs in breast cancer

Adult female mammary growth and development are controlled by tightly regulated cross-talk between a group of steroid hormones and growth factors. Estrogen, along with other growth factors such as IGF-I (insulin-like growth factor-1), stimulates epithelial cell proliferation and growth in the breast tissue, while androgen has opposite effects. Together, these hormones act synergistically to regulate complex pathways which are critical to mammary gland growth and differentiation. It is well known that perturbation of these hormones can lead to the development of hyperplastic breast lesions and can increase the risk of developing breast cancer (Dimitrakakis and Bondy, 2009; Renehan *et al.*, 2004).

A majority of KLKs are known to be regulated by steroid hormones, but the most widely studied example is KLK3. KLK3 is well-known to be regulated by androgens and it has been a well-established model for studying androgen-regulated gene expression.

Three major functional hormone response elements (HRE) have been identified in the *KLK3* gene proximal promoter and the distant enhancer region (Cleutjens *et al.*, 1997; Riegman *et al.*, 1991; Schuur *et al.*, 1996). Similar to *KLK3*, *KLK2* also displays strong androgen responsiveness and restricted prostatic expression. Functional androgen response elements (ARE) have also been discovered in similar regions in the *KLK2* gene (Mitchell *et al.*, 2000; Murtha *et al.*, 1993; Young *et al.*, 1992). In breast tissues, the expression of *KLK2* and *KLK3* is mainly androgen-dependent (Hsieh *et al.*, 1997; Magklara *et al.*, 1999). To a lesser extent, progestins also upregulate *KLK2* and *KLK3*.

For the other KLKs, regulation by multiple steroid hormones has been demonstrated by hormone stimulation experiments, using various breast cancer cell lines. These investigations have shown that KLK4-6 and 8–15 are all regulated by estrogens, androgens, or progestins (Lawrence *et al.*, 2010; Yousef *et al.*, 1999). Although DNA sequence analysis detected one putative progesterone response element (PRE) and one ARE upstream of the *KLK4* promoter region, further experiments have demonstrated that only the PRE seems to be functional (Lai *et al.*, 2009). A search of HREs in the *KLK10* promoter region, however, failed to identify any (Luo *et al.*, 2003). Although a putative ARE is also indicated in the *KLK14* gene, its functionality awaits experimental definition (Yousef *et al.*, 2003a). It is noteworthy that these *KLKs* display very similar patterns in their regulation by steroid hormones. Given that they are colocalized in the same locus and coexpressed in the same tissues, it has been postulated that their coordinated hormonal regulation may represent a unique expression "cassette", which utilizes a common hormone-dependent enhancer (Paliouras *et al.*, 2007).

Another question that remains unanswered is whether the modulation of *KLKs* by steroid hormones is the result of a direct or indirect effect. Interestingly, the roles of sex hormones on the expression of *KLKs* have also been studied *in vivo*. These studies, which were mainly performed in female-to-male and male-to-female transsexuals receiving cross-sex hormones, have shown that, in serum and urine, *KLK2* and *KLK3* are upregulated by testosterone and downregulated by antiandrogens and estrogens, respectively. This observation is consistent with data obtained from *in vitro* studies. However, in these studies, the expression levels of all other KLKs are, surprisingly, unaffected by the disturbance of sex hormones (Slagter *et al.*, 2006).

Another observation worth mentioning is that in breast tumors not all *KLK* expression levels correlate with the steroid hormone receptor status. For instance, although *KLK5* is upregulated by estrogens in breast cancer cell lines, higher *KLK5* expression is actually observed in estrogen receptor-negative breast tumors (Shaw *et al.*, 2008; Yousef *et al.*, 2002b). Additionally, the expression levels of most KLKs in the breast, with the exception of *KLK2* and *KLK3*, do not differ by menopausal status (Sauter *et al.*, 2004a). These findings indicate that expression of KLKs is likely to be regulated not only by steroid hormones but also by other growth factors or signal pathways.

As shown by Paliouras *et al.* (2008a and b), the RAS/MEK/ERK, PI3K/AKT, and c-myc signaling pathways are all involved in the induction of *KLK3*, *10*, and *11* by androgens. Androgens act synergistically to enhance estrogen-induced upregulation of *KLK10, 11,* and *14* in breast cancer cells via a membrane-bound androgen receptor (Paliouras *et al.*, 2008c). Therefore, taken together, it becomes clear that most *KLKs* are upregulated by multiple steroid hormones, suggesting that they might be downstream targets for these hormones. But whether and how they convey the growth-regulation effects of these steroid hormones on breast cancer remains to be further elucidated.

6.5 Tumor suppressor role of KLKs in breast cancer

Tumor suppressors, or anti-oncogenes, are those proteins that can protect cells from malignant transformation. They encompass proteins belonging to several functional classes, such as cell cycle control, DNA repair, apoptosis, and cell adhesion. Classical examples of tumor suppressors include the retinoblastoma protein, p53, VHL, APC, CD95, ST5, YPEL3, ST7, and ST14 (Levine *et al.*, 2010). Quite often, during tumor development, the expression of tumor suppressors is reduced. Some *KLKs* display tumor suppressor characteristics, e.g. *KLK3, 6,* and *10.*

KLK3. In breast cancer, the expression of KLK3 is reduced in comparison to healthy and benign breast tissues, which qualifies KLK3 as a protein with tumor suppressor activity. Several lines of evidence support this hypothesis. Women with lower KLK3 concentrations in their nipple aspirate fluid show a higher risk of developing breast cancer, suggesting that downregulation of KLK3 may predispose to breast cancer development (Sauter *et al.*, 2004a). Otherwise, early *in vitro* studies have shown that KLK3 can inhibit the growth of cultured breast cancer cells (Lai *et al.*, 1996). KLK3 can also function as an antiangiogenic molecule. It can inhibit endothelial cell proliferation, migration, and invasion, by blocking their response to FGF-2 and VEGF (Fortier *et al.*, 1999). KLK3 can also digest plasminogen to release an angiostatin-like peptide, which inhibits new blood vessel formation in tumors (Heidtmann *et al.*, 1999).

Paradoxically, not all experimental evidence agrees that KLK3 acts as a tumor suppressor gene. In some other investigations, KLK3 was shown to effect deleterious outcome in cancer. Kaulsay *et al.* (1999) determined serum IGF-binding protein-6 (IGFBP-6) and KLK3 in breast cancer patients. Sutkowski *et al.* (1999) reported that KLK3 can cleave IGFBP-3, thereby releasing the growth factor IGF-1. Additionally, KLK3 can activate latent TGF-β to then stimulate cell detachment and thereby facilitate tumor spread. Furthermore, KLK3 can weaken the basement membrane through proteolysis, in order to ease tumor invasion and metastasis (Pezzato *et al.*, 2004).

KLK6. KLK6 was originally identified as being downregulated in metastatic breast tumors, compared to the primary tumor, suggesting that KLK6 might function as a metastatic inhibitor and tumor suppressor (Anisowicz *et al.*, 1996). In line with this putative role of being a tumor suppressor, overexpression of KLK6 in a cell line derived from metastatic breast tumor tissue cells resulted in a marked reversal of its malignant phenotype (Pampalakis *et al.*, 2009). Further differential proteomic profiling by these authors unveiled that KLK6 overexpression may lead to significant down-regulation of vimentin, which is known as an established marker of EMT of tumor cells, paralled by up-regulation of calreticulin and cytokeratins 8 and 19. These observations indicate that KLK6 may play a protective role against breast cancer progression, mediated by inhibition of EMT (Pampalakis *et al.*, 2009).

KLK6, displaying trypsin-like enzymatic activity, can efficiently degrade high-molecular-weight ECM proteins, such as fibronectin, laminin, vitronectin, and collagen (Ghosh *et al.*, 2004). This was shown for mouse keratinocytes that overexpress KLK6 and for *KLK6*-transgenic mice (Klucky *et al.*, 2007). In these models, enhanced epidermal keratinocyte proliferation and migration were acting in concert with decreased E-cadherin protein levels. Also, breast cancer cells that were treated with a neutralizing antibody, directed to KLK6's proteolytic center, migrated less than untreated control cells (Ghosh *et al.*, 2004). Another study, employing colon cancer cells, has shown that in these cells KLK6 is up-regulated via the K-Ras pathway. This increased KLK6 expression was correlated with enhanced tumor cell migration and invasion (Henkhaus *et al.*, 2008). In addition, in non-small-cell lung cancer cells, ectopic expression of KLK6 resulted in strongly enhanced cell growth, owing to accelerated G1- and S-phase cell cycles (Nathalie *et al.*, 2009). This data suggests that KLK6 is involved in tumor cell progression.

These contradictory observations and paradoxical roles of KLK3 and KLK6 in breast cancer may be related to variations in tumor tissue type and histology, molecular grade, nuclear grade, and cancer stage (Sotiriou and Pusztai, 2009) and/or might depend on the concentration and/or activity state of the respective KLK. Indeed, it has been reported that the tumor-suppressive effects of KLK6 are restricted to normal concentrations of the protein, whereas a marked overexpression of KLK6 was associated with enhanced tumor growth (Diamandis *et al.*, 2000; Sotiropoulou *et al.*, 2009).

KLK10. The tumor suppressor role of *KLK10* in breast cancer is more definitive than that of KLK3 and KLK6. Despite the fact that *KLK10* is expressed in the majority of normal breast tissues, it is mainly absent in *ductal carcinoma in situ* and in invasive ductal carcinoma (Dhar *et al.*, 2001; Sidiropoulos *et al.*, 2005; Zhang *et al.*, 2006). Interestingly, KLK10 is downregulated in the majority of breast cancer cell lines as well, but when overexpressed in such KLK10-negative breast cancer cells, these KLK10-positive cells did suppress anchorage-independent growth and tumor formation in nude mice (Goyal *et al.*, 1998).

6.6 DNA-methylation of KLKs as the basis of KLK downregulation in breast cancer

KLKs are regulated at the transcriptional, translational, and posttranslational level (Emami and Diamandis, 2008; Pasic *et al.*, 2012) but epigenetic changes may play a critical role in *KLK* gene regulation (Emami and Diamandis, 2008; Pasic *et al.*, 2012;

Sotiropoulou *et al.*, 2009). DNA-methylation is one of the most important epigenetic mechanisms, often leading to gene silencing during cancer development and progression (Esteller, 2008). In normal cells, the DNA-methylation patterns are conserved throughout cell division, allowing the expression of a particular set of cellular genes necessary for that cell type. However, in cancer cells, CpG islands, which typically occur at or near the transcription start site of genes, are often methylated, but are mostly unmethylated in normal tissues. Such aberrantly methylated DNA sequences may render them unrecognizable due to transcription factors, alteration of chromatin structure, and modification of histones, leading to inactivation of cancer-related genes.

Several of the CpG islands within the *KLK* gene locus are not located in promoter regions, but it is known that exonic CpG island hypermethylation can occur independently of promoter methylation, which still can result in gene inactivation (El-Naggar *et al.*, 1997; Kioulafa *et al.*, 2009; Li *et al.*, 2001). Recent data supports the notion that epigenetically modified KLKs, such as *KLK6* and *KLK10*, are involved in cancer (Bharaj *et al.*, 2002; Kioulafa *et al.*, 2009; Pasic *et al.*, 2012; Pampalakis and Sotiropoulou, 2006; Sidiropoulos *et al.*, 2005; White *et al.*, 2010).

Loss of *KLK6* expression in breast cancer may be due to hypermethylation of specific CpG islands located in the *KLK6* proximal promoter, thus prompting the *KLK6* gene as a potential target for pharmacological intervention. However, unlike *KLK10*, *KLK6* mRNA expression could also be restored by the vitamin D3 analog EB1089 in certain breast cancer cell lines. Therefore, the mechanisms underlying the transcriptional deregulation of *KLK6* in breast cancer cells seem to be more complicated, and are certainly not uniform. Both epigenetic mechanisms and pathways regulated by nuclear receptors may possibly be involved (Pampalakis and Sotiropoulou, 2006).

In breast cancer, among the KLKs, methylation of the *KLK10* gene is most extensively investigated. The *KLK10* gene spans about 5.5 kb in length and is composed of five introns, five coding exons, and one non-coding exon (Luo *et al.*, 1998). Since no gross deletion, insertion, or rearrangements have been identified in the *KLK10* gene in any of the breast cancer cell lines that lose *KLK10* expression, other mechanisms are likely to contribute (Liu *et al.*, 1996). Sequence analysis of the *KLK10* gene has revealed that, while its promoter does not contain CpG-rich islands, its exon 3 is CpG-rich, and these CpG islands are in fact highly methylated in breast cancer cell lines and primary breast cancer tumor tissues (Sidiropoulos *et al.*, 2005). When these cells were treated with 5-aza-deoxycytidine, *KLK10* mRNA expression was restored (Li *et al.*, 2001; Sidiropoulos *et al.*, 2005). These findings demonstrate that methylation of *KLK10* exon 3 is responsible for the loss of *KLK10* gene expression in breast cancer. Although not yet shown for breast cancer, it is worth mentioning that epigenetic activation of *KLK13* enhances the malignancy of lung adenocarcinoma cells by promoting N-cadherin expression and laminin degradation (Chou *et al.*, 2011). These results reveal the enhancing effects of *KLK13* on tumor cell invasion and migration.

6.7 Conclusions

Here, we provide insights into the functions and clinical relevance of KLKs in breast cancer. It is now known that extracellular matrix proteins, growth factor precursors, and their binding proteins, as well as cell surface receptors, are all potential targets of KLK proteolysis. Such targets may well explain some, but by no means all of the actions of the KLKs during breast cancer development and progression. It is evident that the KLKs' functions are far more complex than initially thought, given that these proteases do more than degrading physical barriers.

Rather, they can act independently of their proteolytic activity and affect multiple signaling pathways modulating the biology of breast cancer cells. Depending on the circumstances, KLKs may either suppress or promote tumorigenesis and metastasis. Preliminary clinical studies have demonstrated that many KLKs potentially have utility as biomarkers. Individually or in combination with other candidate biomarkers, they may aid early detection of breast cancer, helping to predict patient outcome, and to select treatment strategies. These findings underscore the importance of KLKs in breast cancer and warrant further investigations.

Bibliography

Alanen, K.A., Kuopio, T., Collan, Y.U., Kronqvist, P., Juntti, L., and Nevalainen, T.J. (1999). Immuno-histochemical labelling for prostate-specific antigen in breast carcinomas. Breast Cancer Res. Treat. 56, 169–176.

Anisowicz, A., Sotiropoulou, G., Stenman, G., Mok, S.C., and Sager, R. (1996). A novel protease homolog differentially expressed in breast and ovarian cancer. Mol. Med. 2, 624–636.

Avgeris, M., Mavridis, K., and Scorilas, A. (2010). Kallikrein-related peptidase genes as promising biomarkers for prognosis and monitoring of human malignancies. Biol. Chem. 391, 505–511.

Avgeris, M., Papachristopoulou, G., Polychronis, A., and Scorilas, A. (2011). Down-regulation of kallikrein-related peptidase 5 (KLK5) expression in breast cancer patients: a biomarker for the differential diagnosis of breast lesions. Clin. Proteomics 8, 5.

Avgeris, M., Mavridis, K., and Scorilas A. (2012). Kallikrein-related peptidases in prostate, breast, and ovarian cancers: from pathobiology to clinical relevance. Biol. Chem. 393, 301–317.

Batra, J., Nagle, C.M., O'Mara, T., Higgins, M., Dong, Y., Tan, O.L., Lose, F., Skeie, L.M., Srinivasan, S., Bolton, K.L., Song, H., Ramus, S.J., Gayther, S.A., Pharoah, P.D., Kedda, M.A., Spurdle, A.B., and Clements, J.A. (2011). A kallikrein 15 (KLK15) single nucleotide polymorphism located close to a novel exon shows evidence of association with poor ovarian cancer survival. BMC Cancer 11, 119.

Beaufort, N., Debela, M., Creutzburg, S., Kellermann, J., Bode, W., Schmitt, M., Pidard, D., and Magdolen, V. (2006). Interplay of human tissue kallikrein 4 (hK4) with the plasminogen activation system: hK4 regulates the structure and functions of the urokinase-type plasminogen activator receptor (uPAR). Biol. Chem. 387, 217–222.

Bharaj, B., Scorilas, A., Diamandis, E.P., Giai, M., Levesque, M.A., Sutherland, D.J., Hoffman, B.R. (2000). Breast cancer prognostic significance of a single nucleotide polymorphism in the proximal androgen response element of the prostate specific antigen gene promoter. Breast Cancer Res. Treat. 61, 111–119.

Bharaj, B.B., Luo, L.Y., Jung, K., Stephan, C., and Diamandis, E. P. (2002). Identification of single nucleotide polymorphisms in the human kallikrein 10 (KLK10) gene and their association with prostate, breast, testicular, and ovarian cancers. Prostate 51, 35–41.

Bhoola, K.D., and Fink, E. (2006). Kallikrein-kinin cascade. In: Laurent G.J., Shapiro S.D., eds. Encyclopedia of Respiratory Medicine, Vol 2. Oxford, UK: Elsevier Ltd, 483–493.

Black, M.H., Magklara, A., Obiezu, C., Levesque, M.A., Sutherland, D.J., Tindall, D.J., Young, C.Y., Sauter, E.R., and Diamandis, E.P. (2000a). Expression of a prostate-associated protein, human glandular kallikrein (hK2), in breast tumours and in normal breast secretions. Br. J. Cancer 82, 361–367.

Black, M.H., Giai, M., Ponzone, R., Sismondi, P., Yu, H., and Diamandis, E.P. (2000b). Serum total and free prostate-specific antigen for breast cancer diagnosis in women. Clin. Cancer Res. 6, 467–473.

Black, M.H., and Diamandis, E.P. (2000c). The diagnostic and prognostic utility of prostate-specific antigen for diseases of the breast. Breast Cancer Res. Treat. 59, 1–14.

Boire, A., Covic, L., Agarwal, A., Jacques, S., Sherifi, S., and Kuliopulos, A. (2005). PAR1 is a matrix metalloprotease-1 receptor that promotes invasion and tumorigenesis of breast cancer cells. Cell 120, 303–313.

Booden, M.A., Eckert, L.B., Der, C.J., and Trejo, J. (2004). Persistent signaling by dysregulated thrombin receptor trafficking promotes breast carcinoma cell invasion. Mol. Cell. Biol. 24, 1990–1999.

Borchert, G.H., Melegos, D.N., Tomlinson, G., Giai, M., Roagna, R., Ponzone, R., Sgro, L., and Diamandis, E.P. (1997). Molecular forms of prostate-specific antigen in the serum of women with benign and malignant breast diseases. Br. J. Cancer 76, 1087–1094.

Borgoño, C.A., Fracchioli, S., Yousef, G.M., Rigault de la Longrais, I.A., Luo, L.Y., Soosaipillai, A., Puopolo, M., Grass, L., Scorilas, A., Diamandis, E.P., and Katsaros, D. (2003a). Favorable prognostic value of tissue human kallikrein 11 (hK11) in patients with ovarian carcinoma. Int. J. Cancer 106, 605–610.

Borgoño, C.A., Grass, L., Soosaipillai, A., Yousef, G.M., Petraki, C.D., Howarth, D.H., Fracchioli, S., Katsaros, D., and Diamandis, E.P. (2003b). Human kallikrein 14: a new potential biomarker for ovarian and breast cancer. Cancer Res. 63, 9032–9041.

Borgoño, C.A., and Diamandis, E.P. (2004). The emerging roles of human tissue kallikreins in cancer. Nat Rev Cancer 4, 876–890.

Borgoño, C.A., Kishi, T., Scorilas, A., Harbeck, N., Dorn, J., Schmalfeldt, B., Schmitt, M., Diamandis, E.P. (2006). Human kallikrein 8 protein is a favorable prognostic marker in ovarian cancer. Clin. Cancer Res. 12, 1487–1493.

Borgoño, C.A., Michael, I.P., Shaw, J.L., Luo, L.Y., Ghosh, M.C., Soosaipillai, A., Grass, L., Katsaros, D., and Diamandis, E.P. (2007). Expression and functional characterization of the cancer-related serine protease, human tissue kallikrein 14. J. Biol. Chem. 282, 2405–2422.

Brattsand, M, and Egelrud, T. (1999). Purification, molecular cloning, and expression of a human stratum corneum trypsin-like serine protease with possible function in desquamation. J. Biol. Chem. 274, 30033–30040.

Briot, A., Deraison, C., Lacroix, M., Bonnart, C., Robin, A., Besson, C., Dubus, P., and Hovnanian, A. (2009). Kallikrein 5 induces atopic dermatitis-like lesions through PAR2-mediated thymic stromal lymphopoietin expression in Netherton syndrome. J. Exp. Med. 206, 1135–1147.

Chang, A., Yousef, G.M., Scorilas, A., Grass, L., Sismondi, P., Ponzone, R., and Diamandis, E.P. (2002). Human kallikrein gene 13 (KLK13) expression by quantitative RT-PCR: an independent indicator of favourable prognosis in breast cancer. Br. J. Cancer 86, 1457–1464.

Chee, J., Naran, A., Misso, N.L., Thompson, P.J., and Bhoola, K.D. (2008). Expression of tissue and plasma kallikreins and kinin B1 and B2 receptors in lung cancer. Biol. Chem. 389, 1225–1233.

Chou, R.H., Lin, S.C., Wen, H.C., Wu, C.W., and Chang, W.S. (2011). Epigenetic activation of human kallikrein 13 enhances malignancy of lung adenocarcinoma by promoting N-cadherin expression and laminin degradation. Biochem. Biophys. Res. Commun. 409, 442–447.

Cleutjens, K.B., van der Korput, H.A., van Eekelen, C.C., van Rooij, H.C., Faber, P.W., and Trapman, J. (1997). An androgen response element in a far upstream enhancer region is essential for high, androgen-regulated activity of the prostate-specific antigen promoter. Mol. Endocrinol. 11, 148–161.

Collier, M.E.W., Li, C., and Ettelaie, C. (2008). Influence of exogenous tissue factor on estrogen receptor alpha expression in breast cancer cells: involvement of beta1-integrin, PAR2, and mitogen-activated protein kinase activation. Mol. Cancer Res. 6, 1807–1818.

Cupić, D.F., Tesar, E.C., Ilijas, K.M., Nemrava, J., and Kovacević, M. (2011). Expression of matrix metalloproteinase 9 in primary and recurrent breast carcinomas. Coll. Antropol. 35 (Suppl. 2), 7–10.

Dash, P., Pati, S., Mangaraj, M., Sahu, P.K., and Mohapatra, P.C. (2011). Serum total PSA and free PSA in breast tumors. Indian J. Clin. Biochem. 26, 182–186.

Dhar, S., Bhargava, R., Yunes, M., Li, B., Goyal, J., Naber, S.P., Wazer, D.E., and Band, V. (2001). Analysis of normal epithelial cell specific-1 (NES1)/kallikrein 10 mRNA expression by in situ hybridization, a novel marker for breast cancer. Clin. Cancer Res. 7, 3393–3398.

Diamandis, E.P., Helle, S.I., Yu, H., Melegos, D.N., Lundgren, S., and Lonning, P.E. (1999). Prognostic value of plasma prostate specific antigen after megestrol acetate treatment in patients with metastatic breast carcinoma. Cancer 85, 891–898.

Diamandis, E.P., Yousef, G.M., Soosaipillai, A.R., Grass, L., Porter, A., Little, S., and Sotiropoulou, G. (2000). Immunofluorometric assay of human kallikrein 6 (zyme/protease M/neurosin) and preliminary clinical applications. Clin. Biochem. 33, 369–375.

Diamandis, E.P., Borgoño, C.A., Scorilas, A., Harbeck, N., Dorn, J., and Schmitt, M. (2004). Human kallikrein 11: an indicator of favorable prognosis in ovarian cancer patients. Clin Biochem. 37, 823–829.

Dimitrakakis, C., and Bondy, C. (2009). Androgens and the breast. Breast Cancer Res. 11, 212.

Dorn, J., Magdolen, V., Gkazepis, A., Gerte, T., Harlozinska, A., Sedlaczek, P., Diamandis, E.P., Schuster, T., Harbeck, N., Kiechle, M., Schmitt, M. (2011). Circulating biomarker tissue kallikrein-related peptidase KLK5 impacts ovarian cancer patients' survival. Ann. Oncol. 22, 1783–1790.

Duffy, M.J. (2006). Serum tumor markers in breast cancer: are they of clinical value? Clin. Chem. 52, 345–351.

El-Naggar, A. K., Lai, S., Clayman, G., Lee, J. K., Luna, M. A., Goepfert, H., and Batsakis, J. G. (1997). Methylation, a major mechanism of p16/CDKN2 gene inactivation in head and neck squamous carcinoma. Am. J. Pathol. 151, 1767–1774.

Emami, N., and Diamandis, E.P. (2008). Utility of kallikrein-related peptidases (KLKs) ascancer biomarkers. Clin. Chem. 54, 1600–1607.

Emami, N., Deperthes, D., Malm, J., and Diamandis, E.P. (2008). Major role of human KLK14 in seminal clot liquefaction. J. Biol. Chem. 283, 19561–19569.

Esteller, M. (2008). Epigenetics in cancer. N. Engl. J. Med. 358, 1148–1159.

Even-Ram, S., Uziely, B., Cohen, P., Grisaru-Granovsky, S., Maoz, M., Ginzburg, Y., Reich, R., Vlodavsky, I., and Bar-Shavit, R. (1998). Thrombin receptor overexpression in malignant and physiological invasion processes. Nat. Med. 4, 909–914.

Ewan King, L., Li, X., Cheikh Saad Bouh, K., Pedneault, M., and Chu, C.W. (2007). Human kallikrein 10 ELISA development and validation in breast cancer sera. Clin. Biochem. 40, 1057–1062.

Foekens, J A., Diamandis, E.P., Yu, H., Look, M.P., Meijer-van Gelder, M.E., van Putten, W.L., and Klijn, J. (1999). Expression of prostate-specific antigen (PSA) correlates with poor response to tamoxifen therapy in recurrent breast cancer. Br. J. Cancer 79, 888–894.

Fortier, A.H., Nelson, B.J., Grella, D.K., and Holaday, J.W. (1999). Antiangiogenic activity of prostate-specific antigen. J. Natl. Cancer Inst. 91, 1635–1640.

Fritzsche, F., Gansukh, T., Borgoño, C.A., Burkhardt, M., Pahl, S., Mayordomo, E., Winzer, K.J., Weichert, W., Denkert, C., Jung, K., Stephan, C., Dietel, M., Diamandis, E.P., Dahl, E., and Kristiansen, G. (2006). Expression of human kallikrein 14 (KLK14) in breast cancer is associated with higher tumour grades and positive nodal status. Br. J. Cancer 94, 540–547.

Ghosh, M.C., Grass, L., Soosaipillai, A., Sotiropoulou, G., and Diamandis, E.P. (2004). Human kallikrein 6 degrades extracellular matrix proteins and may enhance the metastatic potential of tumour cells. Tumour Biol. 25, 193–199.

Giai, M., Yu, H., Roagna, R., Ponzone, R., Katsaros, D., Levesque, M.A., and Diamandis, E.P. (1995). Prostate-specific antigen in serum of women with breast cancer. Br. J. Cancer 72, 728–731.

Goyal, J., Smith, K.M., Cowan, J.M., Wazer, D.E., Lee, S.W., and Band, V. (1998). The role for NES1 serine protease as a novel tumor suppressor. Cancer Res. 58, 4782–4786.

Guillon-Munos, A., Oikonomopoulou, K., Michel, N., Smith, C.R., Petit-Courty, A., Canepa, S., Reverdiau, P., Heuzé-Vourc'h, N., Diamandis, E.P., and Courty, Y. (2011). Kallikrein-related peptidase 12 hydrolyzes matricellular proteins of the CCN family and modifies interactions of CCN1 and CCN5 with growth factors. J. Biol. Chem. 286, 25505–25518.

Harris, L., Fritsche, H., Mennel, R., Norton, L., Ravdin, P., Taube, S., Somerfield, M.R., Hayes, D.F., and Bast, R.C. Jr. (2007). American Society of Clinical Oncology 2007 update of recommendations for the use of tumor markers in breast cancer. J. Clin. Oncol. 25, 5287–5312.

Heidtmann, H.H., Nettelbeck, D.M., Mingels, A., Jäger, R., Welker, H.G., Kontermann, R.E. (1999). Generation of angiostatin-like fragments from plasminogen by prostate-specific antigen. Br. J. Cancer 81, 1269–1273.

Henkhaus, R.S., Gerner, E.W., and Ignatenko, N.A. (2008). Kallikrein 6 is a mediator of K-RAS-dependent migration of colon carcinoma cells. Biol. Chem. 389, 757–764.

Hermann, A., Buchinger, P., and Rehbock, J. (1995). Visualization of tissue kallikrein in human breast carcinoma by two-dimensional western blotting and immunohistochemistry. Biol. Chem. Hoppe Seyler 376, 365–370.

Heyl, W., Wolff, J.M., Biesterfeld, S., Schröder, W., Zitzelsberger, D., Jakse, G., Rath, W. (1999). Immunohistochemical analysis of prostate-specific antigen does not correlate to other prognostic factors in breast cancer. Anticancer Res. 19(4A):2563–2565.

Holzscheiter, L., Biermann, J.C., Kotzsch, M., Prezas, P., Farthmann, J., Baretton, G., Luther, T., Tjan-Heijnen, V.C., Talieri, M., Schmitt, M., Sweep, F.C., Span, P.N., and Magdolen, V. (2006). Quantitative reverse transcription-PCR assay for detection of mRNA encoding full-length human tissue kallikrein 7: prognostic relevance of KLK7 mRNA expression in breast cancer. Clin. Chem. 52, 1070–1079.

Hsieh, M.L., Charlesworth, M.C., Goodmanson, M., Zhang, S., Seay, T., Klee, G.G., Tindall, D.J., and Young, C.Y. (1997). Expression of human prostate-specific glandular kallikrein protein (hK2) in the breast cancer cell line T47-D. Cancer Res. 57, 2651–2656.

Iakovlev, V., Siegel, E., Tsao, M.S., and Haun, R.S. (2012). Expression of kallikrein-related peptidase 7 predicts poor prognosis in patients with unresectable pancreatic ductal adenocarcinoma. Cancer Epidemiol. Biomarkers Prev., [Epub ahead of print].

Inoue, Y., Yokobori, T., Yokoe, T. Toiyama, Y., Miki, C., Mimori, K., Mori, M., and Kusunoki, M. (2010). Clinical significance of human kallikrein7 gene expression in colorectal cancer. Ann. Surg. Oncol. 17, 3037–3042.

Jiang, R., Shi, Z., Johnson, J.J., Liu, Y., and Stack, M.S. (2011). Kallikrein-5 promotes cleavage of desmoglein-1 and loss of cell-cell cohesion in oral squamous cell carcinoma. J. Biol. Chem. 286, 9127–9135.

Johnson, S.K., Ramani, V.C., Hennings, L., and Haun, R.S. (2007). Kallikrein 7 enhances pancreatic cancer cell invasion by shedding E-cadherin. Cancer 109, 1811–1820.

Kalluri, R. (2009). EMT: when epithelial cells decide to become mesenchymal-like cells. J. Clin. Invest. 119, 1417–1419.

Kapadia, C., Ghosh, M.C., Grass, L., and Diamandis, E.P. (2004). Human kallikrein 13 involvement in extracellular matrix degradation. Biochem. Biophys. Res. Commun. 323, 1084–1090.

Kaulsay, K.K., Ng, E.H., Ji, C.Y., Ho, G.H., Aw, T.C., and Lee, K.O. (1999). Serum IGF-binding protein-6 and prostate specific antigen in breast cancer. Eur. J. Endocrinol. 140, 164–168.

Kim, J.T., Song, E.Y., Chung, K.S., Kang, M.A., Kim, J.W., Kim, S.J., Yeom, Y.I., Kim, J.H., Kim, K.H., and Lee, H.G. (2011). Up-regulation and clinical significance of serine protease kallikrein 6 in colon cancer. Cancer 117, 2608–2619.

Kioulafa, M., Kaklamanis, L., Stathopoulos, E., Mavroudis, D., Georgoulias, V., and Lianidou, E.S. (2009). Kallikrein 10 (KLK10) methylation as a novel prognostic biomarker in early breast cancer. Ann. Oncol. 20, 1020–1025.

Kishi, T., Grass, L, Soosaipillai, A., Shimizu-Okabe, C., and Diamandis, E.P. (2003). Human kallikrein 8: immunoassay development and identification in tissue extracts and biological fluids. Clin. Chem. 49, 87–96.

Kishi, T., Soosaipillai, A., Grass, L., Little, S.P., Johnstone, E.M., and Diamandis, E.P. (2004). Development of an immunofluorometric assay and quantification of human kallikrein 7 in tissue extracts and biological fluids. Clin. Chem. 50, 709–716.

Klokk, T.I., Kilander, A., Xi, Z., Waehre, H., Risberg, B., Danielsen, H.E., and Saatcioglu, F. (2007). Kallikrein 4 is a proliferative factor that is overexpressed in prostate cancer. Cancer Res. 67, 5221–5230.

Klucky, B., Mueller, R., Vogt, I., Teurich, S., Hartenstein, B., Breuhahn, K., Flechtenmacher, C., Angel, P., and Hess, J. (2007). Kallikrein 6 induces E-cadherin shedding and promotes cell proliferation, migration, and invasion. Cancer Res. 67, 8198–8206.

Konstantoudakis, G., Florou, D., Mavridis, K., Papadopoulos, I.N., and Scorilas, A. (2010). Kallikrein-related peptidase 13 (KLK13) gene expressional status contributes significantly in the prognosis of primary gastric carcinomas. Clin. Biochem. 43, 1205–1211.

Kountourakis, P., Psyrri, A., Scorilas, A., Markakis, S., Kowalski, D., Camp, R.L., Diamandis, E.P., and Dimopoulos, M.A. (2009). Expression and prognostic significance of kallikrein-related peptidase 8 protein levels in advanced ovarian cancer by using automated quantitative analysis. Thromb. Haemost. 101, 541–546.

Kraus, T.S., Cohen, C., and Siddiqui, M.T. (2010). Prostate-specific antigen and hormone receptor expression in male and female breast carcinoma. Diagn. Pathol. 5, 63.

Krenzer, S., Peterziel, H., Mauch, C., Blaber, S.I., Blaber, M., Angel, P., and Hess, J. (2011). Expression and function of the kallikrein-related peptidase 6 in the human melanoma microenvironment. J. Invest. Dermatol. 131, 2281–2288.

Kyriakopoulou, L.G., Yousef, G.M., Scorilas, A., Katsaros, D., Massobrio, M., Fracchioli, S., and Diamandis, E.P. (2003). Prognostic value of quantitatively assessed KLK7 expression in ovarian cancer. Clin. Biochem. 36, 135–143.

Lai, J., Myers, S.A., Lawrence, M.G., Odorico, D.M., and Clements, J.A. (2009). Direct progesterone receptor and indirect androgen receptor interactions with the kallikrein-related peptidase 4 gene promoter in breast and prostate cancer. Mol. Cancer Res. 7, 129–141.

Lai, L.C., Erbas, H., Lennard, T.W., and Peaston, R.T. (1996). Prostate-specific antigen in breast cyst fluid: possible role of prostate-specific antigen in hormone-dependent breast cancer. Int. J. Cancer 66, 743–746.

Lawrence, M.G., Lai, J., and Clements, J.A. (2010). Kallikreins on steroids: structure, function, and hormonal regulation of prostate-specific antigen and the extended kallikrein locus. Endocr. Rev. 31, 407–446.

Lee, J.Y., Park, A.K., Lee, K.M., Park, S.K., Han, S., Han, W., Noh, D.Y., Yoo, K.Y., Kim, H., Chanock, S.J., Rothman, N., and Kang, D. (2009). Candidate gene approach evaluates association between innate immunity genes and breast cancer risk in Korean women. Carcinogenesis 30, 1528–1531.

Levine, A.J., and Puzio-Kuter, A.M. (2010). The control of the metabolic switch in cancers by oncogenes and tumor suppressor genes. Science 330, 1340–1344.

Li, B., Goyal, J., Dhar, S., Dimri, G., Evron, E., Sukumar, S., Wazer, D.E., and Band, V. (2001). CpG methylation as a basis for breast tumor-specific loss of NES1/kallikrein 10 expression. Cancer Res. 61, 8014–8021.

Li, X., Liu, J., Wang, Y., Zhang, L., Ning, L., and Feng, Y. (2009). Parallel underexpression of kallikrein 5 and kallikrein 7 mRNA in breast malignancies. Cancer Sci. 100, 601–607.

Liu, X.L., Wazer, D.E., Watanabe, K., and Band, V. (1996). Identification of a novel serine protease-like gene, the expression of which is down-regulated during breast cancer progression. Cancer Res. 56, 3371–3379.

Lu, Y., Papagerakis, P., Yamakoshi, Y., Hu, J.C., Bartlett, J.D., and Simmer, J.P. (2008). Functions of KLK4 and MMP-20 in dental enamel formation. Biol. Chem. 389, 695–700.

Luo, L., Herbrick, J.A., Scherer, S.W., Beatty, B., Squire, J., and Diamandis, E.P. (1998). Structural characterization and mapping of the normal epithelial cell-specific 1 gene. Biochem. Biophys. Res. Commun. 247, 580–586.

Luo, L.Y., Diamandis, E.P., Look, M.P., Soosaipillai, A.P., and Foekens, J.A. (2002). Higher expression of human kallikrein 10 in breast cancer tissue predicts tamoxifen resistance. Br. J. Cancer 86, 1790–1796.

Luo, L.Y., Grass, L., and Diamandis, E.P. (2003). Steroid hormone regulation of the human kallikrein 10 (KLK10) gene in cancer cell lines and functional characterization of the KLK10 gene promoter. Clin. Chim. Acta 337, 115–126.

Magklara, A., Scorilas, A., López-Otín, C., Vizoso, F., Ruibal, A., and Diamandis, E.P. (1999). Human glandular kallikrein in breast milk, amniotic fluid, and breast cyst fluid. Clin. Chem. 45, 1774–1780.

Magklara, A., Grass, L., and Diamandis, E.P. (2000). Differential steroid hormone regulation of human glandular kallikrein (hK2) and prostate-specific antigen (PSA) in breast cancer cell lines. Breast Cancer Res. Treat. 59, 263–270.

Magklara, A., Scorilas, A., Katsaros, D., Massobrio, M., Yousef, G.M., Fracchioli, S., Danese, S., and Diamandis, E.P. (2001). The human KLK8 (neuropsin/ovasin) gene: identification of two novel

Mangé, A., Desmetz, C., Berthes, M.L., Maudelonde, T., and Solassol, J. (2008). Specific increase of human kallikrein 4 mRNA and protein levels in breast cancer stromal cells. Biochem. Biophys. Res. Commun. 375, 107–112.

Marić, P., Ozretić, P., Levanat, S., Oresković, S., Antunac, K., and Beketić-Oresković, L. (2011). Tumor markers in breast cancer--evaluation of their clinical usefulness. Coll. Antropol. 35, 241–247.

Michael, I.P., Sotiropoulou, G., Pampalakis, G., Magklara, A., Ghosh, M., Wasney, G., and Diamandis, E.P. (2005). Biochemical and enzymatic characterization of human kallikrein 5 (hK5), a.novel serine protease potentially involved in cancer progression. J. Biol. Chem. 280, 14628–14635.

Michael, I.P., Pampalakis, G., Mikolajczyk, S.D., Malm, J., Sotiropoulou, G., and Diamandis, E.P. (2006). Human tissue kallikrein 5 is a member of a proteolytic cascade pathway involved in seminal clot liquefaction and potentially in prostate cancer progression. J. Biol. Chem. 281, 12743–12750.

Misek, D.E., and Kim, E.H. (2011). Protein biomarkers for the early detection of breast cancer. Int. J. Proteomics 2011, 343582.

Mitchell, S.H., Murtha, P.E., Zhang, S., Zhu, W., and Young, C.Y. (2000). An androgen response element mediates LNCaP cell dependent androgen induction of the hK2 gene. Mol. Cell Endocrinol. 168, 89–99.

Mitchell, G., Sibley, P.E., Wilson, A.P., Sauter, E., A'Hern, R., and Eeles, R.A. (2002). Prostate-specific antigen in nipple aspiration fluid: menstrual cycle variability and correlation with serum prostate-specific antigen. Tumour Biol. 23, 287–297.

Mize, G.J., Wang, W., and Takayama, T.K. (2008). Prostate-specific kallikreins-2 and -4 enhance the proliferation of DU-145 prostate cancer cells through protease-activated receptors-1 and -2. Mol. Cancer Res. 6, 1043–1051.

Mo, L., Zhang, J., Shi, J., Xuan, Q., Yang, X., Qin, M., Lee, C., Klocker, H., Li, Q.Q., and Mo, Z. (2010). Human kallikrein 7 induces epithelial-mesenchymal transition-like changes in prostate carcinoma cells: a role in prostate cancer invasion and progression. Anticancer Res. 30, 3413–3420.

Monne, M., Croce, C.M., Yu, H., and Diamandis, E.P. (1994). Molecular characterization of prostate-specific antigen messenger RNA expressed in breast tumors. Cancer Res. 54, 6344–6347.

Mukai, S., Fukushima, T., Naka, D., Tanaka, H., Osada, Y., and Kataoka, H. (2008). Activation of hepatocyte growth factor activator zymogen (pro-HGFA) by human kallikrein 1-related peptidases. FEBS J. 275, 1003–1017.

Murtha, P., Tindall, D.J., and Young, C.Y. (1993). Androgen induction of a human prostate-specific kallikrein, hKLK2: characterization of an androgen response element in the 5′ promoter region of the gene. Biochemistry 32, 6459–6464.

Nagahara, H., Mimori, K., Utsunomiya, T., Barnard, G.F., Ohira, M., Hirakawa, K., and Mori, M. (2005). Clinicopathologic and biological significance of kallikrein 6 overexpression in human gastric cancer. Clin. Cancer Res. 11, 6800–6806.

Naidoo, S., and Raidoo, D.M. (2009). Angiogenesis in cervical cancer is mediated by HeLa metabolites through endothelial cell tissue kallikrein. Oncol. Rep. 22, 285–293.

Narița, D., Raica, M., Anghel, A., Suciu, C., and Cîmpean, A. (2005). Immunohistochemical localization of prostate-specific antigen in benign and malignant breast conditions. Rom. J. Morphol. Embryol. 46, 41–45.

Narita, D., Cimpean, A.M., Anghel, A., and Raica, M. (2006a). Prostate-specific antigen value as a marker in breast cancer. Neoplasma 53, 161–167.

Narita, D., Raica, M., Suciu, C., Cîmpean, A., and Anghel, A. (2006b). Immunohistochemical expression of androgen receptor and prostate-specific antigen in breast cancer. Folia Histochem. Cytobiol. 44, 165–172.

Narița, D., Anghel, A., and Motoc, M. (2008). Prostate-specific antigen may serve as a pathological predictor in breast cancer. Rom. J. Morphol. Embryol. 49, 173–180.

Nathalie, H.V., Chris, P., Serge, G., Catherine, C., Benjamin, B., Claire, B., Christelle, P., Briollais, L., Pascale, R., Marie-Lise, J., and Yves, C. (2009). High kallikrein-related peptidase 6 in non-small cell lung cancer cells: an indicator of tumour proliferation and poor prognosis. J. Cell. Mol. Med. 13, 4014–4022.

Nelson, H.D., Tyne, K., Naik, A., Bougatsos, C., Chan, B.K., and Humphrey, L. (2009). Screening for breast cancer: systematic evidence review update for the US preventive services task force. Ann. Intern. Med. 151, 727–737.

Obiezu, C.V., Shan, S.J., Soosaipillai, A., Luo, L.Y., Grass, L., Sotiropoulou, G., Petraki, C.D., Papanastasiou, P.A., Levesque, M.A., and Diamandis, E.P. (2005). Human kallikrein 4:

quantitative study in tissues and evidence for its secretion into biological fluids. Clin. Chem. 51, 1432–1442.

Obiezu, C.V., Michael, I.P., Levesque, M.A., and Diamandis, E.P. (2006). Human kallikrein 4: enzymatic activity, inhibition, and degradation of extracellular matrix proteins. Biol. Chem. 387, 749–759.

Oikonomopoulou, K., Li, L., Zheng, Y., Simon, I., Wolfert, R.L., Valik, D., Nekulova, M., Simickova, M., Frgala, T., Diamandis, E.P. (2008). Prediction of ovarian cancer prognosis and response to chemotherapy by a serum-based multiparametric biomarker panel. Br. J. Cancer 99, 1103–1113.

Paliouras, M. and Diamandis, E.P. (2007). Coordinated steroid hormone-dependent and independent expression of multiple kallikreins in breast cancer cell lines. Breast Cancer Res. Treat. 102, 7–18.

Paliouras, M. and Diamandis, E.P. (2008a). An AKT activity threshold regulates androgen-dependent and androgen-independent PSA expression in prostate cancer cell lines. Biol. Chem. 389, 773–780.

Paliouras, M., and Diamandis, E.P. (2008b). Intracellular signaling pathways regulate hormone-dependent kallikrein gene expression. Tumour Biol. 29, 63–75.

Paliouras, M., and Diamandis, E.P. (2008c). Androgens act synergistically to enhance estrogen-induced upregulation of human tissue kallikreins 10, 11, and 14 in breast cancer cells *via* a membrane bound androgen receptor. Mol. Oncol. 1, 413–424.

Pampalakis, G., and Sotiropoulou, G. (2006). Multiple mechanisms underlie the aberrant expression of the human kallikrein 6 gene in breast cancer. Biol. Chem. 387, 773–782.

Pampalakis, G., Prosnikli, E., Agalioti, T., Vlahou, A., Zoumpourlis, V., and Sotiropoulou, G. (2009). A tumor-protective role for human kallikrein-related peptidase 6 in breast cancer mediated by inhibition of epithelial-to-mesenchymal transition. Cancer Res. 69, 3779–3787.

Papachristopoulou, G., Avgeris, M., and Scorilas, A. (2009). Expression analysis and study of KLK4 in benign and malignant breast tumors. Thromb. Haemost. 101, 381–387.

Papachristopoulou, G., Avgeris, M., Charlaftis, A., and Scorilas, A. (2011). Quantitative expression analysis and study of the novel human kallikrein-related peptidase 14 gene (KLK14) in malignant and benign breast tissues. Thromb. Haemost. 105, 131–137.

Pasic, M.D., Olkhov, E., Bapat, B., and Yousef, G.M. (2012). Epigenetic regulation of kallikrein-related peptidases: there is a whole new world out there. Biol. Chem. 393, 319–330.

Patsis, C., Yiotakis, I., and Scorilas, A. (2012). Diagnostic and prognostic significance of human kallikrein 11 (KLK11) mRNA expression levels in patients with laryngeal cancer. Clin. Biochem., [Epub ahead of print].

Petraki, C.D., Karavana, V.N., Skoufogiannis, P.T., Little, S.P., Howarth, D.J., Yousef, G.M., and Diamandis, E.P. (2001). The spectrum of human kallikrein 6 (zyme/protease M/neurosin) expression in human tissues as assessed by immunohistochemistry. J. Histochem. Cytochem. 49, 1431–1441.

Petraki, C.D., Karavana, V.N., Luo, L.Y., and Diamandis, E.P. (2002). Human kallikrein 10 expression in normal tissues by immunohistochemistry. J. Histochem. Cytochem. 50, 1247–1261.

Petraki, C.D., Gregorakis, A.K., Vaslamatzis, M.M., Papanastasiou, P.A., Yousef, G.M., Levesque, M.A., and Diamandis, E.P. (2006a). Prognostic implications of the immunohistochemical expression of human kallikreins 5, 6, 10 and 11 in renal cell carcinoma. Tumour Biol. 27, 1–7.

Petraki, C.D., Papanastasiou, P.A., Karavana, V.N., and Diamandis, E.P. (2006b). Cellular distribution of human tissue kallikreins: immunohistochemical localization. Biol. Chem. 387, 653–663.

Petraki, C., Dubinski, W., Scorilas, A., Saleh, C., Pasic, M.D., Komborozos, V., Khalil, B., Gabril, M.Y., Streutker, C., Diamandis, E.P., and Yousef, G.M. (2012). Evaluation and prognostic significance of human tissue kallikrein-related peptidase 6 (KLK6) in colorectal cancer. Pathol. Res. Pract. 208, 104–108.

Pezzato, E., Sartor, L., Dell'Aica, I., Dittadi, R., Gion, M., Belluco, C., Lise, M., and Garbisa, S. (2004). Prostate carcinoma and green tea: PSA-triggered basement membrane degradation and MMP-2 activation are inhibited by (-)epigallocatechin-3-gallate. Int. J. Cancer 112, 787–792.

Planque, C., Li, L., Zheng, Y., Soosaipillai, A., Reckamp, K., Chia, D., Diamandis, E.P., and Goodglick L. (2008). A multiparametric serum kallikrein panel for diagnosis of non-small cell lung carcinoma. Clin. Cancer Res. 14, 1355–1362.

Planque, C., Choi, Y.H., Guyetant, S., Heuzé-Vourc'h, N., Briollais, L., and Courty, Y. (2010). Alternative splicing variant of kallikrein-related peptidase 8 as an independent predictor of unfavorable prognosis in lung cancer. Clin. Chem. 56, 987–997.

Poh, B.H., Jayaram, G., Sthaneshwar, P., and Yip, C.H. (2008). Prostate-specific antigen in breast disease. Malays. J. Pathol. 30, 43–51.

Pollak, M. (2008). Insulin and insulin-like growth factor signalling in neoplasia. Nat. Rev. Cancer 8, 915–928.

Prassas, I., Paliouras, M., Datti, A., and Diamandis, E.P. (2008). High-throughput screening identifies cardiac glycosides as potent inhibitors of human tissue kallikrein expression: implications for cancer therapies. Clin. Cancer Res. 14, 5778–5784.

Prezas, P., Arlt, M.J., Viktorov, P., Soosaipillai, A., Holzscheiter, L., Schmitt, M., Talieri, M., Diamandis, E.P., Krüger, A., and Magdolen, V. (2006). Overexpression of the human tissue kallikrein genes KLK4, 5, 6, and 7 increases the malignant phenotype of ovarian cancer cells. Biol. Chem. 387, 807–811.

Rajapakse, S., and Takahashi, T. (2007). Expression and enzymatic characterization of recombinant human kallikrein 14. Zoolog. Sci. 24, 774–780.

Ramani, V.C., Kaushal, G.P., and Haun, R.S. (2011). Proteolytic action of kallikrein-related peptidase 7 produces unique active matrix metalloproteinase-9 lacking the C-terminal hemopexin domains. Biochim. Biophys. Acta 1813, 1525–1531.

Ramsay, A.J., Dong, Y., Hunt, M.L., Linn, M., Samaratunga, H., Clements, J.A., and Hooper, J.D. (2008). Kallikrein-related peptidase 4 (KLK4) initiates intracellular signaling via protease-activated receptors (PARs). KLK4 and PAR-2 are co-expressed during prostate cancer progression. J. Biol. Chem. 283, 12293–12304.

Rehbock, J., Buchinger, P., Hermann, A., and Figueroa, C. (1995). Identification of immunoreactive tissue kallikrein in human ductal breast carcinomas. J. Cancer Res. Clin. Oncol. 121, 64–68.

Renehan, A.G., Zwahlen, M., Minder, C., O'Dwyer, S.T., Shalet, S.M., and Egger, M. (2004). Insulin-like growth factor (IGF)-I, IGF binding protein-3, and cancer risk: systematic review and meta-regression analysis. Lancet 363, 1346–1353.

Riegman, P.H., Vlietstra, R.J., van der Korput, J.A., Brinkmann, A.O., and Trapman, J. (1991). The promoter of the prostate-specific antigen gene contains a functional androgen responsive element. Mol. Endocrinol. 5, 1921–1930.

Rückert, F., Hennig, M., Petraki, C.D., Wehrum, D., Distler, M., Denz, A., Schröder, M., Dawelbait, G., Kalthoff, H., Saeger, H.D., Diamandis, E.P., Pilarsky, C., and Grützmann, R. (2008). Co-expression of KLK6 and KLK10 as prognostic factors for survival in pancreatic ductal adenocarcinoma. Br. J. Cancer 99, 1484–1492.

Saijo, N. (2012). Critical comments for roles of biomarkers in the diagnosis and treatment of cancer. Cancer Treat. Rev. 38, 63–67.

Sano, A., Sangai, T., Maeda, H., Nakamura, M., Hasebe, T., and Ochiai, A. (2007). Kallikrein 11 expressed in human breast cancer cells releases insulin-like growth factor through degradation of IGFBP-3. Int. J. Oncol. 30, 1493–1498.

Sasaki, H., Kawano, O., Endo, K., Suzuki, E., Haneda, H., Yukiue, H., Kobayashi, Y., Yano, M., and Fujii, Y. (2006). Decreased kallikrein 11 messenger RNA expression in lung cancer. Clin. Lung Cancer 8, 45–48.

Sauter, E.R., Daly, M., Linahan, K., Ehya, H., Engstrom, P.F., Bonney, G., Ross, E.A., Yu, H., and Diamandis, E. (1996). Prostate-specific antigen levels in nipple aspirate fluid correlate with breast cancer risk. Cancer Epidemiol. Biomarkers Prev. 5, 967–970.

Sauter, E.R., Klein, G., Wagner-Mann, C., and Diamandis, E.P. (2004a). Prostate-specific antigen expression in nipple aspirate fluid is associated with advanced breast cancer. Cancer Detect. Prev. 28, 27–31.

Sauter, E.R., Lininger, J., Magklara, A., Hewett, J.E., and Diamandis, E.P. (2004b). Association of kallikrein expression in nipple aspirate fluid with breast cancer risk. Int. J. Cancer 108, 588–591.

Schmitt, M., Mengele, K., Schueren, E., Sweep, F.C., Foekens, J.A., Brünner, N., Laabs, J., Malik, A., and Harbeck, N. (2007). European Organisation for Research and Treatment of Cancer (EORTC) Pathobiology Group standard operating procedure for the preparation of human tumour tissue extracts suited for the quantitative analysis of tissue-associated biomarkers. Eur. J. Cancer 43, 835–844.

Schuur, E.R., Henderson, G.A., Kmetec, L.A., Miller, J.D., Lamparski, H.G., and Henderson, D.R. (1996). Prostate-specific antigen expression is regulated by an upstream enhancer. J. Biol. Chem. 271, 7043–7051.

Scorilas, A., Borgoño, C.A., Harbeck, N., Dorn, J., Schmalfeldt, B., Schmitt, M., and Diamandis, E.P. (2004). Human kallikrein 13 protein in ovarian cancer cytosols: a new favorable prognostic marker. J. Clin. Oncol. 22, 678–685.

Seiz, L., Dorn, J., Kotzsch, M., Walch, A., Grebenchtchikov, N.I., Gkazepis, A., Schmalfeldt, B., Kiechle, M., Bayani, J., Diamandis, E.P., Langer, R., Sweep, F.C., Schmitt, M., and Magdolen, V. (2012). Stromal cell-associated expression of kallikrein-related peptidase 6 (KLK6) indicates poor prognosis of ovarian cancer patients. Biol. Chem. 393, 391–401.

Shan, S.J., Scorilas, A., Katsaros, D., Rigault de la Longrais, I., Massobrio, M., and Diamandis, E.P. (2006). Unfavorable prognostic value of human kallikrein 7 quantified by ELISA in ovarian cancer cytosols. Clin. Chem. 52, 1879–1886.

Shaw, J.L., and Diamandis, E.P. (2007). Distribution of 15 human kallikreins in tissues and biological fluids. Clin. Chem. 53, 1423–1432.

Shaw, J.L., and Diamandis, E.P. (2008). Regulation of human tissue kallikrein-related peptidase expression by steroid hormones in 32 cell lines. Biol. Chem. 389, 1409–1419.

Sher, Y.P., Chou, C.C., Chou, R.H., Wu, H.M., Wayne Chang, W.S., Chen, C.H., Yang, P.C., Wu, C.W., Yu, C.L., and Peck, K. (2006). Human kallikrein 8 protease confers a favorable clinical outcome in non-small cell lung cancer by suppressing tumor cell invasiveness. Cancer Res. 66, 11763–11770.

Shigemasa, K., Tian, X., Gu, L., Tanimoto, H., Underwood, L.J., O'Brien, T.J., and Ohama, K. (2004). Human kallikrein 8 (hK8/TADG-14) expression is associated with an early clinical stage and favorable prognosis in ovarian cancer. Oncol. Rep. 11, 1153–1159.

Sidiropoulos, M., Pampalakis, G., Sotiropoulou, G., Katsaros, D., and Diamandis, E.P. (2005). Downregulation of human kallikrein 10 (KLK10/NES1) by CpG island hypermethylation in breast, ovarian and prostate cancers. Tumour Biol. 26, 324–336.

Slagter, M.H., Gooren, L.J., de Ronde, W., Soosaipillai, A., Scorilas, A., Giltay, E.J., Paliouras, M., and Diamandis, E.P. (2006). Serum and urine tissue kallikrein concentrations in male-to-female transsexuals treated with antiandrogens and estrogens. Clin. Chem. 52, 1356–1365.

Sotiriou, C., and Pusztai, L. (2009). Gene-expression signatures in breast cancer. N. Engl. J. Med. 360, 790–800.

Sotiropoulou, G., Pampalakis, G., and Diamandis, E.P. (2009). Functional roles of human kallikrein-related peptidases. J. Biol. Chem. 284, 32989–32994.

Stavropoulou, P., Gregorakis, A.K., Plebani, M., and Scorilas, A. (2005). Expression analysis and prognostic significance of human kallikrein 11 in prostate cancer. Clin. Chim. Acta. 357, 190–195.

Stuckey, A. (2011). Breast cancer: epidemiology and risk factors. Clin. Obstet. Gynecol. 54, 96–102.

Su, S., Li, Y., Luo, Y., Sheng, Y., Su, Y., Padia, R.N., Pan, Z.K., Dong, Z., and Huang, S. (2009). Proteinase-activated receptor 2 expression in breast cancer and its role in breast cancer cell migration. Oncogene 28, 3047–3057.

Sutkowski, D.M., Goode, R.L., Baniel, J., Teater, C., Cohen, P., McNulty, A.M., Hsiung, H.M., Becker, G.W., and Neubauer, B.L. (1999). Growth regulation of prostatic stromal cells by prostate-specific antigen. J. Natl. Cancer Inst. 91, 1663–1669.

Talieri, M., Diamandis, E.P., Gourgiotis, D., Mathioudaki, K., and Scorilas, A. (2004). Expression analysis of the human kallikrein 7 (KLK7) in breast tumors: a new potential biomarker for prognosis of breast carcinoma. Thromb. Haemost. 91, 180–186.

Talieri, M., Li, L., Zheng, Y., Alexopoulou, D.K., Soosaipillai, A., Scorilas, A., Xynopoulos, D., and Diamandis, E.P. (2009a). The use of kallikrein-related peptidases as adjuvant prognostic markers in colorectal cancer. Br. J. Cancer 100, 1659–1665.

Talieri, M., Mathioudaki, K., Prezas, P., Alexopoulou, D.K., Diamandis, E.P., Xynopoulos, D., Ardavanis, A., Arnogiannaki, N., and Scorilas, A. (2009b). Clinical significance of kallikrein-related peptidase 7 (KLK7) in colorectal cancer. Thromb. Haemost. 101, 741–747.

Talieri, M., Devetzi, M., Scorilas, A., Prezas, P., Ardavanis, A., Apostolaki, A., and Karameris, A. (2011). Evaluation of kallikrein- related peptidase 5 expression and its significance for breast cancer patients: association with kallikrein-related peptidase 7 expression. Anticancer Res. 31, 3093–3100.

Talieri, M., Devetzi, M., Scorilas, A., Pappa, E., Tsapralis, N., Missitzis, I., and Ardavanis, A. (2012). Human kallikrein-related peptidase 12 (KLK12) splice variants expression in breast cancer and their clinical impact. Tumour Biol., [Epub ahead of print].

Tham, S.M, Ng, K.H., Pook, S.H., Esuvaranathan, K., and Mahendran, R. (2011). Tumor and microenvironment modification during progression of murine orthotopic bladder cancer. Clin. Dev. Immunol. 2011, 865684.

Thiery, J.P. (2002). Epithelial-mesenchymal transitions in tumour progression. Nat. Rev. Cancer 2, 442–454.

Versteeg, H.H., Schaffner, F., Kerver, M., Ellies, L.G., Andrade-Gordon, P., Mueller, B.M., and Ruf, W. (2008). Protease-activated receptor (PAR) 2, but not PAR1, signaling promotes the development of mammary adenocarcinoma in polyoma middle T mice. Cancer Res. 68, 7219–7127.

Veveris-Lowe, T.L., Lawrence, M.G., Collard, R.L., Bui, L., Herington, A.C., Nicol, D.L., and Clements, J.A. (2005). Kallikrein 4 (hK4) and prostate-specific antigen (PSA) are associated with the loss of E-cadherin and an epithelial-mesenchymal transition (EMT)-like effect in prostate cancer cells. Endocr. Relat. Cancer 12, 631–643.

Wang, S.M., Mao, J., Li, B., Wu, W., and Tang, L.L (2008). Expression of KLK6 protein and mRNA in primary breast cancer and its clinical significance. Xi Bao Yu Fen Zi Mian Yi Xue Za Zhi 24, 1087–1089.

Wen, Y.G., Wang, Q., Zhou, C.Z., Yan, D.W., Qiu, G.Q., Yang, C., Tang, H.M., and Peng, Z.H. (2011). Identification and validation of kallikrein-ralated peptidase 11 as a novel prognostic marker of gastric cancer based on immunohistochemistry. J. Surg. Oncol. 104, 516–524.

Whitbread, A.K., Veveris-Lowe, T.L., Lawrence, M.G., Nicol, D.L., and Clements, J.A. (2006). The role of kallikrein-related peptidases in prostate cancer: potential involvement in an epithelial to mesenchymal transition. Biol. Chem. 387, 707–714.

White, N.M., Mathews, M., Yousef, G.M., Prizada, A., Popadiuk, C., and Doré, J.J. (2009). KLK6 and KLK13 predict tumor recurrence in epithelial ovarian carcinoma. Br. J. Cancer 101, 1107–1113.

White, N. M., Bui, A., Mejia-Guerrero, S., Chao, J., Soosaipillai, A., Youssef, Y., Mankaruos, M., Honey, R. J., Stewart, R., Pace, K. T., Sugar, L., Diamandis, E.P., Doré, J., and Yousef, G.M. (2010). Dysregulation of kallikrein-related peptidases in renal cell carcinoma: potential targets of miRNAs. Biol. Chem. 391, 411–423.

Xi, Z., Klokk, T.I., Korkmaz, K., Kurys, P., Elbi, C., Risberg, B., Danielsen, H., Loda, M., and Saatcioglu, F. (2004a). Kallikrein 4 is a predominantly nuclear protein and is overexpressed in prostate cancer. Cancer Res. 64, 2365–2370.

Xi, Z., Kaern, J., Davidson, B., Klokk, T.I., Risberg, B., Tropé, C., and Saatcioglu, F. (2004b). Kallikrein 4 is associated with paclitaxel resistance in ovarian cancer. Gynecol. Oncol. 94, 80–85.

Yoon, H., Laxmikanthan, G., Lee, J., Blaber, S.I., Rodriguez, A., Kogot, J.M., Scarisbrick, I.A., and Blaber, M. (2007). Activation profiles and regulatory cascades of the human kallikrein-related peptidases. J. Biol. Chem. 282, 31852–31864.

Young, C.Y., Andrews, P.E., Montgomery, B.T., and Tindall, D.J. (1992). Tissue-specific and hormonal regulation of human prostate-specific glandular kallikrein. Biochemistry 31, 818–824.

Yousef, G.M., Obiezu, C.V., Luo, L.Y., Black, M.H., and Diamandis, E.P. (1999). Prostase/KLK-L1 is a new member of the human kallikrein gene family, is expressed in prostate and breast tissues, and is hormonally regulated. Cancer Res. 59, 4252–4256.

Yousef, G.M., Magklara, A., and Diamandis, E.P. (2000). KLK12 is a novel serine protease and a new member of the human kallikrein gene family – differential expression in breast cancer. Genomics 69, 331–341.

Yousef, G.M., Kyriakopoulou, L.G., Scorilas, A., Fracchioli, S., Ghiringhello, B., Zarghooni, M., Chang, A., Diamandis, M., Giardina, G., Hartwick, W.J., Richiardi, G., Massobrio, M., Diamandis, E.P., and Katsaros, D. (2001a). Quantitative expression of the human kallikrein gene 9 (KLK9) in ovarian cancer: a new independent and favorable prognostic marker. Cancer Res. 61, 7811–7818.

Yousef, G.M., Magklara, A., Chang, A., Jung, K., Katsaros, D., and Diamandis, E.P. (2001b). Cloning of a new member of the human kallikrein gene family, KLK14, which is down-regulated in different malignancies. Cancer Res. 61, 3425–3431.

Yousef, G.M., Scorilas, A., Magklara, A., Memari, N., Ponzone, R., Sismondi, P., Biglia, N., Abd Ellatif, M., and Diamandis, E.P. (2002a). The androgen-regulated gene human kallikrein 15 (KLK15) is an independent and favourable prognostic marker for breast cancer. Br. J. Cancer 87, 1294–1300.

Yousef, G.M., Scorilas, A., Kyriakopoulou, L.G., Rendl, L., Diamandis, M., Ponzone, R., Biglia, N., Giai, M., Roagna, R., Sismondi, P., and Diamandis, E.P. (2002b). Human kallikrein gene 5 (KLK5) expression by quantitative PCR: an independent indicator of poor prognosis in breast cancer. Clin. Chem. 48, 1241–1250.

Yousef, G.M., Borgoño, C.A., Scorilas, A., Ponzone, R., Biglia, N., Iskander, L., Polymeris, M.E., Roagna, R., Sismondi, P., and Diamandis, E.P. (2002c). Quantitative analysis of human kallikrein gene 14 expression in breast tumours indicates association with poor prognosis. Br. J. Cancer 87, 1287–1293.

Yousef, G.M., Fracchioli, S., Scorilas, A., Borgoño, C.A., Iskander, L., Puopolo, M., Massobrio, M., Diamandis, E.P., and Katsaros, D. (2003a). Steroid hormone regulation and prognostic value of the human kallikrein gene 14 in ovarian cancer. Am. J. Clin. Pathol. 119, 346–355.

Yousef, G.M., Polymeris, M.E., Grass, L., Soosaipillai, A., Chan, P.C., Scorilas, A., Borgoño, C., Harbeck, N., Schmalfeldt, B., Dorn, J., Schmitt, M., and Diamandis, E.P. (2003b). Human kallikrein 5: a potential novel serum biomarker for breast and ovarian cancer. Cancer Res. 63, 3958–3965.

Yousef, G.M., Scorilas, A., Nakamura, T., Ellatif, M.A., Ponzone, R., Biglia, N., Maggiorotto, F., Roagna, R., Sismondi, P., and Diamandis, E.P. (2003c). The prognostic value of the human kallikrein gene 9 (KLK9) in breast cancer. Breast Cancer Res. Treat. 78, 149–158.

Yousef, G.M., Yacoub, G.M., Polymeris, M.E., Popalis, C., Soosaipillai, A., and Diamandis, E.P. (2004a). Kallikrein gene downregulation in breast cancer. Br. J. Cancer 90, 167–172.

Yousef, G.M., White, N.M., Kurlender, L., Michael, I., Memari, N., Robb, J.D., Katsaros, D., Stephan, C., Jung, K., and Diamandis, E.P. (2004b). The kallikrein gene 5 splice variant 2 is a new biomarker for breast and ovarian cancer. Tumour Biol. 25, 221–227.

Yousef, G.M., Borgoño, C.A., White, N.M., Robb, J.D., Michael, I.P., Oikonomopoulou, K., Khan, S., and Diamandis, E.P. (2004c). *In silico* analysis of the human kallikrein gene 6. Tumour Biol. 25, 282–289

Yu, H., Diamandis, E.P., Monne, M., and Croce, C.M. (1995a). Oral contraceptive-induced expression of prostate-specific antigen in the female breast. J. Biol. Chem. 270, 6615–6618.

Yu, H., Giai, M., Diamandis, E.P., Katsaros, D., Sutherland, D.J., Levesque, M.A., Roagna, R., Ponzone, R., and Sismondi, P. (1995b). Prostate-specific antigen is a new favorable prognostic indicator for women with breast cancer. Cancer Res. 55, 2104–2110.

Yu, H., Levesque, M.A., Clark, G.M., and Diamandis, E.P. (1998). Prognostic value of prostate-specific antigen for women with breast cancer: a large United States cohort study. Clin. Cancer Res. 4, 1489–1497.

Yu, X., Tang, H.Y., Li, X.R., He, X.W., and Xiang, K.M. (2010). Over-expression of human kallikrein 11 is associated with poor prognosis in patients with low rectal carcinoma. Med. Oncol. 27, 40–44.

Yunes, M.J., Neuschatz, A.C., Bornstein, L.E., Naber, S.P., Band, V., and Wazer, D.E. (2003). Loss of expression of the putative tumor suppressor NES1 gene in biopsy-proven ductal carcinoma *in situ* predicts for invasive carcinoma at definitive surgery. Int. J. Radiat. Oncol. Biol. Phys. 56, 653–657.

Zarghami, N., Grass, L., Sauter, E.R., and Diamandis, E.P. (1997). Prostate-specific antigen in serum during the menstrual cycle. Clin. Chem. 43, 1862–1867.

Zhang, Y., Bhat, I., Zeng, M., Jayal, G., Wazer, D.E., Band, H., and Band, V. (2006). Human kallikrein 10, a predictive marker for breast cancer. Biol. Chem. 387, 715–721.

Zhao, H., Dong, Y., Quan, J., Smith, R., Lam, A., Weinstein, S., Clements, J., Johnson, N.W., and Gao, J. (2011). Correlation of the expression of human kallikrein-related peptidases 4 and 7 with the prognosis in oral squamous cell carcinoma. Head Neck 33, 566–572.

Zhao, Y., Verselis, S.J., Klar, N., Sadowsky, N.L., Kaelin, C.M., Smith, B., Foretova, L., and Li, F.P. (2001). Nipple fluid carcinoembryonic antigen and prostate-specific antigen in cancer-bearing and tumor-free breasts. J. Clin. Oncol. 19, 1462–1467.

Zheng, Y., Katsaros, D., Shan, S.J., de la Longrais, I.R., Porpiglia, M., Scorilas, A., Kim, N.W., Wolfert, R.L., Simon, I., Li, L., Feng, Z., and Diamandis, E.P. (2007). A multiparametric panel for ovarian cancer diagnosis, prognosis, and response to chemotherapy. Clin. Cancer Res. 13, 6984–6992.

Julia Dorn, Valentina Milou, Vathany Kulasingam,
Barbara Schmalfeldt, Eleftherios P. Diamandis,
and Manfred Schmitt

7 Clinical Relevance of Kallikrein-related Peptidases in Ovarian Cancer

7.1 Introduction

The family of kallikrein-related peptidases (KLK) consists of 15 members abundantly expressed in various tissues, some including the ovary (Shaw and Diamandis, 2007). KLKs are known to be involved in neurodegenerative diseases and skin disorders but also in cancer with hormone-dependent malignancies of the breast, prostate, and the ovary and tumors of the brain, skin, kidney, and the gastrointestinal tract (Borgoño and Diamandis, 2004; Krenzer et al., 2011; Pampalakis et al., 2007; Sidiropoulos et al., 2005). Remarkably, the clinical impact of KLK family members as novel biomarkers for screening, diagnosis, prognosis, or therapy response prediction of cancer patients has been explored most extensively for patients afflicted with cancer of the ovary (Avgeris et al., 2010; Borgoño and Diamandis, 2004; Clements et al., 2004; Clements, 2008; Emami and Diamandis, 2008; Mavridis and Scorilas, 2010; Oikonomopoulou et al., 2010; Paliouras and Diamandis, 2006; Schmitt and Magdolen, 2009; Yousef and Diamandis, 2009).

7.2 Ovarian cancer pathology, diagnosis, and therapy

Prognosis of the relatively rare tumors of the ovary is poor, owing to late diagnosis and often inefficient primary debulking surgery, but also because of rapidly developing chemoresistance. The general term "ovarian cancer" denominates those tumors of the ovary that originate from the epithelial surface (ovarian carcinoma) and accounts for more than 80% of all solid ovarian tumors. Others, such as sex cord-stromal tumors, germ cell tumors, and metastases (for example from gastrointestinal tumors) are less common. For the year 2012, more than 22,000 new cases and 15,500 deaths from ovarian cancer are estimated for the USA, which makes ovarian cancer the second most common neoplasm of the female reproductive tract, after uterine/endometrial cancer, but the leading cause of death from a gynecological malignancy (http://www.cancer.gov/cancertopics/types/ovarian). Early ovarian cancer detection could potentially decrease mortality, but still, due to often only unspecific symptoms and the lack of a valid screening tool, 75% of the cancer patients present themselves with an advanced stage of the disease. Two-third of patients will relapse and develop chemoresistance within the first 5 years after diagnosis and initial treatment. Therefore, for ovarian cancer, valid tumor markers for screening, diagnosis, prognosis, and therapy response prediction are urgently needed.

Ovarian carcinomas are categorized according to their predominant epithelial cells and molecular type, as recently reviewed by Lengyel (2010). The most common histopathology is serous (or serous-papillary) cystadenocarcinoma, followed by endometrioid carcinoma (often associated with endometriosis) and mucinous cystadenocarcinomas (resembling tumors originating from the gastrointestinal tract). Epithelial ovarian cancers are molecularly classified as Type I or Type II (McCluggage, 2011). Type I ovarian cancers include low-grade serous-papillary, endometrioid, and borderline tumors of low malignant potential, with better patient outcome, even though this group of patients shows resistance to chemotherapy. Type I ovarian tumors are less common and contain mutations in *KRAS* and/or *BRAF*, *CTNNB1*, *PTEN*, and *PIK3CA* and, in low grade mucinous tumors, of *TP53*. Type II ovarian cancers are more frequent and include high-grade serous carcinomas, as well as undifferentiated carcinomas and carcinosarcomas. Type II ovarian cancers have a poor prognosis, carry *TP53* mutations, as well as LOH on chromosomes 7q and 9p, and are believed to originate from the fallopian tubes and the peritoneum.

Risk factors for ovarian cancer include cancer of the breast, uterus, colon, or rectum in the patient's family and personal history, as well as advanced age. Other risk factors are infertility, obesity, and menopausal hormone therapy for longer than 10 years. Ovarian cancer patients with a familiy history and BRCA1/2 mutations (who constitute approx. 5% of all patients) usually carry Type II tumors (Bast *et al.*, 2009; Lengyel *et al.*, 2010).

There is no effective screening procedure: neither vaginal ultrasonography, nor CA125 analysis in serum, nor other protein or gene expression analyses are sufficiently specific (Baggerly *et al.*, 2005; Lu *et al.*, 2004; Skates *et al.*, 2003) to achieve the generally claimed positive predictive value (PPV) of 10%, meaning at maximum ten surgeries will be required to diagnose one ovarian cancer case. Given the incidence of one ovarian cancer per 2,500 women over 50 years of age, sensitivity to the early stage disease must exceed 75%, with a specificity of 99.6%.

Ovarian cancer is usually diagnosed during surgery. In the case of an adnexal mass of unknown dignity, CA125 in serum as well as OVA1, a blood test approved by the FDA recently, can support the clinician's decision to perform at least a biopsy. With many false-positive cases for OVA1, this test is only useful in preventing a malignant mass to be falsely classified as benign (Yip *et al.*, 2011). CA125 levels, combined with transvaginal ultrasound, could help to detect early stage ovarian cancer, as elaborated on in early reports from the UK Collaborative Trial of Ovarian Cancer Screening clinical trial (UKCTOCS), but the influence on ovarian cancer mortality is not yet clear (Menon *et al.*, 2009). Otherwise, approximately 20% of ovarian carcinomas have little or no CA125 expression. Yet, for those patients who are CA125-positive, it is a useful biomarker to monitor response to ovarian cancer therapy (Bast *et al.*, 2005). CA125 and another serum marker, HE4 (human epididymis protein 4), can also be used to diagnose recurrent patients before symptoms occur. As this earlier diagnosis does not lead to prolonged survival, but into shorter therapy-free time, follow-up with CA125 and HE4 currently is not recommended, albeit often implemented in clinical routine (Rustin *et al.*, 2010; Yip *et al.*, 2011).

Staging at the time of diagnosis is performed intraoperatively, according to the guidelines of the International Federation of Gynecology and Obstetrics, FIGO. FIGO early stage I tumors are limited to one or both ovaries, FIGO early stage II tumors include pelvic extensions or cells in peritoneal washings or ascites, FIGO late stage III tumors present peritoneal metastases outside the pelvis or lymph node involvement, and FIGO late stage IV indicates cases with distant metastases. Staging laparotomy consists of a longitudinal abdominal section with bilateral salpingo-oophorectomy, hysterectomy, infragastric omentectomy, and paraaortic and pelvic lymphadenectomy, as well as appendectomy, if indicated. Cytologic washings, as well as random biopsies of the peritoneum at the diaphragm, the colon rings, and all suspect regions are required for appropriate staging. The goal of primary surgery is the removal of as much of the visible tumor as possible (debulking surgery), and may include sigmoidectomy, hemicolectomy, peritonectomy, or even resection of parts of the liver and the lung.

The stage of the cancer at the time of diagnosis represents the major traditional prognostic factor, as the 5-year survival of FIGO stage I patients is more than 90%, while survival of patients with FIGO stage III and IV is approximately 25% (Bethesda, 2011). Another very important traditional prognostic factor is the size of residual disease after cytoreductive surgery, which is the only parameter to be influenced by therapy for now. This is why it is of prime importance to leave no residual tumor during the primary ovarian cancer surgery (Polterauer *et al.*, 2012). Other established clinical prognostic factors are histology, tumor grade, and presence of ascites (Elattar *et al.*, 2011).

After diagnosis, staging, and cyto-reductive therapy by primary surgery, patients receive adjuvant therapy, usually consisting of 6 cycles of paclitaxel and carboplatin, every 21 days. Bevacizumab (Avastin®), a humanized monoclonal antibody to vascular endothelial growth factor (VEGF), has recently been approved by the EMA for standard first-line treatment of FIGO stage IIIB to IV. Bevacizumab is administered concomitant with chemotherapy and, subsequently, as monotherapy every 3 weeks for 15 months in total.

7.3 KLKs in ovarian cancer

As mentioned earlier, blood-borne or tissue-based biomarkers for screening and risk-group subclassification of early and advanced ovarian cancer patients are urgently needed. In this respect, the expression of members of the KLK gene family has been studied extensively at the gene and protein level in a variety of normal and diseased human tissues and biological fluids, in the last decade (Avgeris *et al.*, 2010; Borgoño and Diamandis, 2004; Shaw and Diamandis, 2007).

In the normal human ovary, *KLK* expression at the mRNA level is highest for *KLK6-8,* and *10,* whereas low expression was noted for *KLK1, 9, 11,* and *14,* and no expression for *KLK2-5, 12,* and *15.* At the protein level, moderate to high amounts were found for KLK1, 6, 7, 10, and 11 and low concentrations for KLK8 and 14. KLK2-5, 9, 12, 13, and 15 proteins are not expressed by the normal ovary (Shaw and Diamandis, 2007). Interestingly, compared to the normal ovary, concurrent upregulation of twelve (3–11 and 13–15) of the fifteen KLKs, reported for the mRNA and/or protein expression level, is a very unique feature, but characteristic for ovarian cancer (**Tab. 7.1, 7.2**).

7.3.1 Circulating KLKs as screening/diagnostic and/or prognostic ovarian cancer biomarkers

Most importantly, for ovarian cancer, early diagnosis is vital. Therefore, apart from sufficient specificity and sensitivity, an ideal screening or diagnostic marker must distinguish between early-stage malignancy and the benign/healthy state. Yet, regarding the screening/diagnostic potential of KLKs in ovarian cancer, most results have been acquired by analyzing blood serum specimens of advanced stage ovarian cancer patients due to wider sample availability. Otherwise, validation of the findings in a multicenter setting, also including early-stage ovarian cancer patients, is a prerequisite, before any KLK can be recommended as a cancer biomarker for diagnostic purposes in order to identify patients at risk.

In early and/or advanced ovarian cancer patients, seven KLKs (KLK5-8, 10, 11, and 14) are released into the blood (serum) and are elevated in comparison to serum levels of healthy individuals, patients with benign diseases of the ovary, or cancer of other origins (Bandiera *et al.*, 2009; Borgoño *et al.*, 2003b; Diamandis *et al.*, 2003a and b; Dorn *et al.*, 2011a; Kishi *et al.*, 2003; Koh *et al.*, 2011; Luo *et al.*, 2003) (**Tab. 7.1, 7.2**). Likewise, six of these KLKs are also released into peritoneal ascites (KLK5, 7, 8, 10, 11, and 14), and eight were detected in pleural effusions (KLK5-8, 10, 11, 13, and 14) of ovarian cancer patients. So far, the scientific literature provides no detailed information regarding secretion of KLKs by ovarian cancer cells into the patient's blood, peritoneum, or pleura, concerning the other seven KLKs (KLK1-4, 9, 12, 15).

KLK proteins present in sera, ascites, or pleural effluents may also serve as valuable prognostic cancer biomarkers in order to predict the course of early and/or late stage ovarian cancer (**Tab. 7.1**). For five of the KLKs, their prognostic value was evaluated (KLK5, 6, 8, 10, and 11). Elevated KLK5 serum and ascites levels are associated with an unfavorable prognosis, and a rapid decrease of KLK5 serum concentration, when measured serially before and after the first cycle of chemotherapy, is associated with favorable response (Dorn *et al.*, 2011a; Oikonompoulou *et al.*, 2008). Presence of KLK8 in serum indicates a favorable prognosis (Kishi *et al.*, 2003), while elevated

Tab. 7.1 Clinical utility of KLKs present in bodily fluids (serum, ascites, and/or pleural effusion) of ovarian cancer patients.

	KLK expression in	Elevated KLK discriminates	Tumor stage	Tumor grade	PFS	OS	Clinical outcome	Marker for response to chemotherapy	Reference
KLK5	Serum, ascites	cancer from benign	Late	n.d.	Short	n.s.	Unfavorable prognosis (serum and ascites)	n.d.	Dorn et al., 2011a
	Serum	cancer from normal, benign, borderline	n.d.	n.d.	n.d.	n.d.	Stage related to KLK5 content	n.d.	Bandiera et al., 2009
	Serum		Late	n.d.	Short	Short	Unfavorable prognosis	Yes	Oikonomopoulo et al., 2008
	Pleural effusions and ascites	cancer from normal, benign	n.d.	n.d.	n.d.	n.d.	n.d.	n.d.	Shih le et al., 2007
	Serum, ascites	cancer from normal	n.d.	n.d.	n.d.	n.d.	n.d.	n.d.	Yousef et al., 2003b
KLK6	Serum	cancer from normal, benign	Late	n.d.	n.d.	Short	Unfavorable prognosis	n.d.	Koh et al., 2011
	Serum	cancer from normal, benign	Late	High	Short	Short	Unfavorable prognosis. High KLK6 is associated with presence of residual tumor and suboptimal debulking.	Yes	Diamandis et al., 2003b
	Serum	n.d.	Late	n.d.	Short	Short	Unfavorable prognosis	Yes	Oikonomopoulo et al., 2008
	Serum	n.d.	n.d.	n.d.	n.d.	n.d.	n.d.	n.d.	Hutchinson et al., 2003
	Serum	cancer from normal	n.d.	n.d.	n.d.	n.d.	n.d.	n.d.	Diamandis et al., 2000
	Pleural effusions and ascites	cancer from normal, benign	n.d.	n.d.	n.d.	n.d.	n.d.	n.d.	Shih le et al., 2007
KLK7	Serum	n.d.	n.s	n.d.	n.s.	n.s.	n.s.	n.d.	Yip et al., 2011
	Serum	n.d.	n.s	n.d.	n.s.	n.s.	n.s.	Yes	Oikonomopoulo et al., 2008
	Pleural effusions and ascites	cancer from normal, benign	n.d.	n.d.	n.d.	n.d.	n.d.	n.d.	Shih le et al., 2007

Tab. 7.1 (continued)

	KLK expression in	Elevated KLK discriminates	Tumor stage	Tumor grade	PFS	OS	Clinical outcome	Marker for response to chemotherapy	Reference
KLK8	Serum	n.d.	n.s.		n.s.	n.s.	n.s.	Yes	Oikonomopoulo et al., 2008
	Serum, ascites	cancer from normal	Early	n.d.	Long	n.s.	Favorable prognosis (ascites)	n.d.	Kishi et al., 2003
	Pleural effusions and ascites	cancer from normal, benign	n.d.	n.d.	n.d.	n.d.	n.d.	n.d.	Shih Ie et al., 2007
KLK10	Serum	cancer from benign	Late	n.d.	n.s.	n.s.	Unfavorable prognosis	n.d.	Koh et al., 2011
	Serum	n.d.	Late	n.d.	Short	Short	Unfavorable prognosis	Yes	Oikonomopoulo et al., 2008
	Serum	cancer from normal	Late	High	Short	Short	Unfavorable prognosis High KLK10 is associated with presence of residual tumor and suboptimal debulking	Yes	Luo et al., 2003
	Pleural effusions and ascites	cancer from normal, benign	n.d.	n.d.	n.d.	n.d.	n.d.	n.d.	Shih Ie et al., 2007
	Serum	n.d.	n.d.	n.d.	n.d.	n.d.	n.d.	n.d.	Welsh et al., 2003
KLK11	Pleural effusions and ascites	cancer from normal, benign	n.d.	n.d.	n.d.	n.d.	n.d.	n.d.	Shih Ie et al., 2007
	Serum	n.d.	Late	n.d.	Short	n.s.	Unfavorable prognosis	n.s.	Oikonomopoulo et al., 2008
	Serum	Normal, benign from cancer	n.d.	n.d.	n.d.	n.d.	n.d.	n.d.	McIntosh et al., 2007
KLK13	Pleural effusions and ascites	cancer from normal, benign	n.d.	n.d.	n.d.	n.d.	n.d.	n.d.	Shih Ie et al., 2007
KLK14	Pleural effusions and ascites	cancer from normal, benign	n.d.	n.d.	n.d.	n.d.	n.d.	n.d.	Shih Ie et al., 2007
	Serum	cancer from normal	n.d.	n.d.	n.d.	n.d.	n.d.	n.d.	Borgoño et al., 2003b

Tab. 7.2 Clinical utility of KLKs present in tumor tissues of ovarian cancer patients.

	Normal		Cancerous		Clinical relevance of KLK in ovarian cancer (technique applied)	Reference
	mRNA	Protein	mRNA	Protein		
KLK1	Moderate	Moderate	62% up, 18% down$ n.d.	n.d.	n.d	Girgis et al., 2012
KLK2	Absent	Absent	3% up, 1% down$ n.d.	n.d.	n.d.	Girgis et al., 2012
KLK3	Absent	Absent	1% up, 13% down$ Increased	Increased	Not relevant (ELISA). Presence of rs1108403 allele correlated with poor survival (Sequenom iplex mass array)	Gilks et al., 2005 Girgis et al., 2012 Kucera et al., 1997 O'Mara et al., 2011 Yu et al., 1995
KLK4	Absent	Absent	6% up, 4% down$ Increased	Increased	Poor prognosis (RT-PCR, IHC). Response prediction to paclitaxel chemo-therapy (IHC)	Davidson et al., 2005 Dong et al., 2001 Girgis et al., 2012 Obiezu et al., 2001 Xi et al., 2004
KLK5	Absent	Absent	4% up, 91% down$ Increased	Increased	Poor prognosis (RT-PCR, ELISA)	Diamandis et al., 2003a Dong et al., 2003 Girgis et al., 2012 Hibbs et al., 2004 Kim et al., 2001 Yousef et al., 2003c Yousef et al., 2002 Zheng et al., 2007

Tab. 7.2 (continued)

	Normal		Cancerous		Clinical relevance of KLK in ovarian cancer (technique applied)	Reference
	mRNA	Protein	mRNA	Protein		
KLK6	High	Moderate	13% up, 75% down$ Increased	Increased	Poor prognosis (qRT-PCR, Affymetrix microarray, ELISA, IHC)	Adib et al., 2004 Anisowicz et al., 1996 Bignotti et al., 2006 Gilks et al., 2005 Girgis et al., 2012 Gyorffy et al., 2012 Hibbs et al., 2004 Hoffman et al., 2002 Hu et al., 2009 Kontourakis et al., 2008 Lu et al., 2004 Ni et al., 2004 Seiz et al., 2012 Shan et al., 2007 Tanimoto et al., 2001 Welsh et al., 2003 White et al., 2009 Yousef et al., 2003c Yousef et al., 2004 Zheng et al., 2007
KLK7	High	Moderate	6% up, 85% down$ Increased	Increased	Poor prognosis (qRT-PCR, ELISA, IHC, IHC-AQUA) Response prediction to paclitaxel chemotherapy (IHC)	Adib et al., 2004 Bignotti et al., 2006 Dong et al., 2003 Dong et al., 2010 Girgis et al., 2012 Hibbs et al., 2004

				References
KLK8	High	5% up, 77% down$	Favorable prognosis (qRT-PCR, IHC, ELISA)	Kyriakopoulou et al., 2003; Psyrri et al., 2008; Shan et al., 2006; Tanimoto et al., 1999; Yousef et al., 2003c
	Low	Increased	Poor prognosis (IHC-AQUA)	Adib et al., 2004; Bignotti et al., 2006; Borgoño et al., 2006; Gilks et al., 2005; Girgis et al., 2012; Hibbs et al., 2004; Kishi et al., 2003; Kontourakis et al., 2009; Magklara et al., 2001; Shigemasa et al., 2004a; Yousef et al., 2003c
KLK9	Low	1% up, 17% down$	Favorable prognosis (qRT-PCR)	Girgis et al., 2012; Yousef et al., 2001a
KLK10	High	Absent	Poor prognosis (ELISA: Luo)	Adib et al., 2004; Bignotti et al., 2006; Cheng et al., 2010; Dorn et al., 2007; Gilks et al., 2005; Girgis et al., 2012; Hibbs et al., 2004; Lu et al., 2004; Luo et al., 2001b; Shvartsman et al., 2003; Welsh et al., 2003; Yousef et al., 2003c; Zheng et al., 2007
	Increased	n.d.	Favorable prognosis (ELISA: Dorn)	
			Response prediction to platinum-based chemotherapy	

Tab. 7.2 (continued)

	Normal		Cancerous		Clinical relevance of KLK in ovarian cancer (technique applied)	Reference
	mRNA	Protein	mRNA	Protein		
KLK11	Moderate	Moderate	36% up, 38% down[$] Increased	Increased	Poor prognosis (qRT-PCR: Shigemasa). Favorable prognosis (ELISA: Diamandis, Borgoño)	Adib et al., 2004 Borgoño et al., 2003a Diamandis et al., 2004 Girgis et al., 2012 Shigemasa et al., 2004b Yousef et al., 2003c Zheng et al., 2007 Girgis et al., 2012
KLK12	Absent	Absent	14% up, 45% down[$] n.d.	n.d.	n.d.	
KLK13	Low	Absent	7% up, 63% down[$] Increased	Increased	Poor prognosis (qRT-PCR) Favorable prognosis (ELISA) Response prediction to platinum-based chemotherapy.	Dorn et al., 2007 Girgis et al., 2012 Scorilas et al., 2004 Zheng et al., 2007 White et al., 2009
KLK14	Moderate	Low	<1% up, 4% down[$] Decreased	Increased	Favorable prognosis (qRT-PCR)	Borgoño et al., 2003b Girgis et al., 2012 Yousef et al., 2001b Yousef et al., 2003a Yousef et al., 2003c
KLK15	Absent	Absent	13% up, 5% down[$] Increased	n.d.	Poor prognosis (qRT-PCR). SNP rs266851 associated with poor overall survival (in silico and DNA sequencing)	Batra et al., 2011 Girgis et al., 2012 Yousef et al., 2003d

$ Girgis et al., 2012: in silico analysis of 516 serous ovarian cancer patients compared to non-malignant counterparts.

n.d.: not determined

protein levels of KLK6, 10, and 11 are markers of poor prognosis for the ovarian cancer patient (Diamandis *et al.*, 2003b; Koh *et al.*, 2011; Luo *et al.*, 2003; Oikonomopoulo *et al.*, 2008).

7.3.2 Serum ovarian cancer biomarkers CA125 and KLKs

To enhance the percentage of ovarian cancers detected at an early stage, screening strategies have been devised that utilize quantification of serum CA125 protein. CA125 (or mucin 16; MUC16) is a member of the mucin glycoprotein family, encoded by the MUC16 gene (Yin *et al.*, 2001). CA125 is especially useful as a monitoring cancer bio-marker for detecting recurrence of ovarian cancer. It also is clinically approved for monitoring the efficacy of chemotherapy.

Several ovarian cancer studies addressed the question as to whether certain KLKs, or combinations thereof, are better diagnostic/screening and/or monitoring biomark-ers than CA125. Kishi *et al.* (2003), for instance, reported a direct correlation between CA125 and KLK8, for both serum and ascites. Diamandis *et al.* (2003b) found that serum KLK6 concentration correlates moderately with CA125 concentration and is higher in late-stage ovarian cancer patients with higher-grade disease and in patients with serous histotype. El Sherbini *et al.* (2011) described that, for diagnostic purposes, serum KLK6 and KLK10 alone have lower overall sensitivities than serum CA125, but a combination with serum KLK6 improves the sensitivity of CA125. Koh *et al.* (2011), on the other hand, were interested in assessing the clinical value of CA125, compared to KLK6 and KLK10, with regard to overall survival probability. Patients with late stage FIGOIII/IV cancers showed upregulation of CA125, KLK6, and KLK10. Mortality within 3 years of disease was associated with older age and upregulation of CA125 and KLK6 only.

Interestingly, when KLK6 and CA125 were determined in combination with KLK13, White *et al.* (2009) demonstrated that the combination of these three biomarkers is a more sensitive test to detect early stage ovarian cancer than CA125 alone. Sera from ovarian cancer patients were also assessed for expression of CA125 and KLK5-8, 10, and 11, in combination with three other biomarkers (B7-H4, regenerating protein IV, Spondin-2) (Oikonomopoulou *et al.*, 2008). All markers examined (except KLK7 and regenerating protein IV), predicted the time-to-progression among chemotherapy responders.

It is worth mentioning, however, that up to 20% of ovarian cancers do not express CA125. Thus, other cancer biomarkers that can be detected in sera of ovarian cancer patients lacking CA125 protein expression have to be considered. For this purpose, Rosen *et al.* (2005) devised a panel consisting of ten potential serum markers (amongst others including KLK6 and KLK10), being present in sera of ovarian cancer patients with no or low expression of CA125.

7.3.3 Tumor tissue-associated KLKs as prognostic ovarian cancer biomarkers

In ovarian cancer tissue, both mRNA and protein levels of KLKs have been reported to exceed levels found in physiological situations (**Tab. 7.2**). Ten *KLK* mRNAs (*KLK4-8, 10, 11,* and *13–15)* were found to be elevated in ovarian cancer tissue when compared to tissue obtained from normal ovaries, benign ovarian tumors and, in some of the analyses, when compared to tumors of low malignant potential, or between late and early stage or low and high grade tumors (Adib *et al.*, 2004; Bignotti *et al.*, 2006; Dong *et al.*, 2001; 2003; Gilks *et al.*, 2005; Shvartsman *et al.*, 2003; Tanimoto *et al.*, 1999; 2001; White *et al.*, 2009; Yousef *et al.*, 2001b; 2003a, c and d). When assessing KLKs at the protein level by immunohistochemistry or ELISA, seven KLKs (KLK5-8, 10, 11, and 14) were found to be upregulated in ovarian cancer, compared to normal ovarian tissues or other non-malignant gynecological diseases (**Tab. 7.2**) (Dorn *et al.*, 2006; 2007; Shan *et al.*, 2006; Yousef *et al.*, 2003c; Zheng *et al.*, 2007).

Most KLK family members are associated with a poor prognosis for ovarian cancer patients, but some are linked to a favorable course of the cancer, and for some, the findings are different for mRNA expression, compared to protein content: *KLK4-7, 10, 11, 13,* and *15* mRNA overexpression is associated with late-stage, high-grade disease, and shorter disease-free and overall survival. The opposite is true for elevated *KLK8, 9, 11,* and *14* mRNA (**Tab. 7.2**). Analogous to this, KLK5-7 proteins were reported as predictors of poor and KLK8 of good patient outcome. KLK10 protein expression, by different studies, was found to be a predictor of either good or poor prognosis. KLK11 and KLK13 protein content represent favorable disease outcome, which is opposite to the results reported for *KLK11* and *KLK13* mRNA expression. More specifically, *KLK4* mRNA expression in ovarian tumor tissue is associated with earlier relapse and death (only in univariate analysis) (Obiezu *et al.*, 2001), similar to KLK5 protein levels in tissue, serum, or ascites, which are correlated with disease-free and overall survival (Diamandis *et al.*, 2003a; Dorn *et al.*, 2011a; Yousef *et al.*, 2003).

Analysis of KLK6 in tumor cytosols, using ELISA, indicates that KLK6-positive tumors are more likely attached to advanced disease, serous histology, and suboptimal debulking. KLK6 has an impact on overall and progression-free survival as well, especially in ovarian cancer subgroups who show, at first sight, a good prognosis, e.g. with low-grade tumors and optimally debulked patients (Hoffman *et al.*, 2002). Stromal-cell-associated overexpression of KLK6, as assessed by immunohistochemistry, is associated with shorter overall and progression-free survival (Seiz *et al.*, 2012).

High concentrations of KLK7 and KLK10 protein in ovarian cancer tissue extracts are associated with advanced stage and larger residual tumor size (Shan *et al.*, 2006; Shvartsman *et al.*, 2003), as well as reduced overall and progression-free survival (Luo *et al.*, 2001b), which is even more accurate when used in a serum-based multiparametric biomarker panel including B7-H4 and Spondin-2 (Oikonomopoulou *et al.*, 2008). Concomitant with these findings, higher level differences of KLK5-7, and 10 between primary tumor and omentum metastastases are associated with suboptimal debulking, as well as with disease progression (Dorn *et al.*, 2011b). KLK8 protein is elevated in tumor cytosols and serum of ovarian cancer patients, as well as in ascitic fluid, and higher levels indicate a better prognosis (Borgoño *et al.*, 2006; Kishi *et al.*, 2003). The prognostic impact of KLK9 has been evaluated by Yousef *et al.* (2001a), where KLK9 gene expression in tumor cytosols is a favorable factor.

KLK10 in tumor cytosols and in serum of ovarian cancer patients is associated with advanced stage, higher FIGO stage, serous histology, higher nuclear grade, sub-optimal debulking, and low response to chemotherapy (Koh *et al.*, 2011; Luo *et al.*, 2001a; 2001b; 2003). As already mentioned, residual tumor mass after primary debulking is one of the most important prognostic factors for ovarian cancer. A score, based on clinical factors including nuclear grade and ascites volume, in addition to KLK6 and KLK13, indicates the outcome of surgery and could help to identify those patients who might be spared the burden of surgery (Dorn *et al.*, 2007).

KLK11 and KLK13 protein levels in ovarian cancer tissue are correlated with early stage and favorable outcome for overall survival (Borgoño *et al.*, 2003a; Diamandis *et al.*, 2004; Scorilas *et al.*, 2004). This is in contrast to the findings by Shigemasa *et al.* (2004b) and White *et al.* (2009), who reported an association with high grade tumors and worse prognosis for patients having higher *KLK11* and *KLK13* mRNA expression quantified by PCR. Thus, the prognostic impact of KLK11 and KLK13 could here be different, when regarding the mRNA level and the protein level.

KLK14 mRNA in ovarian tumor tissue is associated with longer overall and progression-free survival (OS and PFS) (Yousef *et al.*, 2003a), whereas *KLK15* mRNA in tissue of ovarian cancer patients is associated with shorter OS and PFS (Yousef *et al.*, 2003d). Taken as a whole, KLKs are promising prognostic factors for ovarian cancer. But due to the lack of sufficiently effective alternative therapy models in ovarian cancer, the clinical use of this knowledge for now is small and should be a subject of further investigations.

7.4 Tumor tissue-associated and blood-borne KLKs as predictive ovarian cancer biomarkers

Standard of care in ovarian cancer is cytoreductive surgery, followed by platinum-containing chemotherapy. However, up to 20% of patients turn out to be platinum-resistant and relapse within the first 6 months after start of therapy. Even among the group of responders at first sight, more than 60% relapse in the years following diagnosis (Rustin *et al.*, 2010). To date, there is no approved approach to foresee the response to chemotherapy fairly accurately, but several clinical studies have demonstrated the potential clinical use of some of the KLK to act as novel predictive cancer biomarkers.

For this purpose, tumor tissues obtained at primary surgery are screened. Sera and ascites of ovarian cancer patients are collected at baseline and during chemotherapy, to be analyzed by ELISA for KLK protein expression levels. Oikonomopou-lou *et al.* (2008) determined serum protein levels of KLK5-8, 10, and 11, CA125, B7-H4 (a member of the B7 family of immune costimulatory proteins), regenerating protein IV, and spondin-2 at baseline, or after the first chemotherapy cycle, to predict the patients' response to carboplatin and/or taxol-based (CBDCA/CFA; taxol/CBDCA) chemotherapy. The authors reported that a panel of these serum markers, including KLK5, among others KLK5, KLK7, and CA125, predicted chemotherapy response. All markers, including KLK5, KLK7, and regenerating protein IV, also were powerful pre-dictors of time-to-progression among the chemotherapy responders (Oikonomopou-lou *et al.*, 2008).

The clinical utility of certain KLKs, besides KLK5 and KLK7 (Oikonomopoulou *et al.*, 2008), released into the blood of ovarian cancer patients in order to predict response or failure to chemotherapy was also shown for three other KLKs (KLK6, 8, and 10) (Diamandis *et al.*, 2003b; Luo *et al.*, 2003; Oikonomopoulou *et al.*, 2008). Oikonomopoulou *et al.* (2008) stated that KLK8 is a candidate predictive marker as well, especially in the group of patients with no or low expression of CA125. The clinical relevance of the KLK6 protein to predict chemotherapy response was already shown by Diamandis *et al.* (2003b), who demonstrated that high serum KLK6 protein concentration is associated with the presence of residual tumor, suboptimal debulk-ing, and poor response of ovarian cancer patients to platinum-based chemotherapy. Luo *et al.* (2003) reported that elevation of KLK10 protein in serum of ovarian cancer patients treated with platinum-based chemotherapy will indicate failure of the patient to respond to this chemotherapy. High serum KLK10 was also strongly associ-ated with serous epithelial type, late-stage, advanced grade, suboptimal debulking, and residual tumor mass.

With regard to KLK expression in tumor tissues, the study by Xi *et al.* (2004), for a cohort of ovarian cancer patients with recurrent disease showed that KLK4 protein expression, as assessed by immunohistochemistry, was elevated in the group of patients progressing under taxane (paclitaxel) treatment but not in the group of responders. In line with that, higher protein levels of KLK7 were detected in tumor tissues of non-responding ovarian cancer patients treated with carboplatin ± taxol (Dong *et al.*, 2010).. Combining KLK6 protein levels with that of KLK8 and KLK13, plus cancer stage and debulking status, discriminates between responder (complete or partial) and non-responder ovarian cancer patients treated with platinum-based chemotherapy (Zheng *et al.*, 2007).

7.5 Conclusion

The KLK transcriptome is altered during neoplastic progression and this makes it an ideal molecular candidate biomarker for cancer screening/diagnosis and for monitor-ing of the course of the disease, but also for predicting patient outcome and response to cancer therapy. KLKs are fifteen different secreted proteins, several of which are also made by ovarian cancer cells and released into tumor tissue, blood, and other bodily fluids, such as ascites, pleura effusion, or urine. In ovarian cancer, deregulated gene and protein expression occurs for the majority of the KLKs. KLK5, 6, 8, 10, and 11 are regarded as the most promising KLKs for ovarian cancer screening and diagnosis: expression of KLK4-7, 10, and 15 indicates poor prognosis and KLK8, 9, 11, 13, and 14 are markers of a favorable prognosis. KLK5-8, 10, 11, 13 hold promise as predictive cancer biomarkers.

Most of the KLKs are highly expressed by serous epithelial ovarian tumors. Non-serous tumors do express KLK5, 11, and 13. KLK expression does not correlate with CA125 expression. Hence, KLKs are considered novel promising biomarkers to complement CA125, especially for early- and late-stage ovarian cancer detection, although further studies that include early-stage ovarian cancer patients are needed.

Since all these reports describe single, non-validated findings, one has to acknowledge that, although promising, these results are at present not strong enough to permit a change in the clinical management of early- or late-stage ovarian cancer patients. One would hope that multicenter validation studies will support and refine the published findings, in order to select, by determination of certain serum KLK proteins, those patients who would not benefit from standard chemotherapy. Such patients should be spared the adverse impact of potentially toxic chemotherapy.

We envision that selective inhibitors to certain KLKs will be developed for future therapeutic application, that aim at blocking their enzymatic activity, in order to interfere with KLK-mediated degradation or activation of other proteins. Nonetheless, one has to bear in mind that KLKs may exist in different enzymatic active and inactive molecular forms. Since reports about the enzymatic state of the various KLKs in different healthy and malignant tissues are scarce at present, the clinical utility of such new synthetic or biological therapeutics is not yet apparent.

Abbreviations

BRAF	v-Raf murine sarcoma viral oncogene homolog B1
BRCA1/2	breast cancer 1/2
CA125	carbohydrate antigen 125 or cancer antigen 125
CBDCA	cis-diammine(1,1-cyclobutanedicarboxylato)platinum(II)
CFA	cyclophosphamide
CTNNB1	catenin (cadherin-associated protein), beta 1
DFS	disease-free survival
EMA	European Medicines Agency
FIGO	International Federation of Gynecology and Obstetrics
KLK	kallikrein-related peptidase
KRAS	V-Ki-ras2 Kirsten rat sarcoma viral oncogene homolog

mRNA   messenger RNA
n.d.   not done
n.s.   not significant
OS   overall survival
PIK3CA   phosphoinositide-3-kinase, catalytic, alpha-polypeptide
PPV   positive predictive value
PTEN   phosphatase and tensin homolog
RT-PCR   reverse transcription polymerase chain reaction
TP53   tumor protein p53
uPA   urokinase-type plasminogen activator

Bibliography

Adib, T.R., Henderson, S., Perrett, C., Hewitt, D., Bourmpoulia, D., Ledermann, J., and Boshoff, C. (2004). Predicting biomarkers for ovarian cancer using gene-expression microarrays. Br. J. Cancer 90, 686–692.

Anisowicz, A., Sotiropoulou, G., Stenman, G., Mok, S.C., and Sager, R. (1996). A novel protease homolog differentially expressed in breast and ovarian cancer. Mol. Med. 2, 624–636.

Avgeris, M., Mavridis, K., and Scorilas, A. (2010). Kallikrein-related peptidase genes as promising biomarkers for prognosis and monitoring of human malignancies. Biol. Chem. 391, 505–511.

Baggerly, K.A., Morris, J.S., Edmonson, S.R., and Coombes, K.R. (2005). Signal in noise: evaluating reported reproducibility of serum proteomic tests for ovarian cancer. J. Natl. Cancer. Inst. 97, 307–309.

Bandiera, E., Zanotti, L., Bignotti, E., Romani, C., Tassi, R., Todeschini, P., Tognon, G., Ragnoli, M., Santin, A.D., Gion, M., Pecorelli, S., and Ravaggi, A. (2009). Human kallikrein 5: an interesting novel biomarker in ovarian cancer patients that elicits humoral response. Int. J. Gynecol. Cancer 19, 1015–1021.

Bast, R.C. Jr., Badgwell, D., Lu, Z., Marquez, R., Rosen D., Liu, J., Baggerly K.A., Atkinson E.N., Skates S., Zhang Z., Lokshin, A., and Menon, U., Jacobs, I., and Lu, K. (2005). New tumor markers: CA125 and beyond. Int. J. Gynecol. Cancer. 15 (Suppl. 3), 274–281.

Bast, R.C. Jr., Hennessy, B., and Mills, G.B. (2009). The biology of ovarian cancer: new opportunities for translation. Nat. Rev. Cancer 9, 415–428.

Batra, J., Nagle, C.M., O'Mara, T., Higgins, M., Dong, Y., Tan, O.L., Lose, F., Skeie, L.M., Srinivasan, S., Bolton, K.L., Song, H., Ramus, S.J., Gayther, S.A., Pharoah ,P.D., Kedda, M.A., Spurdle, A.B., and Clements, J.A. (2011). A kallikrein 15 (KLK15) single nucleotide polymorphism located close to a novel exon shows evidence of association with poor ovarian cancer survival. BMC Cancer 11, 119.

Bethesda, MD, http://seer.cancer.gov/csr/1975_2008/, based on November 2010 SEER data submission, posted to the SEER web site, 2011.

Bignotti, E., Tassi, R.A., Calza, S., Ravaggi, A., Romani, C., Rossi, E., Falchetti, M., Odicino, F.E., Pecorelli, S., and Santin, A.D. (2006). Differential gene expression profiles between tumor biopsies and short-term primary cultures of ovarian serous carcinomas: identification of novel molecular biomarkers for early diagnosis and therapy. Gynecol. Oncol. 103, 405–416.

Borgoño, C.A., Fracchioli, S., Yousef, G.M., Rigault de la Longrais, I.A., Luo, L.Y., Soosaipillai, A., Puopolo, M., Grass, L., Scorilas, A., Diamandis, E.P., and Katsaros, D. (2003a). Favorable prognostic value of tissue human kallikrein 11 (hK11) in patients with ovarian carcinoma. Int. J. Cancer 106, 605–610.

Borgoño, C.A., Grass, L., Soosaipillai, A., Yousef, G.M., Petraki, C.D., Howarth, D.H., Fraccioli, S., Katsaros, D., and Diamandis, E.P. (2003b). Human kallikrein 14: a new potential biomarker for ovarian and breast cancer. Cancer Res. 63, 9032–9041.

Borgoño, C.A., and Diamandis, E.P. (2004). The emerging roles of human tissue kallikreins in cancer. Nat. Rev. Cancer 4, 876–890.

Borgoño, C.A., Kishi, T., Scorilas, A., Harbeck, N., Dorn, J., Schmalfeldt, B., Schmitt, M., and Diamandis, E.P. (2006). Human kallikrein 8 protein is a favorable prognostic marker in ovarian cancer. Clin. Cancer Res. 12, 1487–1493.

Cheng, L., Lu, W., Kulkarni, B., Pejovic, T., Yan, X., Chiang, J.H., Hood, L., Odunsi, K., and Lin, B. (2010). Analysis of chemotherapy response programs in ovarian cancers by the next-generation sequencing technologies. Gynecol. Oncol. 117, 159–169.

Clements, J.A., Willemsen, N.M., Myers, S.A., and Dong, Y. (2004). The tissue kallikrein family of serine proteases: functional roles in human disease and potential as clinical biomarkers. Crit. Rev. Clin. Lab. Sci. 41, 265–312.

Clements, J.A. (2008). Reflections on the tissue kallikrein and kallikrein-related peptidase family – from mice to men – what have we learnt in the last two decades? Biol. Chem. 389, 1447–1454.

Davidson, B., Xi, Z., Klokk, T.I., Trope, C.G., Dorum, A., Scheistroen, M., and Saatcioglu, F. (2005). Kallikrein 4 expression is up-regulated in epithelial ovarian carcinoma cells in effusions. Am. J. Clin. Pathol. 123, 360–368.

Diamandis, E.P., Borgoño, C.A., Scorilas, A., Harbeck, N., Dorn, J., and Schmitt, M. (2004). Human kallikrein 11: an indicator of favorable prognosis in ovarian cancer patients. Clin. Biochem. 37, 823–829.

Diamandis, E.P., Borgoño, C.A., Scorilas, A., Yousef, G.M., Harbeck, N., Dorn, J., Schmalfeldt, B., and Schmitt, M. (2003a). Immunofluorometric quantification of human kallikrein 5 expression in ovarian cancer cytosols and its association with unfavorable patient prognosis. Tumour Biol. 24, 299–309.

Diamandis, E.P., Scorilas, A., Fracchioli, S., Van Gramberen, M., De Bruijn, H., Henrik, A., Soosaipillai, A., Grass, L., Yousef, G.M., Stenman, U.H., Massobrio M, van der Zee, A.G., Vergote, I., and Katsaros, D. (2003b). Human kallikrein 6 (hK6): a new potential serum biomarker for diagnosis and prognosis of ovarian carcinoma. J. Clin. Oncol. 21, 1035–1043.

Diamandis, E.P., Yousef, G.M., Soosaipillai, A.R., and Bunting, P. (2000). Human kallikrein 6 (zyme/protease M/neurosin): a new serum biomarker of ovarian carcinoma. Clin. Biochem. 33, 579–583.

Dong, Y., Kaushal, A., Brattsand, M., Nicklin, J., and Clements, J.A. (2003). Differential splicing of KLK5 and KLK7 in epithelial ovarian cancer produces novel variants with potential as cancer biomarkers. Clin. Cancer Res. 9, 1710–1720.

Dong, Y., Kaushal, A., Bui, L., Chu, S., Fuller, P.J., Nicklin, J., Samaratunga, H., and Clements, J.A. (2001). Human kallikrein 4 (KLK4) is highly expressed in serous ovarian carcinomas. Clin. Cancer Res. 7, 2363–2371.

Dong, Y., Tan, O.L., Loessner, D., Stephens, C., Walpole, C., Boyle, G.M., Parsons, P.G., and Clements, J.A. (2010). Kallikrein-related peptidase 7 promotes multicellular aggregation via the alpha(5)beta(1) integrin pathway and paclitaxel chemoresistance in serous epithelial ovarian carcinoma. Cancer Res. 70, 2624–2633.

Dorn, J., Harbeck, N., Kates, R., Magdolen, V., Grass, L., Soosaipillai, A., Schmalfeldt, B., Diamandis, E.P., and Schmitt, M. (2006). Disease processes may be reflected by correlations among tissue kallikrein proteases but not with proteolytic factors uPA and PAI-1 in primary ovarian carcinoma. Biol. Chem. 387, 1121–1128.

Dorn, J., Schmitt, M., Kates, R., Schmalfeldt, B., Kiechle, M., Scorilas, A., Diamandis, E.P., and Harbeck, N. (2007). Primary tumor levels of human tissue kallikreins impact surgical success and survival in ovarian cancer patients. Clin. Cancer Res. 13, 1742–1748.

Dorn, J., Magdolen, V., Gkazepis, A., Gerte, T., Harlozinska, A., Sedlaczek, P., Diamandis, E.P., Schuster, T., Harbeck, N., Kiechle, M., and Schmitt, M. (2011a). Circulating biomarker tissue kallikrein-related peptidase KLK5 impacts ovarian cancer patients' survival. Ann. Oncol. 22, 1783–1790.

Dorn, J., Harbeck, N., Kates, R., Gkazepis, A., Scorilas, A., Soosaipillai, A., Diamandis, E., Kiechle, M., Schmalfeldt, B., and Schmitt, M. (2011b). Impact of expression differences of KLK and of uPA and PAI-1 between primary tumor and omentum metastasis in advanced ovarian cancer. Ann. Oncol. 22, 877–883.

Elattar, A., Bryant, A., Winter-Roach, B.A., Hatem, M., and Naik, R. (2011). Optimal primary surgical treatment for advanced epithelial ovarian cancer. Cochrane Database Syst. Rev. CD007565.

El Sherbini, M.A., Sallam, M.M., Shaban, E.A., and El-Shalakany, A.H. (2011). Diagnostic value of serum kallikrein-related peptidases 6 and 10 versus CA125 in ovarian cancer. Int. J. Gynecol. Cancer 21, 625–632.

Emami, N., and Diamandis, E.P. (2008). Utility of kallikrein-related peptidases (KLKs) as cancer biomarkers. Clin. Chem. 54, 1600–1607.

Gilks, C.B., Vanderhyden B.C., Zhu, S., van de Rijn, M., and Longacre, T.A. (2005). Distinction between serous tumors of low malignant potential and serous carcinomas based on global mRNA expression profiling. Gynecol. Oncol. 96, 684–694.

Girgis, A.H., Bui, A., White, N.M., and Yousef, G.M. (2012). Integrated genomic characterization of the kallikrein gene locus in cancer. Anticancer Res. 32, 957–963.

Gyorffy, B., Lanczky, A., and Szallasi, Z. (2012). Implementing an online tool for genome-wide validation of survival-associated biomarkers in ovarian-cancer using microarray data from 1287 patients. Endocr. Rel. Cancer, [Epub ahead of print].

Hibbs, K., Skubitz, K.M., Pambuccian, S.E., Casey, R.C., Burleson, K.M., Oegema, T.R. Jr., Thiele, J.J., Grindle, S.M., Bliss, R.L., and Skubitz, A.P. (2004). Differential gene expression in ovarian carcinoma: identification of potential biomarkers. Am. J. Pathol. 165, 397–414.

Hoffman, B.R., Katsaros, D., Scorilas, A., Diamandis, P., Fracchioli, S., Rigault de la Longrais, I.A., Colgan, T., Puopolo, M., Giardina, G., Massobrio, M., and Diamandis, E.P. (2002). Immunoflu-orometric quantitation and histochemical localisation of kallikrein 6 protein in ovarian cancer tissue: a new independent unfavourable prognostic biomarker. Br. J. Cancer 87, 763–771.

Hu, C.J., Zhang, F., Chen, Y.J., Sun, X.M., and Zheng, J.F. (2009). Correlation of hK6 expression with clinicopathological features and prognosis in epithelial ovarian cancer. Zhonghua Zhong Liu Za Zhi. 31, 520–523.

Hutchinson, S., Luo, L.Y., Yousef, G.M., Soosaipillai, A., and Diamandis, E.P. (2003). Purification of human kallikrein 6 from biological fluids and identification of its complex with alpha(1)-antichy-motrypsin. Clin. Chem. 49, 746–751.

Kim, H., Scorilas, A., Katsaros, D., Yousef, G.M., Massobrio, M., Fracchioli, S., Piccinno, R., Gordini, G., and Diamandis, E.P. (2001). Human kallikrein gene 5 (KLK5) expression is an indicator of poor prognosis in ovarian cancer. Br. J. Cancer 84, 643–650.

Kishi, T., Grass, L., Soosaipillai, A., Scorilas, A., Harbeck, N., Schmalfeldt, B., Dorn, J., Mysliwiec, M., Schmitt, M., and Diamandis, E.P. (2003). Human kallikrein 8, a novel biomarker for ovarian carcinoma. Cancer Res. 63, 2771–2774.

Koh, S.C., Razvi, K., Chan, Y.H., Narasimhan, K., Ilancheran, A., Low, J.J., Choolani, M., and the Ovarian Cancer Research Consortium of SE Asia (2011). The association with age, human tissue kallikreins 6 and 10 and hemostatic markers for survival outcome from epithelial ovarian cancer. Arch. Gynecol. Obstet. 284, 183–190.

Kountourakis, P., Psyrri, A., Scorilas, A., Camp, R., Markakis, S., Kowalski, D., Diamandis, E.P., and Dimopoulos, M.A. (2008). Prognostic value of kallikrein-related peptidase 6 protein expression levels in advanced ovarian cancer evaluated by automated quantitative analysis (AQUA). Cancer Sci. 99, 2224–2229.

Kountourakis, P., Psyrri, A., Scorilas, A., Markakis, S., Kowalski, D., Camp, R.L., Diamandis, E.P., and Dimopoulos, M.A. (2009). Expression and prognostic significance of kallikrein-related peptidase 8 protein levels in advanced ovarian cancer by using automated quantitative analysis. Thromb. Haemost. 101, 541–546.

Krenzer, S., Peterziel, H., Mauch, C., Blaber, S.I., Blaber, M., Angel, P., and Hess, J. (2011). Expression and function of the kallikrein-related peptidase 6 in the human melanoma microenvironment. J. Invest. Dermatol. 131, 2281–2288.

Kucera, E., Kainz, C., Tempfer, C., Zeillinger, R., Koelbl, H., and Sliutz, G. (1997). Prostate specific antigen (PSA) in breast and ovarian cancer. Anticancer Res. 17, 4735–4737.

Kyriakopoulou, L.G., Yousef, G.M., Scorilas, A., Katsaros, D., Massobrio, M., Fracchioli, S., and Diamandis, E.P. (2003). Prognostic value of quantitatively assessed KLK7 expression in ovarian cancer. Clin. Biochem. 36, 135–143.

Lengyel, E. (2010). Ovarian cancer development and metastasis. Am. J. Pathol. 177, 1053–1064.

Lu, K.H., Patterson, A.P., Wang, L., Marquez, R.T., Atkinson, E.N., Baggerly, K.A., Ramoth, L.R., Rosen, D.G., Liu, J., Hellstrom, I., Smith, D., Hartmann, L., Fishman, D., Berchuck, A., Schmandt, R., Whitaker, R., Gershenson, D.M., Mills, G.B., and Bast, R.C. Jr. (2004). Selection of potential markers for epithelial ovarian cancer with gene expression arrays and recursive descent partition analysis. Clin. Cancer Res. 10, 3291–3300.

Luo, L.Y., Bunting, P., Scorilas, A., and Diamandis, E.P. (2001a). Human kallikrein 10: a novel tumor marker for ovarian carcinoma? Clin. Chim. Acta 306, 111–118.

Luo, L.Y., Katsaros, D., Scorilas, A., Fracchioli, S., Piccinno, R., Rigault de la Longrais, I.A., Howarth, D.J., and Diamandis, E.P. (2001b). Prognostic value of human kallikrein 10 expression in epithelial ovarian carcinoma. Clin. Cancer Res. 7, 2372–2379.

Luo, L.Y., Katsaros, D., Scorilas, A., Fracchioli, S., Bellino, R., van Gramberen, M., de Bruijn, H., Henrik, A., Stenman, U.H., Massobrio, M., van der Zee A.G., Vergote I., and Diamandis, E.P. (2003). The serum concentration of human kallikrein 10 represents a novel biomarker for ovarian cancer diagnosis and prognosis. Cancer Res. 63, 807–811.

Magklara, A., Scorilas, A., Katsaros, D., Massobrio, M., Yousef, G.M., Fracchioli, S., Danese, S., and Diamandis, E.P. (2001). The human KLK8 (neuropsin/ovasin) gene: identification of two novel splice variants and its prognostic value in ovarian cancer. Clin. Cancer Res. 7, 806–811.

Mavridis, K., and Scorilas, A. (2010). Prognostic value and biological role of the kallikrein-related peptidases in human malignancies. Future Oncol. 6, 269–285.

McCluggage, W.G. (2011). Morphological subtypes of ovarian carcinoma: a review with emphasis on new developments and pathogenesis. Pathology. 43, 420–432.

McIntosh, M.W., Liu, Y., Drescher, C., Urban N., and Diamandis, E.P. (2007). Validation and Characterization of human kallikrein 11 as a serum marker for diagnosis of ovarian carcinoma. Clin. Cancer Res. 13, 4422–4428.

Menon, U., Gentry-Maharaj, A., Hallett, R., Ryan, A., Burnell, M., Sharma, A., Lewis, S., Davies, S., Philpott, S., Lopes, A., Godfrey, K., Oram, D., Herod, J., Williamson, K., Seif, M.W., Scott, I., Mould, T., Woolas, R., Murdoch, J., Dobbs, S., Amso, N.N., Leeson, S., Cruickshank, D., McGuire, A., Campbell, S., Fallowfield, L., Singh, N., Dawnay, A., Skates, S.J., Parmar, M., and Jacobs, I. (2009). Sensitivity and specificity of multimodal and ultrasound screening for ovarian cancer, and stage distribution of detected cancers: results of the prevalence screen of the UK Collaborative Trial of Ovarian Cancer Screening (UKCTOCS). Lancet Oncol. 10, 327–340.

Ni, X., Zhang, W., Huang, K.C., Wang, Y., Ng, S.K., Mok, S.C., Berkowitz, R.S., and Ng, S.W. (2004). Characterisation of human kallikrein 6/protease M expression in ovarian cancer. Br. J. Cancer 91, 725–731.

Obiezu, C.V., Scorilas, A., Katsaros, D., Massobrio, M., Yousef, G.M., Fracchioli, S., Rigault de la Longrais, I.A., Arisio, R., and Diamandis, E.P. (2001). Higher human kallikrein gene 4 (KLK4) expression indicates poor prognosis of ovarian cancer patients. Clin. Cancer Res. 7, 2380–2386.

Oikonomopoulou, K., Li, L., Zheng, Y., Simon, I., Wolfert, R.L., Valik, D., Nekulova, M., Simickova, M., Frgala, T., and Diamandis, E.P. (2008). Prediction of ovarian cancer prognosis and response to chemotherapy by a serum-based multiparametric biomarker panel. Br. J. Cancer 99, 1103–1113.

Oikonomopoulou, K., Diamandis, E.P., and Hollenberg, M.D. (2010). Kallikrein-related peptidases: proteolysis and signaling in cancer, the new frontier. Biol. Chem. 391, 299–310.

O'Mara, T.A., Nagle, C.M., Batra, J., Kedda ,M.A., Clements, J.A., and Spurdle, A.B. (2011). Kallikrein-related peptidase 3 (KLK3/PSA) single nucleotide polymorphisms and ovarian cancer survival. Twin Res. Hum. Genet. 14, 323–327.

Paliouras, M., and Diamandis, E.P. (2006). The kallikrein world: an update on the human tissue kallikreins. Biol. Chem. 387, 643–562.

Pampalakis, G., and Sotiropoulou, G. (2007). Tissue kallikrein proteolytic cascade pathways in normal physiology and cancer. Biochim. Biophys. Acta 1776, 22–31.

Polterauer, S., Vergote, I., Concin, N., Braicu, I., Chekerov, R., Mahner, S., Woelber, L., Cadron, I., Gorp, T.V., Zeillinger, R., Castillo-Tong, D.C., and Sehouli, J. (2012). Prognostic value of residual tumor size in patients with epithelial ovarian cancer international FIGO Stages IIA-IV: analysis of the OVCAD data. Int. J. Gynecol. Cancer 22, 380–385.

Psyrri, A., Kountourakis, P., Scorilas, A., Markakis, S., Camp, R., Kowalski, D., Diamandis, E.P., and Dimopoulos, M.A. (2008). Human tissue kallikrein 7, a novel biomarker for advanced ovarian carcinoma using a novel in situ quantitative method of protein expression. Ann. Oncol. 19, 1271–1277.

Rosen, D.G., Wang, L., Atkinson, J.N., Yu, Y., Lu, K.H., Diamandis, E.P., Hellstrom, I., Mok, S.C.,Liu, J., and Bast, R.C. Jr. (2005). Potential markers that complement expression of CA125 in epithelial ovarian cancer. Gynecol. Oncol. 99, 267–277.

Rustin, G.J., van der Burg, M.E., Griffin, C.L., Guthrie, D., Lamont, A., Jayson, G.C., Kristensen, G., Mediola, C., Coens, C., Qian, W., Parmar, M.K., and Swart, A.M. (2010). Early versus delayed treatment of relapsed ovarian cancer (MRC OV05/EORTC 55955): a randomized trial. Lancet 376, 1155–1163.

Schmitt, M., and Magdolen, V. (2009). Using kallikrein-related peptidases (KLK) as novel cancer biomarkers. Thromb. Haemost. 101, 222–224.

Scorilas, A., Borgoño, C.A., Harbeck, N., Dorn, J., Schmalfeldt, B., Schmitt, M., and Diamandis, E.P. (2004). Human kallikrein 13 protein in ovarian cancer cytosols: a new favorable prognostic marker. J. Clin. Oncol. 22, 678–685.

Seiz, L., Dorn, J., Kotzsch, M., Walch, A., Grebenchtchikov, N.I., Gkazepis, A., Schmalfeldt, B., Kiechle, M., Bayani, M., Diamandis, E.P., Langer, R., Sweep, F.C.G.J., Schmitt, M., and Magdolen, V. (2012) Stromal cell-associated expression of kallikrein-related peptidase 6 (KLK6) indicates poor prognosis of ovarian cancer patients. Biol. Chem. 393, 391–401.

Shan, S.J., Scorilas, A., Katsaros, D., Rigault de la Longrais, I., Massobrio, M., and Diamandis, E.P. (2006). Unfavorable prognostic value of human kallikrein 7 quantified by ELISA in ovarian cancer cytosols. Clin. Chem. 52, 1879–1886.

Shan, S.J., Scorilas, A., Katsaros, D., and Diamandis, E.P. (2007). Transcriptional upregulation of human tissue kallikrein 6 in ovarian cancer: clinical and mechanistic aspects. Br. J. Cancer 96, 362–372.

Shaw, J.L., and Diamandis, E.P. (2007). Distribution of 15 human kallikreins in tissues and biological fluids. Clin. Chem. 53, 1423–1432.

Sidiropoulos, M., Pampalakis, G., Sotiropoulou, G., Katsaros, D., and Diamandis, E.P. (2005). Downregulation of human kallikrein 10 (KLK10/NES1) by CpG island hypermethylation in breast, ovarian and prostate cancers. Tumour Biol. 26, 324–336.

Shigemasa, K., Tian, X., Gu, L., Tanimoto, H., Underwood, L.J., O'Brien, T.J., and Ohama, K. (2004a). Human kallikrein 8 (hK8/TADG-14) expression is associated with an early clinical stage and favorable prognosis in ovarian cancer. Oncol. Reports 11, 1153–1159.

Shigemasa, K., Gu, L., Tanimoto, H., O'Brien T.J., and Ohama, K. (2004b). Human kallikrein gene 11 (KLK11) mRNA overexpression is associated with poor prognosis in patients with epithelial ovarian cancer. Clin. Cancer Res. 10, 2766–2770.

Shih I, M., Salani, R., Fiegl, M., Wang, T.L., Soosaipillai, A., Marth, C., Muller-Holzner, E., Gastl, G., Zhang, Z., and Diamandis, E.P. (2007). Ovarian cancer specific kallikrein profile in effusions. Gynecol. Oncol. 105, 501–507.

Shvartsman, H.S., Lu, K.H., Lee, J., Lillie, J., Deavers, M.T., Clifford, S., Wolf, J.K., Mills, G.B., Bast, R.C. Jr., Gershenson, D.M., and Schmandt, R. (2003). Overexpression of kallikrein 10 in epithelial ovarian carcinomas. Gynecol. Oncol. 90, 44–50.

Skates, S.J., Menon, U., MacDonald, N., Rosenthal, A.N., Oram, D.H., Knapp, R.C., and Jacobs, I.J. (2003). Calculation of the risk of ovarian cancer from serial CA-125 values for preclinical detection in postmenopausal women. J. Clin- Oncol. 21 (10 Suppl.), 206s-210s.

Tanimoto, H., Underwood, L.J., Shigemasa, K., Yan Yan, M.S., Clarke, J., Parmley, T.H., and O'Brien, T.J. (1999). The stratum corneum chymotryptic enzyme that mediates shedding and desquamation of skin cells is highly overexpressed in ovarian tumor cells. Cancer 86, 2074–2082.

Tanimoto, H., Underwood, L.J., Shigemasa, K., Parmley, T.H., and O'Brien, T.J. (2001). Increased expression of protease M in ovarian tumors. Tumour Biol. 22, 11–18.

Welsh, J.B., Sapinoso, L.M., Kern, S.G., Brown, D.A., Liu ,T., Bauskin, A.R., Ward, R.L., Hawkins, N.J., Quinn, D.I., Russell, P.J., Sutherland, R.L., Breit, S.N., Moskaluk, C.A., Frierson, H.F. Jr., and Hampton, G.M. (2003). Large-scale delineation of secreted protein biomarkers overexpressed in cancer tissue and serum. Proc. Natl. Acad. Sci. USA 100, 3410–3415.

White, N.M., Mathews, M., Yousef, G.M., Prizada, A., Popadiuk, C., and Doré, J.J. (2009). KLK6 and KLK13 predict tumor recurrence in epithelial ovarian carcinoma. Br. J. Cancer 101, 1107–1113.

Xi, Z., Kaern, J., Davidson, B., Klokk, T.I., Risberg, B., Tropé, C., Saatcioglu, and F. (2004). Kallikrein 4 is associated with paclitaxel resistance in ovarian cancer. Gynecol. Oncol., 94, 80–85.

Yin, B.W., and Lloyd, K.O. (2001). Molecular cloning of the CA125 ovarian cancer antigen: identi-fication as a new mucin, MUC16. J. Biol. Chem. 276, 27371–27375.

Yip, P., Chen, T.H., Seshaiah, P., Stephen, L.L., Michael-Ballard K.L., Mapes J.P., Mansfield B.C., and Bertenshaw G.P. (2011). Comprehensive serum profiling for the discovery of epithelial ovarian cancer biomarkers. PLoS One 6, e29533.

Yousef, G.M., Kyriakopoulou, L.G., Scorilas, A., Fracchioli, S., Ghiringhello, B., Zarghooni, M., Chang, A., Diamandis, M., Giardina, G., Hartwick, W.J., Richiardi, G., Massobrio, M., Diamandis, E.P., and Katsaros, D. (2001a). Quantitative expression of the human kallikrein gene 9 (KLK9) in ovarian cancer: a new independent and favorable prognostic marker. Cancer Res. 61, 7811–7818.

Yousef, G.M., Magklara, A., Chang, A., Jung, K., Katsaros, D., and Diamandis, E.P. (2001b). Cloning of a new member of the human kallikrein gene family, KLK14, which is down-regulated in different malignancies. Cancer Res. 61, 3425–3431.

Yousef, G.M., Scorilas, A., Kyriakopoulou, L.G., Rendl, L., Diamandis, M., Ponzone, R., Biglia, N., Giai, M., Roagna, R., Sismondi, P., and Diamandis, E.P. (2002). Human kallikrein gene 5 (KLK5) expression by quantitative PCR: an independent indicator of poor prognosis in breast cancer. Clin. Chem. 48, 1241–1250.

Yousef, G.M., Fracchioli, S., Scorilas, A., Borgoño, C.A., Iskander, L., Puopolo, M., Massobrio, M., Diamandis, E.P., and Katsaros, D. (2003a). Steroid hormone regulation and prognostic value of the human kallikrein gene 14 in ovarian cancer. Am. J. Clin. Pathol. 119, 346–355.

Yousef, G.M., Polymeris, M.E., Grass, L., Soosaipillai, A., Chan, P.C., Scorilas, A., Borgoño, C., Harbeck, N., Schmalfeldt, B., Dorn, J., Schmitt, M, and Diamandis, E.P. (2003b). Human kallikrein 5: a potential novel serum biomarker for breast and ovarian cancer. Cancer Res. 63, 3958–3965.

Yousef, G.M., Polymeris, M.E., Yacoub, G.M., Scorilas, A., Soosaipillai, A., Popalis, C., Fracchioli, S., Katsaros, D., and Diamandis, E.P. (2003c). Parallel overexpression of seven kallikrein genes in ovarian cancer. Cancer Res. 63, 2223–2227.

Yousef, G.M., Scorilas, A., Katsaros, D., Fracchioli, S., Iskander, L., Borgoño, C., Rigault de la Longrais, I.A., Puopolo, M., Massobrio, M., and Diamandis, E.P. (2003d). Prognostic value of the human kallikrein gene 15 expression in ovarian cancer. J. Clin. Oncol. 21, 3119–3126.

Yousef, G.M., Borgoño, C.A., White, N.M., Robb, J.D., Michael, I.P., Oikonomopoulou, K., Khan, S., and Diamandis, E.P. (2004). In silico analysis of the human kallikrein gene 6. Tumour Biol. 25, 282–289.

Yousef, G.M., and Diamandis, E.P. (2009). The human kallikrein gene family: new biomarkers for ovarian cancer. Cancer Treat. Res. 149, 165–187.

Yu, H., Diamandis, E.P., Levesque, M., Asa, S.L., Monne, M., and Croce, C.M. (1995). Expression of the prostate-specific antigen gene by a primary ovarian carcinoma. Cancer Res. 55, 1603–1606.

Zheng, Y., Katsaros, D., Shan, S.J., de la Longrais, I.R., Porpiglia, M., Scorilas, A., Kim, N.W., Wolfert, R.L., Simon, I., Li, L., Feng, Z., and Diamandis, E.P. (2007). A multiparametric panel for ovarian cancer diagnosis, prognosis, and response to chemotherapy. Clin. Cancer Res. 13, 6984–6992.

George M. Yousef, and Nicole M.A. White

8 microRNAs: A New Control Mechanism for Kallikrein-related Peptidases in Kidney and Other Cancers

8.1 Introduction

8.1.1 KLK expression in normal kidney tissue

Kallikrein-related peptidases (KLK) have been shown to be expressed in a wide range of normal tissues, including the breast, prostate, kidney, and skin to name a few (Shaw and Diamandis, 2007). The expression of KLKs in the normal kidney is shown in **Tab. 8.1**. The highest level of messenger RNA (mRNA) expression in the kidney has been reported for KLK1, 6, and 7 (Shaw and Diamandis, 2007; Yousef and Diamandis, 2001). At the protein level, Shaw and Diamandis (Shaw and Diamandis, 2007) reported highest expression levels for KLK1 and KLK9, as determined by enzyme-linked immunosorbent assay (ELISA). Immunohistochemical analyses showed positive immunoexpression for 10 of the 15 KLKs in the kidney (Gabril *et al.*, 2010; Petraki *et al.*, 2006b). In some cases, there exists a discrepancy between reported *KLK* gene and KLK protein expression levels. For example *KLK4*, which was reported to have an undetectable level of mRNA, showed low protein expression in the kidney. These inconsistencies may be accounted for by post-transcriptional modifications such as microRNAs (miRNAs). Also, there may exist some cross-reactivity in these studies, due to a high degree of similarity among the KLK family.

Tab. 8.1 KLK expression in normal kidney tissue.

Mole-cule	Technique	High expression	Moderate expression	Low expression	Undetectable	Reference
mRNA	PCR[1]	KLK1, 6, 7	KLK8, 14	KLK2, 3, 5, 9–13, 15	KLK4	Shaw and Diamandis, 2007
	PCR[1]	KLK1, 6, 7	KLK14, 15			Yousef and Diamandis, 2001
Protein	ELISA[2]	KLK1, 9	KLK3, 6, 7, 11, 15	KLK2, 4, 5, 8, 13, 14	KLK10, 12	Shaw and Diamandis, 2007
	IHC[3]	Positive for KLK5-7, 10–14				Petraki *et al.*, 2006b
	IHC[3]	Positive for KLK1, 6, 7, 15				Gabril *et al.*, 2010

1 polymerase chain reaction
2 enzyme-linked immunosorbent assay
3 immunohistochemistry

8.1.2 KLK dysregulation in kidney cancer

A number of studies has shown that KLKs are differentially expressed in pathologi-cal conditions, including psoriasis (Komatsu *et al.*, 2007), prostate cancer (Ahn *et al.*, 2008), ovarian cancer (Bayani *et al.*, 2011), and kidney cancer (Petraki *et al.*, 2006a). Renal cell carcinoma (RCC) is the most common form of kidney cancer in adults. The incidence of RCC has increased steadily over the past twenty years (Chow *et al.*, 1999). The most common subtype of RCC is the clear cell subtype (ccRCC), which accounts for approximately 80% of all RCCs. About 15% of the patients are diagnosed with the papillary subtype of RCC (pRCC) and the remaining 5% are classified as other subtypes which include chromophobe RCC (chRCC) (McLaughlin *et al.*, 2006). Chro-mosomal abnormalities that have been documented in ccRCC include deletion, muta-tion, or silencing by epigenetic mechanisms in the von Hippel-Lindau (VHL) tumor suppressor gene (Kovacs and Frisch, 1989; Moore *et al.*, 2011; Yoshida *et al.*, 1986). On the other hand, other histological types have distinct patterns of genetic changes, which imply different mechanisms of tumorigenesis.

Different prognostic outcomes are associated with different histological subtypes of RCC. Benign renal tumors (e.g., oncocytoma) have a much better prognosis, com-pared to ccRCC. Among RCC subtypes, collecting duct and medullary carcinomas are much more aggressive (Tokuda *et al.*, 2006). Therefore, accurately distinguishing between benign and malignant tumors and between the different subtypes of RCC is of high clinical importance. This is currently based on histopathology, but unfor-tunately it is not always accurate. Several potentially useful immunohistochemistry markers are used for this purpose, but they lack sensitivity and/or specificity. The presence of mixed tumors, where different subtypes of RCC exist in the same tumor can add to this clinical challenge.

The relationship between KLKs and kidney cancer was hypothesized by Bhoola *et al.* (2001). Also, analysis of an RCC cDNA library by Rae *et al.* identified *KLK1-3* and a novel *KLK1* mRNA transcript (Rae *et al.*, 1999). Petraki *et al.* (2006a) reported that KLK5, 6, 10, and 11 expressions were decreased in RCC, compared to normal kidney tissue, when assessed by immunohistochemistry (IHC). They also found a statisti-cally significant positive correlation between the immunohistochemical expression of KLKs 5, 6, 10, and 11. In addition, more aggressive RCC tumors expressed KLKs in a higher percentage of cases than less aggressive ones, but a statistically significant difference was only observed for KLK6 and KLK10 (55 *vs.* 27%, p = 0.016, and 79 *vs.* 56%). Moreover, both KLK6 and KLK11 showed a positive correlation with pathologi-cal stage. Immunohistochemical expression of KLK6 showed a negative correlation with disease-specific survival.

More recently, Gabril *et al.* (2010) reported differential immunostaining of KLKs in RCC, when compared to normal kidney tissue. There was an overall decrease in KLK1 expression in ccRCC when compared to adjacent normal kidney tissue, although there was a stronger KLK1 signal in higher-grade tumors (including grades III and IV and sarcomatoid carcinoma), when compared with lower grades (grades I and II). KLK1 showed weak expression in pRCC, and negative to weak focal expression in chRCC and oncocytoma.

In ccRCC, KLK6 showed moderate diffuse cytoplasmic granular staining with membranous accentuation and focal moderate nuclear staining in low nuclear grade tumors (Gabril *et al.*, 2010). Also, there was decreased KLK6 expression in higher-grade compared with lower-grade tumors. Papillary RCC showed strong diffuse KLK6 cytoplasmic expression, with focal apical accentuation. Oncocytoma and chRCC showed no expression in tumor cells and only highlighted KLK6 expression in the blood vessels of the stroma.

Both nuclear and cytoplasmic KLK7 expression, with focal membranous accentuation, were reported in ccRCC (Gabril *et al.*, 2010). There was decreased KLK7 expression in higher-grade compared to lower-grade tumors. Oncocytoma showed diffuse, strong granular cytoplasmic expression of KLK7. By contrast, chRCC showed focal weak homogeneous cytoplasmic staining. Papillary RCC showed strong, apical cytoplasmic KLK7 staining. In the sarcomatoid variant, KLK7 expression was decreased, and urothelial carcinoma displayed weak KLK7 expression in tumor cells, with positive expression in the umbrella cells. There was no KLK7 expression in collecting duct carcinoma. There was weak cytoplasmic staining of KLK15 in ccRCC, weak to negative expression in pRCC and no expression in chromophobe RCC, urothelial carcinoma, or oncocytoma (Gabril *et al.*, 2010). Taken together, the above findings indicate the potential use of KLKs as adjuvant tissue markers, in order to distinguish RCC subtypes.

8.2 microRNAs (miRNA)

8.2.1 Biogenesis

The biogenesis of microRNAs (miRNA) has been studied extensively (Bartel, 2004; Du and Zamore, 2005; Kim, 2005). First, miRNAs are transcribed by RNA polymerase II, producing a long primary miRNA (pri-miRNA). The pri-miRNA is then modified in the nucleus through capping and polyadenylation, and subsequently cleaved into smaller segments by the RNase III enzyme Drosha, to form a 60–70 nucleotide precursor (pre-miRNA). The pre-miRNA is then transported to the cytoplasm by the enzyme exportin 5, where it is modified by Dicer to result in a 19–24 nucleotide miRNA duplex. One of the two miRNA strands (mature miRNA) is integrated into a large protein complex called RNA-Induced Silencing Complex (RISC), which leads to either messenger RNA (mRNA) degradation or repression of the target protein translation. Depending on the annealing position and the homology of the miRNA to its substrate, mRNA is either degraded or stored in p-bodies (Konecna *et al.*, 2009; Zhao and Liu, 2009), from which they can be released in high copy numbers, e.g. during cell cycle or response to environmental stimuli.

There are a number of biologically diverse functions of miRNAs and they are known to be involved in many normal processes, including cell mobility, differentiation, development, proliferation, and apoptosis (Ambros, 2004). They are predicted to control the expression of up to two thirds of all human genes (Lewis *et al.*, 2005). It is not surprising then that miRNAs have been shown to be involved in cancer.

8.2.2 miRNAs and cancer

The link between miRNAs and cancer is well-established in the literature. miRNA dysregulation has been reported in almost every known cancer, e.g. of the breast, ovarian, prostate, kidney, lung, and the blood (Iorio *et al.*, 2005; 2007; Lu *et al.*, 2012; Porkka *et al.*, 2007; White *et al.*, 2011a). Also, miRNAs have been shown to frequently be located in cancer-associated genomic regions, including fragile sites, regions of loss of heterozygosity or amplification, or common break points (Calin and Croce, 2006b). Moreover, miRNAs have been shown to act as either oncogenes or tumor suppressor genes, depending on their targets (Esquela-Kerscher and Slack, 2006). For example, if a miRNA targets and subsequently negatively regulates a tumor suppressor gene, it functions as an oncogene, e.g. miR-21 targets the PTEN tumor suppressor gene in a number of different cancers. Also, some miRNAs have been shown to have a tumor-suppressive effect, e.g. miR-15/16. On the other hand, miRNAs can also be downstream effectors of oncogenes or tumor suppressor genes and have also been shown to be regulated epigenetically (Formosa *et al.*, 2012), and to be able to epigenetically silence genes (Wang *et al.*, 2012).

Potential roles for miRNAs in the pathogenesis and spread of different cancers have been shown (White *et al.*, 2011b). Accumulating reports highlight the potential utility of miRNAs as cancer biomarkers (Blenkiron and Miska, 2007; Calin and Croce, 2006a; Lu *et al.*, 2005; Mattie *et al.*, 2006; Metias *et al.*, 2009). Differential miRNA expression in cancer can be a useful clinical tool for diagnosis, accurate classification of tumor subtypes, prognosis, and prediction of treatment efficiency. More recently, the potential use of miRNAs as therapeutic targets is being explored (Bhardwaj *et al.*, 2010; Esau and Monia, 2007; Zhang *et al.*, 2010).

8.2.3 miRNA dysregulation in renal cell carcinoma

miRNA dysregulation in RCC has been examined by a number of groups, with a cumulative 237 miRNAs reported to be dysregulated in ccRCC (Chow *et al.*, 2009; Gottardo *et al.*, 2007; Huang *et al.*, 2009b; Jung *et al.*, 2009; Nakada *et al.*, 2008; Naylor *et al.*, 2009; White *et al.*, 2011a). Many of these miRNAs were reported by more than one group, with the same direction of dysregulation, i.e. up- or downregulation. For example, multiple studies reported the upregulation of miR-210, which is known to be involved in hypoxia (Camps *et al.*, 2008; Chan and Loscalzo, 2010; Huang *et al.*, 2009a), and the downregulation of miR-200c, which is well-known for its role in epithelial-to-mesenchymal transition (Gregory *et al.*, 2008; Li *et al.*, 2009). Recently, the involvement of miRNAs in the pathogenesis of ccRCC has been proven experimentally. An effect of miRNAs on cell proliferation in kidney cancer cell lines has also been shown (Chow *et al.*, 2010; White *et al.*, 2010a).

In addition, Neal *et al.* showed that miRNAs are regulated by VHL (Neal *et al.*, 2010), while Lichner *et al.* showed that miRNAs can target multiple members of RCC-related signaling pathways, including VHL and the HIF1-α (Lichner Z *et al.*, 2012). Potential mechanisms for the involvement of miRNAs in RCC have been proposed (White and Yousef, 2010; 2011). Depending on their target, miRNAs can act as either oncogenes or tumor suppressor genes. They can have a direct effect on key molecules involved in the pathogenesis of RCC, including VHL and hypoxia-inducible factors (HIF). It has been shown that the downregulation of VHL by miRNAs promotes the HIF/vascular endothelial growth factor axis (Ghosh *et al.*, 2009). Finally, miRNAs can be downstream effectors of the hypoxia response or effectors of VHL deletion, through a non-hypoxia-mediated pathway (Neal *et al.*, 2010).

It was possible by miRNA profiling to identify miRNA signatures that have potential for diagnostic and prognostic applications to kidney cancer patients. Specific miRNA signatures can accurately distinguish between normal kidney tissue and kidney cancer (White *et al.*, 2011a), as well as between kidney cancer subtypes (Youssef *et al.*, 2011). In addition, White *et al.* identified a miRNA signature that can accurately distinguish between primary and metastatic RCC (White *et al.*, 2011c). The prognostic value of these miRNAs is currently being explored.

8.2.4 The miRNA-KLK interaction

KLK expression and function are regulated by multiple mechanisms, including steroid hormones, hypermethylation of CpG islands, and modifications of the chromosome structure of the KLK locus. KLKs have also been shown to be regulated at the protein level by activation of the zymogen and inhibition of protease activity by serpins or zinc ions (Pampalakis and Sotiropoulou, 2007). Evidence suggests that KLKs can be controlled by miRNAs at the post-transcriptional level. This hypothesis is supported by several reports showing discrepancies between mRNA and protein levels for many KLKs (Yousef *et al.*, 2005a). Other interesting observations are that a number of KLKs are reported to be co-expressed in the same tissues, and that KLKs are similarly dysregulated under certain pathological conditions. This can be explained, at least in part, by the fact that the same miRNA can simultaneously target multiple KLKs.

Chow *et al.* (2008) were the first to report a link between miRNAs and KLKs. Through target prediction analysis, they identified a total of 550 miRNA-KLK interactions, predicted by at least one program, including miRBase Targets V4, miRanda,

TargetScan 4.0, and Pic Tar (available at the time of analysis). Applying a more stringent cut-off, there were 96 miRNA-KLK "strong interactions" predicted by two or more target prediction programs (**Tab. 8.2**). KLK10 was thereby reported to be the most-targeted KLK, with 19 targeting miRNAs, followed by KLK5 and KLK13. There also was a correlation between locations of KLK-targeting miRNAs and cancer fragile sites. This study also validated the effect of miRNAs on KLK protein expression. Transfection of let-7f resulted in a decrease in KLK6 and KLK10 protein levels.

Tab. 8.2 Predicted KLK-miRNA interactions and the number of predicted target sites on KLK transcripts.

KLK	miRNA	Programs[a]	Target sites[b]
KLK2	hsa-miR-211	2	1
KLK2	hsa-miR-324	2	1
KLK2	hsa-miR-337	2	1
KLK2	hsa-miR-502	2	4
KLK4	hsa-miR-422	2	1
KLK4	hsa-miR-548d	2	1
KLK4	hsa-miR-637	2	5
KLK4	hsa-miR-765	2	10
KLK5	hsa-miR-106a	2	1
KLK5	hsa-miR-106b	2	1
KLK5	hsa-miR-122a	2	1
KLK5	hsa-miR-125a	2	2
KLK5	hsa-miR-125b	2	2
KLK5	hsa-miR-143	2	1
KLK5	hsa-miR-17	2	1
KLK5	hsa-miR-185	3	1
KLK5	hsa-miR-20b	2	1
KLK5	hsa-miR-299	2	1
KLK5	hsa-miR-326	2	1
KLK5	hsa-miR-491	2	2
KLK5	hsa-miR-519d	2	1
KLK5	hsa-miR-519e	2	1
KLK5	hsa-miR-625	2	1
KLK6	hsa-let-7f	3	1
KLK7	hsa-miR-199a	2	1
KLK7	hsa-miR-199b	2	1
KLK7	hsa-miR-30a	2	1
KLK7	hsa-miR-30e	2	1
KLK7	hsa-miR-33a	2	1
KLK7	hsa-miR-338	2	1
KLK7	hsa-miR-33b	2	1
KLK7	hsa-miR-369	2	1
KLK7	hsa-miR-509	2	1
KLK7	hsa-miR-519a	2	1
KLK7	hsa-miR-519c	2	1

Tab. 8.2 (continued)

KLK	miRNA	Programs[a]	Target sites[b]
KLK7	hsa-miR-591	2	1
KLK7	hsa-miR-628	2	1
KLK9	hsa-miR-18a	2	1
KLK9	hsa-miR-18b	2	1
KLK9	hsa-miR-198	2	1
KLK9	hsa-miR-33a	2	1
KLK9	hsa-miR-33b	2	1
KLK9	hsa-miR-376a	2	1
KLK9	hsa-miR-376b	2	1
KLK9	hsa-miR-431	2	1
KLK9	hsa-miR-548c	2	1
KLK9	hsa-miR-598	2	1
KLK9	hsa-miR-621	2	1
KLK9	hsa-miR-663	2	1
KLK10	hsa-let-7a	3	1
KLK10	hsa-let-7b	2	1
KLK10	hsa-let-7c	3	1
KLK10	hsa-let-7d	3	1
KLK10	hsa-let-7e	3	1
KLK10	hsa-let-7f	3	1
KLK10	hsa-let-7g	2	1
KLK10	hsa-miR-148a	2	1
KLK10	hsa-miR-148b	2	1
KLK10	hsa-miR-197	2	2
KLK10	hsa-miR-214	2	2
KLK10	hsa-miR-224	2	1
KLK10	hsa-miR-326	2	2
KLK10	hsa-miR-377	2	1
KLK10	hsa-miR-496	2	2
KLK10	hsa-miR-516a	2	3
KLK10	hsa-miR-516b	2	3
KLK10	hsa-miR-598	2	1
KLK10	hsa-miR-98	3	1
KLK11	hsa-miR-186	2	1
KLK11	hsa-miR-409	2	1
KLK11	hsa-miR-495	2	1
KLK11	hsa-miR-511	2	1
KLK11	hsa-miR-542	2	1
KLK13	hsa-miR-1	2	1
KLK13	hsa-miR-141	2	1
KLK13	hsa-miR-181b	2	1
KLK13	hsa-miR-181c	2	1
KLK13	hsa-miR-181d	2	1
KLK13	hsa-miR-206	2	1
KLK13	hsa-miR-34a	3	1
KLK13	hsa-miR-34c	2	1

Tab. 8.2 (continued)

KLK	miRNA	Programs[a]	Target sites[b]
KLK13	hsa-miR-409	2	1
KLK13	hsa-miR-449	3	1
KLK13	hsa-miR-453	2	1
KLK13	hsa-miR-455	2	1
KLK13	hsa-miR-494	2	1
KLK13	hsa-miR-515	2	1
KLK13	hsa-miR-542	2	1
KLK14	hsa-miR-612	2	1
KLK14	hsa-miR-661	2	1
KLK15	hsa-miR-224	3	1
KLK15	hsa-miR-498	2	1
KLK15	hsa-miR-552	2	1
KLK15	hsa-miR-608	2	1
KLK15	hsa-miR-638	2	1
KLK15	hsa-miR-663	2	1

a Target prediction programs that were used include miRBase Targets V4, miRanda, TargetScan 4.0, and PicTar

b The number of predicted target sites on the 3′-untranslated region of the corresponding KLK mRNA

8.3 miRNA control of KLK expression in renal cell carcinoma

White *et al.* (2010a) provided the first evidence for linking KLK dysregulation in RCC to miRNAs. Target prediction analysis revealed 61 of 117 miRNAs that were shown to be dysregulated in RCC, to be predicted to target KLKs. Confirming earlier results, this study reported that each *KLK* can be targeted by more than one miRNA and that a single miRNA can target more than one *KLK*. Interestingly, the most commonly reported dysregulated miRNAs in ccRCC, miR-122, miR-210, and miR-224 were predicted to target seven KLKs including KLK1, 2, 5, 7, 10, 11, and 15. The KLK-miRNA axis of interaction was validated experimentally, using two independent approaches. The first was to compare target protein levels before and after miRNA transfection. The miR-224 transfection resulted in decreased KLK1 protein expression. In addition, the miRNA-KLK interaction was confirmed, using a luciferase assay. Transfection of let-7f resulted in decreased luciferase activity of a luciferase vector containing the 3′-untranslated region (UTR) of *KLK10*, compared to controls.

8.4 miRNA control of KLK expression in other cancers

The *KLK*-miRNA interaction was also validated in ovarian and prostate cancers. *KLK* dysregulation in ovarian cancer is well-documented in the literature. Recent reports have also shown that miRNAs are dysregulated in ovarian cancer (Bearfoot *et al.*, 2008; Dahiya *et al.*, 2008; Iorio *et al.*, 2007; Nam *et al.*, 2008; Resnick *et al.*, 2009; Taylor and Gercel-Taylor, 2008; Zhang *et al.*, 2008). A recent study compiled 99 miRNAs that are reported to be dysregulated in ovarian cancer. Interestingly, 25 of these miRNAs were located in chromosomal hotspots that frequently showed chromosomal imbalance in ovarian cancer, suggesting miRNA dysregulation in ovarian cancer may be partially explained by chromosomal aberrations (White *et al.*, 2010b). Target prediction analysis for the miRNAs that are dysregulated in ovarian cancer identified that 63% (62/99) of these miRNAs are predicted to target KLKs. Each *KLK*, with the exception of *KLK14*, could be targeted by at least one miRNA. Twenty-three miRNAs were predicted to target *KLK6*. Interestingly, 61 (49%) of the dysregulated miRNAs correlated with their *KLK* target expression in ovarian cancer, i.e. increased miRNA expression correlated with decreased *KLK* expression and *vice versa*. The same study validated the miRNA-*KLK* interaction in ovarian cancer. Upon transfection with a luciferase vector containing the *KLK10* 3′-UTR and either of miR-224, miR-516a, or let-7f, there was a decrease in luciferase signal, which demonstrated that these miRNAs can directly target *KLK10*. The authors also provided preliminary indirect evidence that miRNAs have a negative effect on ovarian cancer cell proliferation, at least in part, through regulating KLK protein expression.

The interaction between miRNAs and *KLKs* has also been studied in prostate cancer. Dysregulated miRNAs in prostate cancer were identified by comparing prostate cancer to normal prostate tissue from the same patient. There were a total of 55 miRNAs dysregulated in prostate cancer, that were predicted to target *KLKs*. Seven of these (miR-1, miR-140, miR-143, miR-17-5p, miR-21, miR-24, and miR-331-3p) were predicted to target more than one *KLK* (our unpublished data). When the expression of miRNA and their target *KLKs* were examined in prostate cancer tissues, an inverse correlation pattern was found, i.e. miRNA upregulation in cancer was associated with downregulation of their *KLK* targets and *vice versa*. The miRNA-*KLK* interaction was also validated experimentally in prostate cancer cell lines. Upon transfection of miR-330 and miR-143, expression of target *KLKs* were decreased. Furthermore, this interaction had a negative effect on cellular proliferation. Also, Mavridis *et al.* (unpublished data) state that in prostate tumors *KLK15* mRNA and miR-224 levels are negatively correlated.

The interaction between miRNAs and prostate specific antigen (*PSA; KLK3*) has been explored recently by Sun *et al.* (2011), who showed that there is a direct relationship between members of the miR-99 family, including miR-99a, -100 and -125b, and *PSA* (*KLK3*). Transfection of miR-99a, miR-99b, or miR-100 inhibited the growth of prostate cancer cells and decreased the expression of PSA (KLK3) in prostate cancer cell lines. PSA levels increased after inhibition of the miR-99 family.

The miRNA-KLK axis of interaction was also shown to have potential clinical utility for prostate cancer patients. Hao *et al.* (2011) evaluated the clinical utility of the integration of KLK3 (PSA) blood levels with miRNA profiling to improve prostate cancer detection. They reported that the positive predictive value analysis of prostate cancer was increased from 40 to 87.5% by integrating PSA blood levels with miR-21 and miR-141 profiles.

8.5 Conclusions and outlook

It is now becoming evident that the KLK-miRNA interaction is actively involved in cancer pathogenesis. This has a number of implicated potential clinical uses. Cancer gene therapy based on miRNA offers the theoretical appeal of simultaneously targeting multiple gene networks that are controlled by a single, aberrantly expressed miRNA (Tong and Nemunaitis, 2008; White and Yousef, 2010). Reconstitution of tumor-suppressor miRNAs or sequence-specific knockdown of oncogenic miRNAs has produced favorable anti-tumor outcomes in experimental models and they are now being tested for therapeutic applications in different cancers (Garzon *et al.*, 2010; White *et al.*, 2011b). Also, miRNAs can be utilized to knock out KLK genes and that will facilitate an in-depth understanding of the role of KLKs in carcinogenesis.

The relationship is, however, more complicated than previously understood. On one hand, KLKs can be miRNA targets. As experimentally validated, one KLK can be targeted by more than one miRNA, leading to a mechanism for quantitative control over KLKs involved in cancer pathogenesis. In addition, the same miRNA can target more than one KLK in the same pathway, leading to exponential effects on tumorigenesis. This is further complicated by the fact that KLKs may represent indirect targets of miRNAs. On the other hand, KLKs may potentially exert transcriptional control over miRNAs by affecting the miRNA promoters. Recent reports proposed a transcriptional regulator function of KLKs, as suggested by their nuclear localization (Klokk *et al.*, 2007; Korkmaz *et al.*, 2001).

It has also been suggested that target genes, including KLKs, can control miRNA function (Marson *et al.*, 2008). This can be explained by the consumption of the limited pool of cellular miRNAs by some targets, with consequent release of their inhibitory effect on other targets. This can also provide a potential role for splice variants (Yousef *et al.*, 2004b; 2005b; Kurlender *et al.*, 2005) and pseudogenes (Lu *et al.*, 2006; Yousef *et al.*, 2004a) identified in the KLK family without known functions.

It has also been recently hypothesized that genes such as *KLKs* can communicate with each other using miRNAs, according to the "competing endogenous RNA theory" (Salmena *et al.*, 2011). This is likely to occur for genes with a high degree of homology in their 3′-UTR, such as *KLKs* that contain closely related miRNA response elements (MRE). The dysregulation of one gene can influence the stability and protein translation of the other by altering the amount of available miRNAs. Finally, it should be noted that, contrary to the established role of miRNAs in repressing target proteins, few recent reports showed potential new control mechanisms, including the effect of miRNAs on promoter sequence and protein coding regions.

In advancing the miRNA-KLK axis of research, there are also technical considerations that should be considered. An inherited difficulty is the need for target validation. The presence of an indirect effect should always be considered and may be very difficult to exclude experimentally. Another challenge is the fact that the same miRNA can have multiple targets and a direct one-on-one interaction may not be easy to validate experimentally.

Bibliography

Ahn, J., Berndt, S.I., Wacholder, S., Kraft, P., Kibel, A.S., Yeager, M., Albanes, D., Giovannucci, E., Stampfer, M.J., Virtamo, J., Thun, M.J., Feigelson, H.S., Cancel-Tassin, G., Cussenot, O., Thomas, G., Hunter, D.J., Fraumeni, J.F. Jr., Hoover, R.N., Chanock, S.J., and Hayes, R.B. (2008). Variation in KLK genes, prostate-specific antigen and risk of prostate cancer. Nat. Genet. 40, 1032–1034.

Ambros, V. (2004). The functions of animal microRNAs. Nature 431, 350–355.

Bartel, D.P. (2004). MicroRNAs: genomics, biogenesis, mechanism, and function. Cell 116, 281–297.

Bayani, J., Marrano, P., Graham, C., Zheng,Y., Li, L., Katsaros, D., Lassus, H., Butzow, R., Squire, J.A., and Diamandis, E.P. (2011). Genomic instability and copy-number heterogeneity of chromosome 19q, including the kallikrein locus, in ovarian carcinomas. Mol. Oncol. 5, 48–60.

Bearfoot, J.L., Choong, D.Y., Gorringe, K.L., and Campbell, I.G. (2008). Genetic analysis of cancer-implicated MicroRNA in ovarian cancer. Clin. Cancer Res. 14, 7246–7250.

Bhardwaj, A., Singh, S., and Singh, A.P. (2010). MicroRNA-based cancer therapeutics: big hope from small RNAs. Mol. Cell Pharmacol. 2, 213–219.

Bhoola, K., Ramsaroop, R., Plendl, J., Cassim, B., Dlamini, Z., and Naicker, S. (2001). Kallikrein and kinin receptor expression in inflammation and cancer. Biol. Chem. 382, 77–89.

Blenkiron, C. and Miska, E.A. (2007). miRNAs in cancer: approaches, aetiology, diagnostics and therapy. Hum. Mol. Genet. 16 Spec No 1, R106-R113.

Calin, G.A. and Croce, C.M. (2006a). MicroRNA signatures in human cancers. Nat. Rev. Cancer 6, 857–866.

Calin, G.A. and Croce, C.M. (2006b). MicroRNAs and chromosomal abnormalities in cancer cells. Oncogene 25, 6202–6210.

Camps, C., Buffa, F.M., Colella, S., Moore, J., Sotiriou, C., Sheldon, H., Harris, A.L., Gleadle, J.M., and Ragoussis, J. (2008). hsa-miR-210 Is induced by hypoxia and is an independent prognostic factor in breast cancer. Clin. Cancer Res. 14, 1340–1348.

Chan, S.Y. and Loscalzo, J. (2010). MicroRNA-210: a unique and pleiotropic hypoxamir. Cell Cycle 9, 1072–1083.

Chow, T.F., Crow, M., Earle, T., El-Said, H., Diamandis, E.P., and Yousef, G.M. (2008). Kallikreins as microRNA targets: an in silico and experimental-based analysis. Biol. Chem. 389, 731–738.

Chow, T.F., Youssef, Y.M., Iianidou, E., Romaschin, A.D., Honey, R.J., Stewart, R., Pace, K.T., and Yousef, G.M. (2009). Differential expression profiling of microRNAs and their potential involvement in renal cell carcinoma pathogenesis. Clin. Biochem. 43, 150–158.

Chow, T.F., Mankaruos, M., Scorilas, A., Youssef, Y., Girgis, A., Mossad, S., Metias, S., Rofael, Y., Honey, R.J., Stewart, R., Pace, K.T., and Yousef, G.M. (2010). The miR-17-92 cluster is overexpressed in and has an oncogenic effect on renal cell carcinoma. J. Urol. 183, 743–751.

Chow, W.H., Devesa, S.S., Warren, J.L., and Fraumeni, J.F. Jr. (1999). Rising incidence of renal cell cancer in the United States. JAMA 281, 1628–1631.

Dahiya, N., Sherman-Baust, C.A., Wang, T.L., Davidson, B., Shih, I., Zhang, Y., Wood, W., III, Becker, K.G., and Morin, P.J. (2008). MicroRNA expression and identification of putative miRNA targets in ovarian cancer. PLoS. ONE. 3, e2436.

Du, T., and Zamore, P.D. (2005). microPrimer: the biogenesis and function of microRNA. Development 132, 4645–4652.

Esau, C.C., and Monia, B.P. (2007). Therapeutic potential for microRNAs. Adv. Drug Deliv. Rev. 59, 101–114.

Esquela-Kerscher, A., and Slack, F.J. (2006). Oncomirs – microRNAs with a role in cancer. Nat. Rev. Cancer 6, 259–269.

Formosa, A., Lena, A.M., Markert, E.K., Cortelli, S., Miano, R., Mauriello, A., Croce, N., Vandesompele, J., Mestdagh, P., Finazzi-Agro, E., Levine, A.I., Melino, G., Bernardini, S., and Candi, E. (2012). DNA methylation silences miR-132 in prostate cancer. Oncogene, [Epub ahead of print].

Gabril, M., White, N.M., Moussa, M., Chow, T.F., Metias, S.M., Fatoohi, E., and Yousef, G.M. (2010). Immunohistochemical analysis of kallikrein-related peptidases in the normal kidney and renal tumors: potential clinical implications. Biol. Chem. 391, 403–409.

Garzon, R., Marcucci, G., and Croce, C.M. (2010). Targeting microRNAs in cancer: rationale, strategies and challenges. Nat. Rev. Drug Discov. 9, 775–789.

Ghosh, A.K., Shanafelt, T.D., Cimmino, A., Taccioli, C., Volinia, S., Liu, C.G., Calin, G.A., Croce, C.M., Chan, D.A., Giaccia, A.J., Secreto, C., Wellik, L.E., Lee, Y.K., Mukhopadhyay, D., and Kay, N.E. (2009). Aberrant regulation of pVHL levels by microRNA promotes the HIF/VEGF axis in CLL B cells. Blood 113, 5568–5574.

Gottardo, F., Liu, C.G., Ferracin, M., Calin, G.A., Fassan, M., Bassi, P., Sevignani, C., Byrne, D., Negrini, M., Pagano, F., Gomella, L.G., Croce, C.M., and Baffa, R. (2007). Micro-RNA profiling in kidney and bladder cancers. Urol. Oncol. 25, 387–392.

Gregory, P.A., Bert, A.G., Paterson, E.L., Barry, S.C., Tsykin, A., Farshid, G., Vadas, M.A., Khew-Goodall, Y., and Goodall, G.J. (2008). The miR-200 family and miR-205 regulate epithelial to mesenchymal transition by targeting ZEB1 and SIP1. Nat. Cell Biol. 10, 593–601.

Hao, Y., Zhao, Y., Zhao, X., He, C., Pang, X., Wu, T.C., Califano, J.A., and Gu, X. (2011). Improvement of prostate cancer detection by integrating the PSA test with miRNA expression profiling. Cancer Invest. 29, 318–324.

Huang, X., Ding, L., Bennewith, K.L., Tong, R.T., Welford, S.M., Ang, K.K., Story, M., Le, Q.T., and Giaccia, A.J. (2009a). Hypoxia-inducible mir-210 regulates normoxic gene expression involved in tumor initiation. Mol. Cell 35, 856–867.

Huang, Y., Dai, Y., Yang, J., Chen, T., Yin, Y., Tang, M., Hu, C., and Zhang, L. (2009b). Microarray analysis of microRNA expression in renal clear cell carcinoma. Eur. J. Surg. Oncol. 35, 1119–1123.

Iorio, M.V., Ferracin, M., Liu, C.G., Veronese, A., Spizzo, R., Sabbioni, S., Magri, E., Pedriali, M., Fabbri, M., Campiglio, M., Menard, S., Palazzo, J.P., Rosenberg, A., Musiani, P., Volinia, S., Nenci, I., Calin, G.A., Querzoli, P., Negrini, M., and Croce, C.M. (2005). MicroRNA gene expression deregulation in human breast cancer. Cancer Res. 65, 7065–7070.

Iorio, M.V., Visone, R., Di, L.G., Donati, V., Petrocca, F., Casalini, P., Taccioli, C., Volinia, S., Liu, C.G., Alder, H., Calin, G.A., Menard, S., and Croce, C.M. (2007). MicroRNA signatures in human ovarian cancer. Cancer Res. 67, 8699–8707.

Jung, M., Mollenkopf, H.-J., Grimm, C., Wagner, I., Albrecht, M., Waller, T., Pilarsky, C., Johannsen, M., Stephan, C., Lehrach, H., Nietfeld, W., Rudel, T., Jung, K., and Kristiansen, G. (2009). MicroRNA profiling of clear cell renal cell cancer identifies a robust signature to define renal malignancy. J. Cell. Mol. Med. 19, 3918–3928.

Kim, V.N. (2005). MicroRNA biogenesis: coordinated cropping and dicing. Nat. Rev. Mol. Cell Biol. 6, 376–385.

Klokk, T.I., Kilander, A., Xi, Z., Waehre, H., Risberg, B., Danielsen, H.E., and Saatcioglu, F. (2007). Kallikrein 4 is a proliferative factor that is overexpressed in prostate cancer. Cancer Res. 67, 5221–5230.

Komatsu, N., Saijoh, K., Kuk, C., Shirasaki, F., Takehara, K., and Diamandis, E.P. (2007). Aberrant human tissue kallikrein levels in the stratum corneum and serum of patients with psoriasis: dependence on phenotype, severity and therapy. Br. J. Dermatol. 156, 875–883.

Konecna, A., Heraud, J.E., Schoderboeck, L., Raposo, A.A., and Kiebler, M.A. (2009). What are the roles of microRNAs at the mammalian synapse? Neurosci. Lett. 466, 63–68.

Korkmaz, K.S., Korkmaz, C.G., Pretlow, T.G., and Saatcioglu, F. (2001). Distinctly different gene structure of KLK4/KLK-L1/prostase/ARM1 compared with other members of the kallikrein family: intracellular localization, alternative cDNA forms, and regulation by multiple hormones. DNA Cell Biol. 20, 435–445.

Kovacs, G., and Frisch, S. (1989). Clonal chromosome abnormalities in tumor cells from patients with sporadic renal cell carcinomas. Cancer Res. 49, 651–659.

Kurlender, L., Borgoño, C., Michael, I.P., Obiezu, C., Elliott, M.B., Yousef, G.M., and Diamandis, E.P. (2005). A survey of alternative transcripts of human tissue kallikrein genes. Biochim. Biophys. Acta 1755, 1–14.

Lewis, B.P., Burge, C.B., and Bartel, D.P. (2005). Conserved seed pairing, often flanked by adenosines, indicates that thousands of human genes are microRNA targets. Cell 120, 15–20.

Li, Y., van den Boom, T.G., Kong, D., Wang, Z., Ali, S., Philip, P.A., and Sarkar, F.H. (2009). Up-regulation of miR-200 and let-7 by natural agents leads to the reversal of epithelial-to-mesenchymal transition in gemcitabine-resistant pancreatic cancer cells. Cancer Res. 69, 6704–6712.

Lichner, Z., Mejia-Guerrero, S., Ignacak, M.L., Krizova, A., Bao, T.T., Girgis, A., Youssef, Y., and Yousef, G.M. (2012). Pleiotropic action of renal cell carcinoma-dysregulated miRNAs on hypoxia related signaling pathways. Am. J. Pathol. 180, 1675–1687.

Lu, J., Getz, G., Miska, E.A., Alvarez-Saavedra, E., Lamb, J., Peck, D., Sweet-Cordero, A., Ebert, B.L., Mak, R.H., Ferrando, A.A., Downing, J.R., Jacks, T., Horvitz, H.R., and Golub, T.R. (2005). MicroRNA expression profiles classify human cancers. Nature 435, 834–838.

Lu, W., Zhou, D., Glusman, G., Utleg, A.G., White, J.T., Nelson, P.S., Vasicek, T.J., Hood, L., and Lin, B. (2006). KLK31P is a novel androgen regulated and transcribed pseudogene of kallikreins that is expressed at lower levels in prostate cancer cells than in normal prostate cells. Prostate 66, 936–944.

Lu, Y., Govindan, R., Wang, L., Liu, P.Y., Goodgame, B., Wen, W., Sezhiyan, A., Li, Y.F., Hua, X., Wang, Y., Yang, P., and You, M. (2012). MicroRNA profiling and prediction of recurrence/relapse-free survival in stage I lung cancer. Carcinogenesis 33, 1046–1054.

Marson, A., Levine, S.S., Cole, M.F., Frampton, G.M., Brambrink, T., Johnstone, S., Guenther, M.G., Johnston, W.K., Wernig, M., Newman, J., Calabrese, J.M., Dennis, L.M., Volkert, T.L., Gupta, S., Love, J., Hannett, N., Sharp, P.A., Bartel, D.P., Jaenisch, R., and Young, R.A. (2008). Connecting microRNA genes to the core transcriptional regulatory circuitry of embryonic stem cells. Cell 134, 521–533.

Mattie, M.D., Benz, C.C., Bowers, J., Sensinger, K., Wong, L., Scott, G.K., Fedele, V., Ginzinger, D., Getts, R., and Haqq, C. (2006). Optimized high-throughput microRNA expression profiling provides novel biomarker assessment of clinical prostate and breast cancer biopsies. Mol. Cancer 5, 24.

McLaughlin, J.K., Lipworth, L., and Tarone, R.E. (2006). Epidemiologic aspects of renal cell carcinoma. Semin. Oncol. 33, 527–533.

Metias, S.M., Lianidou, E., and Yousef, G.M. (2009). MicroRNAs in clinical oncology: at the crossroads between promises and problems. J. Clin. Pathol. 62, 771–776.

Moore, L.E., Nickerson, M.L., Brennan, P., Toro, J.R., Jaeger, E., Rinsky, J., Han, S.S., Zaridze, D., Matveev, V., Janout, V., Kollarova, H., Bencko, V., Navratilova, M., Szeszenia-Dabrowska, N., Mates, D., Schmidt, L.S., Lenz, P., Karami, S., Linehan, W.M., Merino, M., Chanock, S., Boffetta, P., Chow, W.H., Waldman, F.M., and Rothman, N. (2011). Von Hippel-Lindau (VHL) inactivation in sporadic clear cell renal cancer: associations with germline VHL polymorphisms and etiologic risk factors. PLoS. Genet. 7, e1002312.

Nakada, C., Matsuura, K., Tsukamoto, Y., Tanigawa, M., Yoshimoto, T., Narimatsu, T., Nguyen, L.T., Hijiya, N., Uchida, T., Sato, F., Mimata, H., Seto, M., and Moriyama, M. (2008). Genome-wide microRNA expression profiling in renal cell carcinoma: significant down-regulation of miR-141 and miR-200c. J. Pathol. 216, 418–427.

Nam, E.J., Yoon, H., Kim, S.W., Kim, H., Kim, Y.T., Kim, J.H., Kim, J.W., and Kim, S. (2008). MicroRNA expression profiles in serous ovarian carcinoma. Clin. Cancer Res. 14, 2690–2695.

Naylor, S.L., Garcia, D., Parekh, D., Troyer, D.A., Grizzle, W.A., and Thompson, I.M. (2009). miRNA profiling in kidney cancer. Keystone Symposia on Molecular and Cellular Biology: microRNA and Cancer. Silverstone, CO., 248, 78.

Neal, C.S., Michael, M.Z., Rawlings, L.H., van der Hoek, M.B., and Gleadle, J.M. (2010). The VHL-dependent regulation of microRNAs in renal cancer. BMC Med. 8, 64.

Pampalakis, G., and Sotiropoulou, G. (2007). Tissue kallikrein proteolytic cascade pathways in normal physiology and cancer. Biochim. Biophys. Acta 1776, 22–31.

Petraki, C.D., Gregorakis, A.K., Vaslamatzis, M.M., Papanastasiou, P.A., Yousef, G.M., Levesque, M.A., and Diamandis, E.P. (2006a). Prognostic implications of the immunohistochemical expression of human kallikreins 5, 6, 10 and 11 in renal cell carcinoma. Tumour Biol. 27, 1–7.

Petraki, C.D., Papanastasiou, P.A., Karavana, V.N., and Diamandis, E.P. (2006b). Cellular distribution of human tissue kallikreins: immunohistochemical localization. Biol. Chem. 387, 653–663.

Porkka, K.P., Pfeiffer, M.J., Waltering, K.K., Vessella, R.L., Tammela, T.L., and Visakorpi, T. (2007). MicroRNA expression profiling in prostate cancer. Cancer Res. 67, 6130–6135.

Rae, F., Bulmer, B., Nicol, D., and Clements, J. (1999). The human tissue kallikreins (KLKs 1–3) and a novel KLK1 mRNA transcript are expressed in a renal cell carcinoma cDNA library. Immunopharmacology 45, 83–88.

Resnick, K.E., Alder, H., Hagan, J.P., Richardson, D.L., Croce, C.M., and Cohn, D.E. (2009). The detection of differentially expressed microRNAs from the serum of ovarian cancer patients using a novel real-time PCR platform. Gynecol. Oncol. 112, 55–59.

Salmena, L., Poliseno, L., Tay, Y., Kats, L., and Pandolfi, P.P. (2011). A ceRNA hypothesis: the Rosetta Stone of a hidden RNA language? Cell 146, 353–358.

Shaw, J.L., and Diamandis, E.P. (2007). Distribution of 15 human kallikreins in tissues and biological fluids. Clin. Chem. 53, 1423–1432.

Sun, D., Lee, Y.S., Malhotra, A., Kim, H.K., Matecic, M., Evans, C., Jensen, R.V., Moskaluk, C.A., and Dutta, A. (2011). miR-99 family of MicroRNAs suppresses the expression of prostate-specific antigen and prostate cancer cell proliferation. Cancer Res. 71, 1313–1324.

Taylor, D.D., and Gercel-Taylor, C. (2008). MicroRNA signatures of tumor-derived exosomes as diagnostic biomarkers of ovarian cancer. Gynecol. Oncol. 110, 13–21.

Tokuda, N., Naito, S., Matsuzaki, O., Nagashima, Y., Ozono, S., and Igarashi, T. (2006). Collecting duct (Bellini duct) renal cell carcinoma: a nationwide survey in Japan. J. Urol. 176, 40–43.

Tong, A.W. and Nemunaitis, J. (2008). Modulation of miRNA activity in human cancer: a new paradigm for cancer gene therapy? Cancer Gene Ther. 15, 341–355.

Wang, Y.S., Chou, W.W., Chen, K.C., Cheng, H.Y., Lin, R.T., and Juo, S.H. (2012). microRNA-152 Mediates DNMT1-Regulated DNA Methylation in the Estrogen Receptor alpha Gene. PLoS One 7, e30635.

White, N.M., and Yousef, G.M. (2010). MicroRNAs: exploring a new dimension in the pathogenesis of kidney cancer. BMC Med. 8, 65.

White, N.M., Bui, A., Mejia-Guerrero, S., Chao, J., Soosaipillai, A., Youssef, Y., Mankaruos, M., Honey, R.J., Stewart, R., Pace, K.T., Sugar, L., Diamandis, E.P., Dore, J., and Yousef, G.M. (2010a). Dysregulation of kallikrein-related peptidases in renal cell carcinoma: potential targets of miRNAs. Biol. Chem. 391, 411–423.

White, N.M., Chow, T.F., Mejia-Guerrero, S., Diamandis, M., Rofael, Y., Faragalla, H., Mankaruous, M., Gabril, M., Girgis, A., and Yousef, G.M. (2010b). Three dysregulated miRNAs control kallikrein 10 expression and cell proliferation in ovarian cancer. Br. J. Cancer 102, 1244–1253.

White, N.M., and Yousef, G.M. (2011). Translating molecular signatures of renal cell carcinoma into clinical practice. J. Urol. 186, 9–11.

White, N.M., Bao, T.T., Grigull, J., Youssef, Y.M., Girgis, A., Diamandis, M., Fatoohi, E., Metias, M., Honey, R.J., Stewart, R., Pace, K., Bjarnason, G.A., and Yousef, G.M. (2011a). miRNA profiling in clear cell renal cell carcinoma: biomarker discovery and the Identification of potential controls and consequences of miRNA dysregulation. J. Urol. 186, 1077–1083.

White, N.M., Fatoohi, E., Metias, M., Jung, K., Stephan, C., and Yousef, G.M. (2011b). Metastamirs: a stepping stone towards improved cancer management. Nat. Rev. Clin. Oncol. 8, 75–84.

White, N.M., Khella, H.W., Grigull, J., Adzovic, A., Youssef, Y.M., Honey, R.J., Stewart, R., Pace, K.T., Bjarnason, G.A., Jewett, M.A., Evans, A.J., Gabril, M., and Yousef, G.M. (2011c). miRNA profiling in metastatic renal cell carcinoma reveals a tumor suppressor effect for miR-215. Br. J. Cancer 105, 1741–1749.

Yoshida, M.A., Ohyashiki, K., Ochi, H., Gibas, Z., Pontes, J.E., Prout, G.R. Jr., Huben, R., and Sandberg, A.A. (1986). Cytogenetic studies of tumor tissue from patients with nonfamilial renal cell carcinoma. Cancer Res. 46, 2139–2147.

Yousef, G.M. and Diamandis, E.P. (2001). The new human tissue kallikrein gene family: structure, function, and association to disease. Endocr. Rev. 22, 184–204.

Yousef, G.M., Borgoño, C.A., Michael, I.P., and Diamandis, E.P. (2004a). Cloning of a kallikrein pseudogene. Clin. Biochem. 37, 961–967.

Yousef, G.M., White, N.M., Kurlender, L., Michael, I., Memari, N., Robb, J.D., Katsaros, D., Stephan, C., Jung, K., and Diamandis, E.P. (2004b). The kallikrein gene 5 splice variant 2 is a new biomarker for breast and ovarian cancer. Tumour Biol. 25, 221–227.

Yousef, G.M., Obiezu, C.V., Luo, L.Y., Magklara, A., Borgoño, C.A., Kishi, T., Memari, N., Michael, P., Sidiropoulos, M., Kurlender, L., Economopolou, K., Kapadia, C., Komatsu, N., Petraki, C., Elliott, M., Scorilas, A., Katsaros, D., Levesque, M.A., and Diamandis, E.P. (2005a). Human tissue kallikreins: from gene structure to function and clinical applications. Adv. Clin. Chem. 39, 11–79.

Yousef, G.M., White, N.M., Michael, I.P., Cho, J.C., Robb, J.D., Kurlender, L., Khan, S., and Diamandis, E.P. (2005b). Identification of new splice variants and differential expression of the human kallikrein 10 gene, a candidate cancer biomarker. Tumour Biol. 26, 227–235.

Youssef, Y., White, N.M., Grigull, J., Krizova, A., Samy, C., Mejia-Guerrero, S., Evans, A., Jewett, M., and Yousef, G.M. (2011). MiRNA profiling in kidney cancer subtypes: accurate molecular classification and correlation with cytogenetic and mRNA data identifies unique and shared biological pathways. Eur. Urol. 59, 721–730.

Zhang, G., Wang, Q., and Xu, R. (2010). Therapeutics based on microRNA: a new approach for liver cancer. Curr. Genomics 11, 311–325.

Zhang, L., Volinia, S., Bonome, T., Calin, G.A., Greshock, J., Yang, N., Liu, C.G., Giannakakis, A., Alexiou, P., Hasegawa, K., Johnstone, C.N., Megraw, M.S., Adams, S., Lassus, H., Huang, J., Kaur, S., Liang, S., Sethupathy, P., Leminen, A., Simossis, V.A., Sandaltzopoulos, R., Naomoto, Y., Katsaros, D., Gimotty, P.A., DeMichele, A., Huang, Q., Butzow, R., Rustgi, A.K., Weber, B.L., Birrer, M.J., Hatzigeorgiou, A.G., Croce, C.M., and Coukos, G. (2008). Genomic and epigenetic alterations deregulate microRNA expression in human epithelial ovarian cancer. Proc. Natl. Acad. Sci. USA 105, 7004–7009.

Zhao, S., and Liu, M.F. (2009). Mechanisms of microRNA-mediated gene regulation. Sci. China C. Life Sci. 52, 1111–1116.

Jane Bayani, and Eleftherios P. Diamandis

9 Genomic Instability of the *KLK*-locus in Cancer

9.1 Introduction

Genomic instability is a broad term that defines a characteristic feature of most cancers which implicates abnormal events during DNA repair, replication, and division. In fact, genomic instability is recognized as an accepted characteristic to the classical hallmarks of cancer (Hanahan and Weinberg, 2011; Negrini *et al.*, 2010). In the majority of solid tumors, genomic instability contributes to aberrant expression of genes. For some genes, the presence of sequence mutations is sufficient to affect expression, while others are dependent on changes in gene dosage. Although the *KLK* locus was fully elucidated more than a decade ago (Harvey *et al.*, 2000; Yousef *et al.*, 1999; Yousef and Diamandis, 2001), very little is known regarding the extent to which the locus is affected by changes precipitated by DNA damaging events, genomic sequence, and DNA architecture, as well as genomic instability. It is also unclear what effect this may have on the observed expression changes seen in cancers. As discussed in Chapter 1, Volume 1, the *KLK* locus is unique in harboring all members of the gene family on chromosome 19q, which could arguably pose both a benefit and a disadvantage. In this chapter, we will discuss the recent concepts of genomic instability in cancers, the methods for detecting these changes, and the issue of how this may be functionally significant to the *KLK* locus.

9.2 Defining genomic instability

The estimated rate of somatic cell mutation in humans is approximately 1.1×10^{-8} *per* site *per* generation (Loeb and Christians, 1996; Roach *et al.*, 2010). Therefore, genetic or genomic instability refers to the overall process that increases this rate of mutation and, more importantly, results in a phenotypic change. In the context of cancer, genomic instability is classically categorized by two main forms: microsatellite instability (MIN) and chromosomal instability (CIN). The prototypic example for defining genomic instability comes from studies in colorectal cancer, where distinct genomic phenotypes are seen between hereditary and sporadic cases (Vilar and Gruber, 2010). The MIN phenotype causes the shortening or expansion of short tandem repeat units that make up microsatellite sequences scattered throughout the genome. Such events are indicative of an underlying mismatch repair (MMR) deficiency associated with hereditary colorectal cancer (Lynch and de la Chapelle, 2003). Indeed, these hereditary nonpolyposis colorectal cancer (HNPCC) patients possess germline mutations in a variety of MMR genes (Lynch and de la Chapelle, 2003), but possess cancer genomes that are predominantly near-diploid. In contrast, sporadic cancers comprise the majority of colorectal cancers and exhibit the CIN phenotype (Lengauer et al., 1997), that is, genomes with gross chromosomal changes. Because the CIN phenotype is seen for the vast majority of solid cancers, this will form the basis for the following discussion.

9.2.1 Defining CIN and its underlying mechanisms

The various forms and mechanisms of CIN have been extensively reviewed in the scientific literature (Geigl et al., 2008; Holland and Cleveland, 2009; Thompson et al., 2010) but will be briefly summarized here. As genomic instability refers to a rate of genomic change, CIN refers to the rate of chromosomal changes. The term aneuploidy is often used interchangeably with CIN. However, it is important to mention that the presence of aneuploidy alone does not constitute CIN. Thus, the level of genomic heterogeneity, that is the genomic differences at a cell-by-cell basis, defines the level of CIN. Moreover, changes at the chromosomal level encompass both numerical CIN (N-CIN) as well as structural CIN (S-CIN) (Bayani et al., 2007), and have a direct effect on the copy-number of the affected genomic regions. Generally, N-CIN describes the on-going gains or losses of whole chromosomes, whole chromosome arms, or changes in DNA ploidy (i.e. multiples of a haploid number – 2n/3n/4n etc). Similarly, S-CIN refers to on-going chromosomal breakage events, which include chromosomal additions, deletions, translocations, and other structural rearrangements.

Changes in chromosome number typically result from errors in chromosomal segregation during cell division, including the abnormal equatorial orientation of chromosomes, improper spindle attachment between centromeres and centrosomes, as well as supernumary centrosomes (Holland and Cleveland, 2009; Gisselsson, 2011). While these factors contribute to the whole or partial gains and losses of individual chromosomes, changes in ploidy reflect the whole gains or losses of multiple haploid complements (Ganem et al., 2007; Gisselsson, 2011; Krajcovic et al., 2011; Storchova and Pellman, 2004). In many epithelial cancers and sarcomas, polyploidy is frequently observed (Weaver and Cleveland, 2008) and provides sufficient genetic material for the generation of virtually limitless combinations that promote (or possibly suppress) progression (Weaver and Cleveland, 2007). In contrast, structural changes of chromosomes appear to occur through different mechanisms (Hastings et al., 2009).

Typical structural changes include chromosomal additions, deletions, and translocations, as well as the formation of specialized amplification structures, such as double minute chromosomes (dms) and homogeneously staining regions (hsrs). Like polyploidy, structural changes can affect gene copy number, but may also facilitate the formation of abnormal gene fusion products. Changes in chromosome structure are primarily influenced by two major DNA repair mechanisms: homologous recombination (HR) and non-homologous recombination (NHR) (reviewed by Hastings *et al.* (2009)). When impaired, these mechanisms are unable to properly resolve DNA breakage events (Lindahl and Wood, 1999). Together with the ability of these transformed cells to escape cell cycle checkpoints, indiscriminate DNA repair occurs and leads to associations between different chromosomes. However, such promiscuous DNA repair can also be potentiated by other factors such as sequence architecture. Indeed, examination of chromosomal regions associated with structural rearrangements (Abeysinghe *et al.*, 2003; Chuzhanova *et al.*, 2003) have shown that regions containing high GC-content are associated with translocations, while high AT-content regions are associated with deletions. Furthermore, repeat elements, including *Alu* motifs (Batzer and Deininger, 2002), *alpha*-satellite DNA (Grady *et al.*, 1992), and telomeric sequences (Gisselsson and Hoglund, 2005; Murnane, 2006), which are highly interspersed throughout the human genome, can act as recipient and donor sites for translocations. With the high proportion of such architectural features, seen on chromosome 19 as well as within the *KLK* locus itself (discussed below), and together with DNA repair and mitotic segregation errors, the *KLK* locus possesses the qualities for instability during tumorigenesis.

9.2.2 Methods for detecting CIN

We have already discussed that CIN can be defined as the rate of on-going numerical changes (N-CIN), or as the rate on-going structural changes (S-CIN) (Bayani *et al.*, 2007). Therefore, to properly ascertain this rate of change, one must directly examine the genome on a cell-by-cell basis. Visualizing both numerical and structural changes ideally requires the preparation of metaphase chromosomes directly from the cancer cells, which is technically challenging (Bayani *et al.*, 2007). Nevertheless, the integration of classical cytogenetics with Fluorescence *in situ* Hybridization (FISH)-based techniques such as Spectral Karyotyping (SKY) (Bayani and Squire, 2002; Schrock *et al.*, 1996) has enabled the elucidation of complex chromosomal rearrangements that could not be seen from G-banding alone. Interphase cytogenetics, while useful for inferring numerical changes that affect ploidy through the FISH-based enumeration of centromeres, fails to determine global structural rearrangements. This is because the interphase nuclei rather than the chromosomes are interrogated. Locus-specific rearrangements, however, may be determined by interphase FISH through the use of locus-specific probe configurations, whose co-localization or "break apart" status indicate the presence or absence of the rearrangement (Bayani and Squire, 2007).

In the past decade, we have witnessed the rapid accumulation of cancer genome data derived from microarray and next-generation sequencing initiatives (Beroukhim *et al.*, 2010; Hudson *et al.*, 2010; Meyerson *et al.*, 2010). The benefit of performing microarrays for determining copy-number changes is its ability to use small amounts of DNA extracted directly from the tissue. The test DNA is then compared to normal

DNA, resulting in copy-number profiles generated at an astounding level of resolution (Tan *et al.*, 2007). However, these copy-number profiles only provide the average copy-number change across those cells from which the DNA was extracted. Moreover, caution must be taken, as one cannot presume that the DNA extracted from the tumor tissue is purely tumor-derived. An admixture of contaminating normal stroma and vascularly-derived cells may skew the findings. Therefore, while microarray analyses of average copy-number changes can be revealed across the entire genome, the level of karyotypic complexity, (aneu)ploidy, or CIN can only be inferred. Establishing levels of CIN has been of increasing interest (Greaves and Maley, 2012), with the association of CIN with cancer progression, and with generally poorer outcome, having been recognized for some time. While increased CIN can be an indication of genomic diversity (Gorringe *et al.*, 2005), laying the foundation for the emergence of more malignant clones, many now observe that too much CIN can be physiologically detrimental to the cell (Weaver and Cleveland, 2008, 2009). Thus, the challenge for investigators is to uncover patterns of CIN that promote and sustain tumorigenesis, *versus* patterns of CIN that suppress it (Weaver and Cleveland, 2008, 2009).

9.3 Chromosome 19 and the *KLK* locus in cancer

9.3.1 Chromosome 19

In 2004, the completed sequence for human chromosome 19 was published in *Nature* (Grimwood *et al.*, 2004) and the reader is referred to this body of work for further details. Remarkably, despite being one of the smallest chromosomes, chromosome 19 possesses many unexpected and unique properties, namely the 26 protein-coding gene loci per megabase average and the finding that 50% of the chromosome is composed of protein coding loci (i.e. exons plus introns). There are at least 1,461 protein-coding genes and 321 pseudogenes. Twenty duplicated gene families, which include the *KLK* locus, make up more than 25% of the chromosome. The majority of gene families is composed of Kruppe-type zinc finger genes, olfactory receptors, cyotochrome P450, and immunoglobulin-like receptor genes. Notable for this high prevalence in the number of gene families, chromosome 19 is, more interestingly, known for the tandemly-clustered or large segmentally-duplicated orientation of these genes. Chromosome 19 has an unusually high G+C content of 48%, which is associated with the high gene and *Alu* repeat density. Indeed, approximately 55% of the chromosome is composed of repeat elements, in contrast to the genome average of 44.8% (Lander *et al.*, 2001). The high G+C content also reflects the finding that two-thirds of genes have at least one CpG island within 2,000 bp upstream and 1,000 bp downstream of the putative transcription start site.

According to the Online Mendelian Inheritance of Man database, (http://www.ncbi.nlm.nih.gov/omim), at least 97 genes are linked with single-gene mendelian traits associated with rare genetic disorders (Grimwood *et al.*, 2004). Constitutional chromosomal aberrations involving chromosome 19 are generally rare (Babic *et al.*, 2007; Puvabanditsin *et al.*, 2009; Salbert *et al.*, 1992; Schluth-Bolard *et al.*, 2008). Therefore, based on the uniqueness of the overall high gene density and the large collection of gene families associated with normal physiological homeostasis, it is not unusual to expect that constitutional aberrations affecting chromosome 19 could have more deleterious effects than other chromosomes. In contrast to these rare constitutional changes, chromosome 19 aberrations are common in cancers, as seen from the compendium of cytogenetic and genomic changes at the Mitelman Database of Chromosome Aberrations and Gene Fusions in Cancer (http://cgap.nci.nih.gov/Chromosomes/Mitelman), the Atlas of Genetics and Cytogenetics in Oncology and Haematology (http://atlasgeneticsoncology.org//index.html), the SKY/M-FISH/CGH database (http://www.ncbi.nlm.nih.gov/sky/), and the Progenetix CGH database (http://www.progenetix.org/cgi-bin/pgHome.cgi). Recurrent rearrangements and copy-number changes have been frequently identified among both hematological malignancies and solid tumors. Due to the observed deleterious effects on fitness, and because of abnormalities resulting in physiological and mental disturbances, it appears that there is a selection against constitutional chromosome 19 aberrations. However, in cancers, the large collection of genes and genomic architectural qualities creates the perfect recipe for contributing to carcinogenesis.

9.3.2 The *KLK* locus and cancer

The reader is referred to the chapters in this book that discuss the normal genomic structure and expression status of the *KLK* genes in various tissues and fluids, as well as the evolution of the locus. However, some salient points will be mentioned here. The *KLK* locus at 19q13.3/13.4 houses all the known human kallikrein-related serine peptidases (Harvey *et al.*, 2000; Yousef *et al.*, 1999; 2000), which map in a contiguous manner. This contiguous mapping provides important insights as to the evolutionary history of this gene family. Very recently, Pavlopoulou *et al.* (2010) reported a comprehensive evolutionary history of the locus through the use of mammalian, avian, reptilian and amphibious genomes. Using comparative sequencing suites for cross-genome analyses, the emergence of the *KLK* genes was traced to some 330 million years ago (mya), which is twice the suggested 150 mya (see Chapter 3, Volume 1). Comparative sequencing across various genomes suggests the selection of these genes for co-clustering. Indeed, while the number of genes and pseudogenes across the various genomes differs, whether mapping to one or more chromosomal loci, the co-clustering of *KLK* family members is generally maintained. Owing to the high density of repeat elements on chromosome 19, this may have precipitated the development and apparently ongoing evolution of the gene cluster (Lu *et al.*, 2006). Indeed, the *KLK* locus itself has a range of 34–52% of repeat elements composed of *Alu*-derived, short interspersed nuclear elements (SINEs), and mammalian-wide interspersed repeats (MIR) (Gan *et al.*, 2000; Yousef *et al.*, 2001). Simple mini-satellite repeats (MSR) have also been identified in the 3' untranslated regions (UTR) of *KLK4* and *KLK14*, namely the introns of *KLK6, 7,* and *14*, in addition to intergenic areas across the locus (Grimwood *et al.*, 2004; Hooper *et al.*, 2001; Nelson *et al.*, 1999; Yousef *et al.*, 2001), begging the question as to whether these regions could be more sensitive to copy-number events under DNA damaging conditions.

Both mRNA and protein studies show the general overexpression of these genes in cancers, when compared to their normal tissues (Borgoño and Diamandis, 2004; Shaw and Diamandis, 2007). The primary mechanisms studied to date that influence such over-expression include the response to hormones (Lawrence *et al.*, 2010), DNA-methylation, and transcriptional binding sites (Borgoño and Diamandis, 2004). As already mentioned, the effects on gene expression caused by chromosomal changes have not been extensively studied for the *KLK* locus. In the following sections, we will review the newest findings regarding the locus in cancer.

9.3.3 *KLK* sequence mutations and single nucleotide polymorphisms in cancer

With the evolutionary evidence to support the expansion and diversity of the *KLK* genes, and in addition to the genomic architectural features unique to chromosome 19, it is unclear how these characteristics precipitate sequence mutations, polymorphism, or genomic instability. In some respects, tumorigenesis can be linked to evolution, with the rapid generation of genomic clones whose survival is tested through various physiological selective pressures (Greaves and Maley, 2012). In uncovering gene expression mechanisms, one often first looks for mutations or polymorphisms in the gene that may influence the observed differential expression. Interrogation of The Cancer Genome Atlas (http://tcga-data.nci.nih.gov/tcga/) shows no identified mutations for any of the *KLK* genes among glioblastoma cases, with no information identified for other cancer groups. However, analyzed cancers from The International Cancer Genome Consortium (http://www.icgc.org/), have identified a very low frequency of mutations among the gene members, many of which were intronic. These findings are similarly reflected by the Catalogue of Somatic Mutations (http://www.sanger.ac.uk/genetics/CGP/cosmic/). These mutations likely reflect Single Nucleotide Polymorphisms (SNP), the most common form of variation. The reader is referred to Chapter 2, Volume 1 for a detailed review of SNPs in the *KLK* locus. Briefly, if defined as common variations at a single nucleotide position, where the least common allele is present in at least 1% of a given population (Collins *et al.*, 1998), SNPs occur approximately once in every 150–300 bp in the human genome (Brookes, 1999; Kruglyak and Nickerson, 2001; Ladiges *et al.*, 2004). Indeed, the number of SNPs within the *KLK* locus now exceeds 1,800 (Goard *et al.*, 2007) and the identification of these SNPs has led to disease association studies, predominantly in prostate cancer for *KLK2*,

3, and *15* (Chavan *et al.*, 2010; Cramer *et al.*, 2008; Lose *et al.*, 2011; Nam *et al.*, 2009; Pal *et al.*, 2007; Parikh *et al.*, 2010a; Park *et al.*, 2010). However, few studies have investigated the functional significance of these variants (Klein *et al.*, 2010; Kote-Jarai *et al.*, 2011; Lai *et al.*, 2007; Parikh *et al.*, 2010b). For *KLK6*, by mutational sequencing, no mutations have been identified in ovarian cancers (Shan *et al.*, 2007), although SNP changes were identified in the 5′UTRs. For *KLK10* (Batra *et al.*, 2010), two tag SNPs and one exonic SNP were studied in 319 ovarian cancer patients, as well as four ovarian cancer cell lines. A potentially functionally relevant SNP was revealed by sequencing the 3.6 Kb region upstream of the *KLK10* start site. Other studies confirm the lack of somatic activating mutations of *KLK10* in prostate, breast, testicular, or ovarian cancers (Bharaj *et al.*, 2002), as well as in colon and gastric cancers (Feng *et al.*, 2006). It is evident from these initial studies on a few of the *KLK* members that the significance of these polymorphisms should continue to be investigated, not only in order to identify the frequency of these genotypes and their associations with clinical categories, but also to determine their functional significance.

9.3.4 *KLK* translocations

The discovery of the Philadelphia chromosome (Ph) in patients with chronic myelogenous leukemia (CML) (Nowell and Hungerford, 1960; Rowley, 1973) marked the search for other recurrent chromosomal translocations in cancer. This translocation results in the fusion of the *ABL1* gene at chromosome 9q34 with the *BCR* gene at 22q11 (Goldman, 2010; Rowley, 1973). The fusion gene product results in a constitutively active oncogenic BCR-ABL fusion protein. Following its discovery, several additional recurrent translocations were identified among the hematological malignancies, and are indexed in the Mitelman Database of Chromosome Aberrations and Gene Fusions in Cancer. The previous notion that recurrent translocations in epithelial cancers are extremely rare has changed, due to the technical advances in molecular biology and sequencing (Newman and Edwards, 2010). This advance has enabled the identification of cryptic translocations and novel gene fusions (Brenner and Chinnaiyan, 2009; Newman and Edwards, 2010). This idea was supported by the discovery of the *TMPRSS2-ERG* translocation in prostate cancer, using Cancer Outlier Profile Analysis (COPA) to the Oncomine (https://www.oncomine.org/resource/login.html) database (Tomlins *et al.*, 2005). *TMPRSS2* is a prostate-specific and androgen-regulated gene (Afar *et al.*, 2001; Lin *et al.*, 1999). Its fusion to *ERG*, *ETV1* (Hermans *et al.*, 2006; Tomlins *et al.*, 2005), or *ETV4* (Tomlins *et al.*, 2006), members of the *ETS*-family of transcription factors (Oikawa and Yamada, 2003), may result in the dysregulated and oncogenic expression of *ETS* gene targets. On the heels of this discovery, it was identified that *KLK2* is also a fusion partner to *ETV4* (Hermans *et al.*, 2008). Hermans *et al.* (2008) performed quantitative PCR (qRT-PCR) on a large set of clinical prostate cancers, where overexpression of ETV4 was detected in less than 2% of the samples,

Fig. 9.1 **(a) *KLK2-ETS* fusion genes identified in prostate cancer.** We show the three published fusion gene products, involving the *KLK2* and *ETS* family members *ETV4* and *ETV1* (Hermans *et al.*, 2008; Pflueger *et al.*, 2011). Nucleotides shown in green denote *KLK2* sequence, while black print denotes the *ETS* gene member sequence. **(b) Genomic instability and copy-number heterogeneity of the *KLK* locus in cancer.** i) Metaphase fluorescence *in situ* hybridization (FISH) of the *KLK* locus (red), and a near centromeric probe at 19q12 (green) to the tetrapoloid CAOV-3 ovarian cancer cell line, as previously described by Bayani *et al.* (2008b). ii) Sequential Spectral Karyotyping (SKY) analysis and FISH of the metaphase spread shown in (i). The CAOV-3 cell line possesses four copies of the *KLK* locus (red): two located on normal chromosomes 19 and two located on duplicate, rearranged chromosomes resulting from a complex translocation with chromosomes 19, 2, and 13. As seen by the FISH images, only the terminal region of 19q which contains the *KLK* locus (red) was translocated to this derivative chromosome. The 19q12 region (green) was not involved in the translocation event. iii) Metaphase FISH analysis of ovarian cancer ascites. Metaphase preparations from the cancer cells derived from ovarian ascites show blocks of amplified regions of the *KLK* locus (arrow head), as compared to the signal size of normal, single copies (arrows) (*J. Bayani – unpublished data*). iv) FISH to a formalin-fixed, paraffin-embedded ovarian cancer identifies copy-number heterogeneity. A multi-color FISH strategy described by Bayani *et al.* (2011) reveals copy-number heterogeneity for genomic loci 19q12 (green), 19q13.2 (blue), and the *KLK* locus (red). Nuclear boundaries have been indicated for three nuclei, revealing cells with amplification of the *KLK* locus compared to neighboring cells containing fewer copies of the locus.

indicating the rarity of this variant. 5′-RACE and subsequent sequencing of the PCR products identified the fusion partner as *KLK2*. The *KLK2-ETV4* fusion gene was shown to be composed of *KLK2* exon 1 fused to a new *ETV4* exon (denoted as exon 4a) followed by exons 5 and 6 (**Fig. 9.1a**). Unlike the *TMPRSS2-ERG* fusion, whose genes map to the same locus on chromosome 21q22 and which results from a deletion event, the *KLK2-ETV4* fusion results from the translocation between ETV4 at 17q21, and the *KLK* locus at 19q. Interestingly, because of the orientation of *KLK2* and *ETV4*, the fusion product cannot be explained by a single chromosome translocation event. Moreover, the authors indicated that a KLK2-ETV4 fusion protein, containing the NH_2-terminal KLK signal peptide, would be secreted, albeit not function as a transcription factor (Hermans *et al.*, 2008).

More recently, Pflueger *et al.* (2011) applied next-generation transcriptome sequencing (RNA-seq) to 25 human prostate cancers, in order to identify novel fusion transcripts. As a result, an additional *KLK2* fusion transcript was identified, *KLK2-ETV1*, of which two different *KLK2-ETV1* transcripts were detected by RT-PCR and Sanger sequencing: one variant with exon 7 of *ETV1* fused to exon 2 of *KLK2*, and the second one with *ETV1* exon 7 fused to exon 1 of *KLK2*. The prostate-specific and restrictive nature of *KLK2* expression is in keeping with the recurrent theme of androgen regulated partners among these prostate cancer fusion events. With improving RNA sequencing capabilities (Ozsolak and Milos, 2011), the next few years will reveal whether other *KLK* gene members participate in fusion events in other cancers.

(a)

KLK2-ETV4

AGTGGGCACTTATCCTTGGTTTCAGGTCTGAGGGGGGGATGCTGAGGCCGATCGCTCGCTCCTTCACACTTTCTCTGC
GGAGACAGATGCAGCTGCCGGGGCCCTGTCCCCCTGCACCATCCCAACACCACCCCAGCCTCCTCTCCTGTCTCTTC
CCACCAGGTCAGTGACCCCGGCAACCCA

ETV4
(17q21.31)

KLK2
(19q13.3/13.4)

ETV1
(7p21.2)

KLK2-ETV1 Variant 1

GTTCTCTCCATCGCCTTGTCTGTCCCCTCGACTGATTTCGCCGCCAGCTTTCTGAACCCTGTAACTCCTTTCCTCCTTTG
CCGACGATGCCAACCCAACCACGTCCTATGTACCAACGCCAGATGTCTGAGCCAAACATCCCCTTCCCACCACAAGGCT
TTAAGCAGGAGTACCACGACCCAGTGTATGAACACAACACCATGGTTGGCAGTGCGGCCAGCCAAAGCTTTCCCCCTC
CTCTGATGATTAAACAGGAACCCAGAGATnTGCATATGACTCAGAAAGTGCCTAGCTGCCACTCCATTTATATGAGGCAAG
AAGGCTCCTGGCTCATCCCAGCAGAACAGAAGGCTGTATGTTTGAAAAGGGCCCCAGGCAGTTTTATGATGACAGCTGT
GTTGTCCCAGAAAAATTTCGATGGAGACATCAAACAAGAGCCAGGAATGTATCGGGAAGGA

KLK2-ETV1 Variant 2

GTTCTCTCCATCGCCTTGTCTGTGGGGTGCATCTGGTGCCGTGCCCCTCATCCAGTCTCGGATTGTGGGAGGCTGGGA
GTCTGAGAAGCATTCCCAACCCTGGCAGGTGGCTGTGTACAGTCATGGATCGGCACACTGTGGGGGTGTCCTGGTGCA
CCCCCAGTGGGTGTTCACAGCTGCCCCTTGCCTAAAGAATTTCGCCGCCAGCTTTCTGAACCCTGTAACTCCTTTCCTCC
TTTGCCGACGATGCCAAGGGAAGGACGTCCTATGTACCAACGCCAGATGTCTGAGCCAAACATCCCCTTCCCACCACAAG
GCTTTAAGCAGGAGTACCACGACCCAGTGTATGAACACAACACCATGGTTGGCAGTGCGGCCAGCCAAAGCTTTCCCCCT
CCTCTGATGATTAAACAGGAACCCAGAGATTTGCATATGACTCAGAAGTGCCTAGCTGCCACTCCATTTATATGAGGCAAG
AAGGCTTCCTGGCTCATCCCAGCAGAACAGAAGGCTGTATGTTTGAAAAGGGCCCCAGGCAGTTTTATGATGACACCTGT
GTTCTCCCAGAAAAATTCGATGGAGACATCAAACAAGAGCCAGGAATGTATCGGGAAGGA

(b)

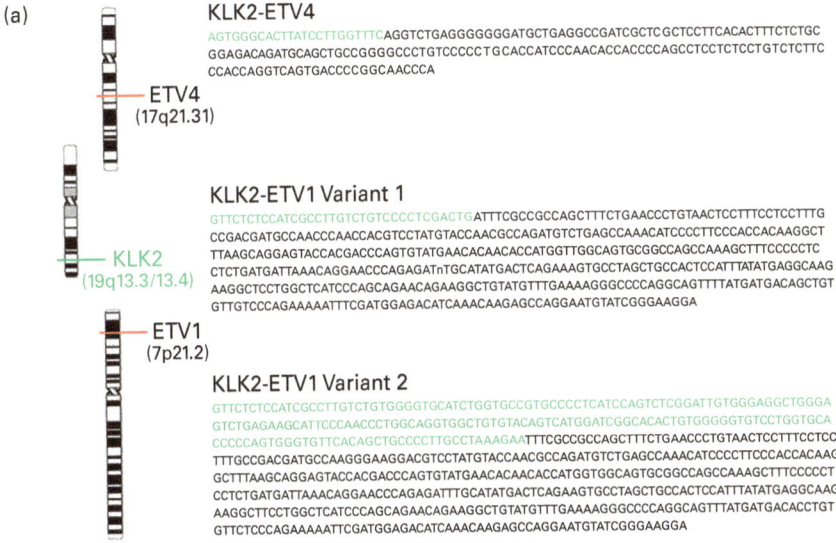

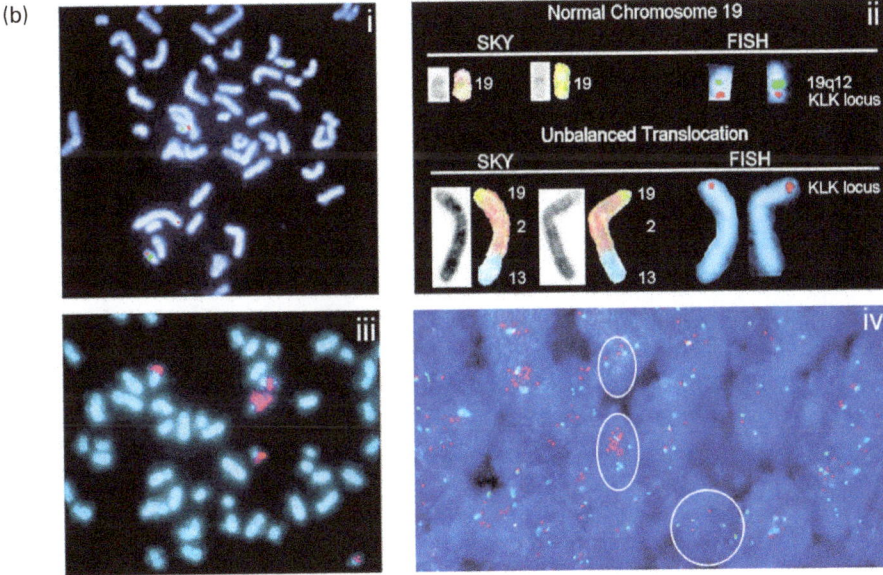

9.3.5 Copy-number changes of the *KLK* locus

We have already mentioned that abnormal mitotic events and failure of DNA repair can facilitate changes in gene copy-number, which result in expression changes at the RNA and protein levels. Chapter 8, Volume 1 describes the reported RNA and protein expression levels associated with normal and diseased tissues. However, the contributions to the copy-number have not been systematically studied. Indeed, Shinoda *et al.* (2007) utilized array-comparative genomic hybridization (aCGH) to detect gains of the *KLK* locus in urinary bladder cancer cells, while Ni *et al.* (2004) detected gains of *KLK6* in ovarian cancers by southern blotting. However, no other studies have examined *KLK* copy-number. We now have access to several publically available databases for which gene copy-number from cancer genomes, cancer cell lines, and normal DNA can be ascertained. These include The Cancer Genome Atlas, The International Cancer Genome Consortium, and the Progenetix Database, as well as published sources such as Beroukhim *et al.* (2010). Based on aCGH methodologies, the resulting copy-number ratios reported among these sources therefore reflect the average copy-number change. As such, it is interesting to note that for each tumor that has been analyzed, and for which there is available copy-number data, the ratio of copy-number change is the same across the *KLK* locus. This suggests that the entire locus experiences the same copy-number change. More specifically, there is no evidence to suggest that individual *KLK*s experience copy-number changes. In the case of the fusion transcripts described by Hermans *et al.* (2008) and Pflueger *et al.* (2011), no accompanying copy-number changes were detected. Indeed, in 2008 this trend was seen by our group among various cancer cell lines and ovarian primary cancer specimens which were analyzed using *KLK*-locus and *KLK*-specific FISH analyses (Bayani *et al.*, 2008b).

In this study, we used bacterial artificial chromosomes (BAC) probes spanning the entire *KLK* locus as FISH probes, in order to interrogate the copy-number and chromosomal mapping status among prostate, breast, and ovarian cancer cell lines. We found that the *KLK* locus was present, in multiple copies, in some of the cancer cell lines and that copy-number imbalances were the result of unbalanced translocations of the *KLK* locus to other chromosomal partners (**Fig. 9.1b**). To determine whether the entire locus was subject to these changes, or whether only some of the *KLK* members were involved, we utilized a multicolor-*KLK*-gene-specific FISH approach whereby *KLK2, 4, 6,* and *13* were differentially labeled and simultaneously hybridized to metaphase preparations of the cell lines. For each cell line, all the probes co-localized with each other, indicating that the entire locus, rather than individual genes, were involved in the observed copy-number changes and/or unbalanced translocations. While the copy-number and pattern of locus translocation were clonal, as expected across the cell lines, primary ovarian carcinomas that were also analyzed showed variations in the number of copies of the locus, as well as in their involvement in structural rearrangements. In most cases, a net gain of the locus was observed when the number of signals was enumerated at a cell-by-cell basis, and when ploidy was accounted for.

The observed heterogeneity in ovarian primary cancers led to a follow-up study, in order to determine whether the locus was subject to genomic instability (Bayani *et al.*, 2011). In our study, we performed interphase FISH analyses on formalin-fixed paraffin-embedded tissues (FFPEs), using multicolor FISH probes that map to the centromeric region at 19q12 as well as at 19q13.2, and to the *KLK* locus at 19q13.4/13.4 (**Fig. 9.1a**). This way, we were able to, first, determine the range of *KLK* copy-number across a number of tumor cells and, second, determine whether the observed copy-number changes may have been precipitated by structural rearrangements along chromosome 19q. Contrary to most aCGH findings for the locus (Beroukhim *et al.*, 2010), we demonstrated that the majority of ovarian cancers shows a net gain and that the copy-number varies from cell to cell. By and large, the locus is subject to a net loss, as is shown by aCGH ratios (Beroukhim *et al.*, 2010). The inconsistencies between the two assays lie in the inability of aCGH to account for the ploidy of the cell. Indeed, the majority of ovarian cancers have genomes that are near-triploid or near-tetraploid (Bayani *et al.*, 2002; 2008a and b), and when these genomes are compared to the normal DNA control (that is diploid), only a (clonal) haploid multiple will be recognized in the aCGH copy-number profile.

Therefore, two copies of a locus in a triploid or near-tetraploid cancer genome may be deemed "lost", when compared to the normal diploid control. This can be further exacerbated when copies vary from cell to cell. We found that 51% of cases had gains, 30% showed a net loss, and 2% were amplified for the *KLK* locus. We also demonstrated that the locus was much more chromosomally unstable than the other loci probed on 19q ($p < 0.001$). When KLK6-specific immunohistochmical staining intensity was associated with copy-number, no strong correlation was shown. However, univariate and multivariate analyses showed a trend for an association between the net loss of the *KLK* locus and longer disease-free survival. Interestingly, FISH analyses indicated that chromosome 19q was subject to structural rearrangement in 62% of cases and was significantly correlated with tumor grade ($p < 0.001$). We speculate that the unique genomic architectural qualities of chromosome 19 play a role in precipitating such breakage events. To our knowledge, these are the only studies that have specifically enumerated the range of *KLK* copy-number in (ovarian) cancer. It is clear, however, that the observed overexpression of the *KLK* genes in various cancers is not influenced solely by copy-number and that other mechanisms may participate in the regulation of such expression.

9.4 Closing remarks

The *KLK* locus resides on a unique chromosome in the human genome. The severely detrimental consequences of constitutional aberrations of chromosome 19 exemplify the importance of development. Evolutionary evidence suggests a physiological need for the (ongoing) expansion of the *KLK* gene family. This is exemplified by the exquisite tissue-specific expression of some family members, and the relatively restrictive coordinate expression of others. In contrast, cancers often show changes in chromosome 19, leveraging this gene-rich chromosome for promoting tumorigenesis. Dysregulated expression of the *KLK* genes has been identified among cancers, and the extent to which chromosomal changes contribute to this change has not been extensively explored. Our studies, as well as others, have shown that the locus is subject to frequent copy-number changes, and also to instability. The significance of such instability in the context of cancer initiation, progression, or its significance as a genomic biomarker, requires further study in additional, clinically annotated specimens.

Bibliography

Abeysinghe, S.S., Chuzhanova, N., Krawczak, M., Ball, E.V., and Cooper, D.N. (2003). Translocation and gross deletion breakpoints in human inherited disease and cancer I: Nucleotide composition and recombination-associated motifs. Hum. Mutat. 22, 229–244.

Afar, D.E., Vivanco, I., Hubert, R.S., Kuo, J., Chen, E., Saffran, D.C., Raitano, A.B., and Jakobovits, A. (2001). Catalytic cleavage of the androgen-regulated TMPRSS2 protease results in its secretion by prostate and prostate cancer epithelia. Cancer Res. 61, 1686–1692.

Babic, I., Brajenovic-Milic, B., Petrovic, O., Mustac, E., and Kapovic, M. (2007). Prenatal diagnosis of complete trisomy 19q. Prenat. Diagn. 27, 644–647.

Batra, J., Tan, O.L., O'Mara, T., Zammit, R., Nagle, C.M., Clements, J.A., Kedda, M.A., and Spurdle, A.B. (2010). Kallikrein-related peptidase 10 (KLK10) expression and single nucleotide polymorphisms in ovarian cancer survival. Int. J. Gyn. Cancer 20, 529–536.

Batzer, M.A., and Deininger, P.L. (2002). Alu repeats and human genomic diversity. Nat. Rev. Genet. 3, 370–379.

Bayani, J., Brenton, J.D., Macgregor, P.F., Beheshti, B., Albert, M., Nallainathan, D., Karaskova, J., Rosen, B., Murphy, J., Laframboise, S., Zanke, B., and Squire, J.A. (2002). Parallel analysis of sporadic primary ovarian carcinomas by spectral karyotyping, comparative genomic hybridization, and expression microarrays. Cancer Res. 62, 3466–3476.

Bayani, J., and Squire, J.A. (2007). Application and interpretation of FISH in biomarker studies. Cancer Lett. 249, 97–109.

Bayani, J., Selvarajah, S., Maire, G., Vukovic, B., Al-Romaih, K., Zielenska, M., and Squire, J.A. (2007). Genomic mechanisms and measurement of structural and numerical instability in cancer cells. Semin. Cancer Biol. 17, 5–18.

Bayani, J., Paderova, J., Murphy, J., Rosen, B., Zielenska, M., and Squire, J.A. (2008a). Distinct patterns of structural and numerical chromosomal instability characterize sporadic ovarian cancer. Neoplasia 10, 1057–1065.

Bayani, J., Paliouras, M., Planque, C., Shan, S.J., Graham, C., Squire, J.A., and Diamandis, E.P. (2008b). Impact of cytogenetic and genomic aberrations of the kallikrein locus in ovarian cancer. Mol. Oncol. 2, 250–260.

Bayani, J., Marrano, P., Graham, C., Zheng, Y., Li, L., Katsaros, D., Lassus, H., Butzow, R., Squire, J.A., and Diamandis, E.P. (2011). Genomic instability and copy-number heterogeneity of chromosome 19q, including the kallikrein locus, in ovarian carcinomas. Mol. Oncol. 5, 48–60.

Bayani, J.M., and Squire, J.A. (2002). Applications of SKY in cancer cytogenetics. Cancer Invest. 20, 373–386.

Beroukhim, R., Mermel, C.H., Porter, D., Wei, G., Raychaudhuri, S., Donovan, J., Barretina, J., Boehm, J.S., Dobson, J., Urashima, M., *et al.*, and Meyerson, M. (2010). The landscape of somatic copy-number alteration across human cancers. Nature 463, 899–905.

Bharaj, B.B., Luo, L.Y., Jung, K., Stephan, C., and Diamandis, E.P. (2002). Identification of single nucleotide polymorphisms in the human kallikrein 10 (KLK10) gene and their association with prostate, breast, testicular, and ovarian cancers. Prostate 51, 35–41.

Borgoño, C.A., and Diamandis, E.P. (2004). The emerging roles of human tissue kallikreins in cancer. Nat. Rev. Cancer 4, 876–890.

Brenner, J.C., and Chinnaiyan, A.M. (2009). Translocations in epithelial cancers. Biochim. Biophys. Acta 1796, 201–215.

Brookes, A.J. (1999). The essence of SNPs. Gene 234, 177–186.

Chavan, S.V., Maitra, A., Roy, N., Patwardhan, S., and Chavan, P.R. (2010). Genetic variants in the distal enhancer region of the PSA gene and their implication in the occurrence of advanced prostate cancer. Mol. Med. Report 3, 837–843.

Chuzhanova, N., Abeysinghe, S.S., Krawczak, M., and Cooper, D.N. (2003). Translocation and gross deletion breakpoints in human inherited disease and cancer II: Potential involvement of repetitive sequence elements in secondary structure formation between DNA ends. Hum. Mutat. 22, 245–251.

Collins, F.S., Brooks, L.D., and Chakravarti, A. (1998). A DNA polymorphism discovery resource for research on human genetic variation. Genome Res. 8, 1229–1231.

Cramer, S.D., Sun, J., Zheng, S.L., Xu, J., and Peehl, D.M. (2008). Association of prostate-specific antigen promoter genotype with clinical and histopathologic features of prostate cancer. Cancer Epidemiol. Biomarkers Prev. 17, 2451–2457.

Feng, B., Xu, W.B., Zheng, M.H., Ma, J.J., Cai, Q., Zhang, Y., Ji, J., Lu, A.G., Qu, Y., Li, J.W., Wang, M.L., Hu, W.G., Liu, B.Y., and Zhu, Z.G. (2006). Clinical significance of human kallikrein 10 gene expression in colorectal cancer and gastric cancer. J. Gastroenterol. Hepatol. 21, 1596–1603.

Gan, L., Lee, I., Smith, R., Argonza-Barrett, R., Lei, H., McCuaig, J., Moss, P., Paeper, B., and Wang, K. (2000). Sequencing and expression analysis of the serine protease gene cluster located in chromosome 19q13 region. Gene 257, 119–130.

Ganem, N.J., Storchova, Z., and Pellman, D. (2007). Tetraploidy, aneuploidy and cancer. Curr. Opin. Genet. Dev. 17, 157–162.

Geigl, J.B., Obenauf, A.C., Schwarzbraun, T., and Speicher, M.R. (2008). Defining ‚chromosomal instability'. Trends Genet. 24, 64–69.

Gisselsson, D., and Hoglund, M. (2005). Connecting mitotic instability and chromosome aberrations in cancer – can telomeres bridge the gap? Semin. Cancer Biol. 15, 13–23.

Gisselsson, D. (2011). Mechanisms of whole chromosome gains in tumors – many answers to a simple question. Cytogenet. Genome Res. 133, 190–201.

Goard, C.A., Bromberg, I.L., Elliott, M.B., and Diamandis, E.P. (2007). A consolidated catalogue and graphical annotation of dbSNP polymorphisms in the human tissue kallikrein (KLK) locus. Mol. Oncol. 1, 303–312.

Goldman, J.M. (2010). Chronic myeloid leukemia: a historical perspective. Semin. Hematol. 47, 302–311.

Gorringe, K.L., Chin, S.F., Pharoah, P., Staines, J.M., Oliveira, C., Edwards, P.A., and Caldas, C. (2005). Evidence that both genetic instability and selection contribute to the accumulation of chromosome alterations in cancer. Carcinogenesis 26, 923–930.

Grady, D.L., Ratliff, R.L., Robinson, D.L., McCanlies, E.C., Meyne, J., and Moyzis, R.K. (1992). Highly conserved repetitive DNA sequences are present at human centromeres. Proc. Natl. Acad. Sci. USA 89, 1695–1699.

Greaves, M., and Maley, C.C. (2012). Clonal evolution in cancer. Nature 481, 306–313.

Grimwood, J., Gordon, L.A., Olsen, A., Terry, A., Schmutz, J., Lamerdin, J., Hellsten, U., Goodstein, D., Couronne, O., Tran-Gyamfi, M., *et al.*, and Lucas, S.M. (2004). The DNA sequence and biology of human chromosome 19. Nature 428, 529–535.

Hanahan, D., and Weinberg, R.A. (2011). Hallmarks of cancer: the next generation. Cell 144, 646–674.

Harvey, T.J., Hooper, J.D., Myers, S.A., Stephenson, S.A., Ashworth, L.K., and Clements, J.A. (2000). Tissue-specific expression patterns and fine mapping of the human kallikrein (KLK) locus on proximal 19q13.4. J. Biol. Chem. 275, 37397–37406.

Hastings, P.J., Lupski, J.R., Rosenberg, S.M., and Ira, G. (2009). Mechanisms of change in gene copy number. Nat. Rev. Genet. 10, 551–564.

Hermans, K.G., van Marion, R., van Dekken, H., Jenster, G., van Weerden, W.M., and Trapman, J. (2006). TMPRSS2:ERG fusion by translocation or interstitial deletion is highly relevant in androgen-dependent prostate cancer, but is bypassed in late-stage androgen receptor-negative prostate cancer. Cancer Res. 66, 10658–10663.

Hermans, K.G., Bressers, A.A., van der Korput, H.A., Dits, N.F., Jenster, G., and Trapman, J. (2008). Two unique novel prostate-specific and androgen-regulated fusion partners of ETV4 in prostate cancer. Cancer Res. 68, 3094–3098.

Holland, A.J., and Cleveland, D.W. (2009). Boveri revisited: chromosomal instability, aneuploidy and tumorigenesis. Nat. Rev. Mol. Cell. Biol. 10, 478–487.

Hooper, J.D., Bui, L.T., Rae, F.K., Harvey, T.J., Myers, S.A., Ashworth, L.K., and Clements, J.A. (2001). Identification and characterization of KLK14, a novel kallikrein serine protease gene located on human chromosome 19q13.4 and expressed in prostate and skeletal muscle. Genomics 73, 117–122.

Hudson, T.J., Anderson, W., Artez, A., Barker, A.D., Bell, C., Bernabe, R.R., Bhan, M.K., Calvo, F., Eerola, I., Gerhard, D.S., et al., and Yang, H. (2010). International network of cancer genome projects. Nature 464, 993–998.

Klein, R.J., Hallden, C., Cronin, A.M., Ploner, A., Wiklund, F., Bjartell, A.S., Stattin, P., Xu, J., Scardino, P.T., Offit, K., Vickers, A.J., Grönberg, H., and Lilja, H. (2010). Blood biomarker levels to aid discovery of cancer-related single-nucleotide polymorphisms: kallikreins and prostate cancer. Cancer Prev. Res. (Phila) 3, 611–619.

Kote-Jarai, Z., Amin Al Olama, A., Leongamornlert, D., Tymrakiewicz, M., Saunders, E., Guy, M., Giles, G.G., Severi, G., Southey, M., Hopper, J.L., et al., and Eeles, R.A. (2011). Identification of a novel prostate cancer susceptibility variant in the KLK3 gene transcript. Hum. Genet. 129, 687–694.

Krajcovic, M., Johnson, N.B., Sun, Q., Normand, G., Hoover, N., Yao, E., Richardson, A.L., King, R.W., Cibas, E.S., Schnitt, S.J., Brugge, J.S., and Overholtzer, M. (2011). A non-genetic route to aneuploidy in human cancers. Nat. Cell Biol. 13, 324–330.

Kruglyak, L., and Nickerson, D.A. (2001). Variation is the spice of life. Nat. Genet. 27, 234–236.

Ladiges, W., Kemp, C., Packenham, J., and Velazquez, J. (2004). Human gene variation: from SNPs to phenotypes. Mutat. Res. 545, 131–139.

Lai, J., Kedda, M.A., Hinze, K., Smith, R.L., Yaxley, J., Spurdle, A.B., Morris, C.P., Harris, J., and Clements, J.A. (2007). PSA/KLK3 AREI promoter polymorphism alters androgen receptor binding and is associated with prostate cancer susceptibility. Carcinogenesis 28, 1032–1039.

Lander, E.S., Linton, L.M., Birren, B., Nusbaum, C., Zody, M.C., Baldwin, J., Devon, K., Dewar, K., Doyle, M., FitzHugh, W., et al., and Chen, Y.J. (2001). Initial sequencing and analysis of the human genome. Nature 409, 860–921.

Lawrence, M.G., Lai, J., and Clements, J.A. (2010). Kallikreins on steroids: structure, function, and hormonal regulation of prostate-specific antigen and the extended kallikrein locus. Endocr. Rev. 31, 407–446.

Lengauer, C., Kinzler, K.W., and Vogelstein, B. (1997). Genetic instability in colorectal cancers. Nature 386, 623–627.

Lin, B., Ferguson, C., White, J.T., Wang, S., Vessella, R., True, L.D., Hood, L., and Nelson, P.S. (1999). Prostate-localized and androgen-regulated expression of the membrane-bound serine protease TMPRSS2. Cancer Res. 59, 4180–4184.

Lindahl, T., and Wood, R.D. (1999). Quality control by DNA repair. Science 286, 1897–1905.

Loeb, L.A., and Christians, F.C. (1996). Multiple mutations in human cancers. Mutat. Res. 350, 279–286.

Lose, F., Batra, J., O'Mara, T., Fahey, P., Marquart, L., Eeles, R.A., Easton, D.F., Al Olama, A.A., Kote-Jarai, Z., Guy, M., *et al.*, and Kedda, M.A. (2011). Common variation in Kallikrein genes KLK5, KLK6, KLK12, and KLK13 and risk of prostate cancer and tumor aggressiveness. Urol. Oncol., [Epub ahead of print].

Lu, W., Zhou, D., Glusman, G., Utleg, A.G., White, J.T., Nelson, P.S., Vasicek, T.J., Hood, L., and Lin, B. (2006). KLK31P is a novel androgen regulated and transcribed pseudogene of kallikreins that is expressed at lower levels in prostate cancer cells than in normal prostate cells. Prostate 66, 936–944.

Lynch, H.T., and de la Chapelle, A. (2003). Hereditary colorectal cancer. N. Engl. J. Med. 348, 919–932.

Meyerson, M., Gabriel, S., and Getz, G. (2010). Advances in understanding cancer genomes through second-generation sequencing. Nat. Rev. Genet. 11, 685–696.

Murnane, J.P. (2006). Telomeres and chromosome instability. DNA Repair (Amst) 5, 1082–1092.

Nam, R.K., Zhang, W.W., Trachtenberg, J., Seth, A., Klotz, L.H., Stanimirovic, A., Punnen, S., Venkateswaran, V., Toi, A., Loblaw, D.A., Sugar, L., Siminovitch, K.A., and Narod, S.A. (2009). Utility of incorporating genetic variants for the early detection of prostate cancer. Clin. Cancer Res. 15, 1787–1793.

Negrini, S., Gorgoulis, V.G., and Halazonetis, T.D. (2010). Genomic instability – an evolving hallmark of cancer. Nat. Rev. Mol. Cell. Biol. 11, 220–228.

Nelson, P.S., Gan, L., Ferguson, C., Moss, P., Gelinas, R., Hood, L., and Wang, K. (1999). Molecular cloning and characterization of prostase, an androgen-regulated serine protease with prostate-restricted expression. Proc. Natl. Acad. Sci. USA 96, 3114–3119.

Newman, S., and Edwards, P.A. (2010). High-throughput analysis of chromosome translocations and other genome rearrangements in epithelial cancers. Genome Med. 2, 19.

Ni, X., Zhang, W., Huang, K.C., Wang, Y., Ng, S.K., Mok, S.C., Berkowitz, R.S., and Ng, S.W. (2004). Characterisation of human kallikrein 6/protease M expression in ovarian cancer. Br. J. Cancer 91, 725–731.

Nowell, P.C., and Hungerford, D.A. (1960). A minute chromosome in human chronic granulocytic leukemia. Science 132, 1497.

Oikawa, T., and Yamada, T. (2003). Molecular biology of the Ets family of transcription factors. Gene 303, 11–34.

Ozsolak, F., and Milos, P.M. (2011). RNA sequencing: advances, challenges and opportunities. Nat. Rev. Genet. 12, 87–98.

Pal, P., Xi, H., Sun, G., Kaushal, R., Meeks, J.J., Thaxton, C.S., Guha, S., Jin, C.H., Suarez, B.K., Catalona, W.J., and Deka, R. (2007). Tagging SNPs in the kallikrein genes 3 and 2 on 19q13 and their associations with prostate cancer in men of European origin. Hum. Genet. 122, 251–259.

Parikh, H., Deng, Z., Yeager, M., Boland, J., Matthews, C., Jia, J., Collins, I., White, A., Burdett, L., Hutchinson, A., Qi, L., Bacior, J.A., Lonsberry, V., Rodesch, M.J., Jeddeloh, J.A., Albert, T.J., Halvensleben, H.A., Harkins, T.T., Ahn, J., Berndt, S.I., Chatterjee, N., Hoover, R., Thomas, G., Hunter, D.J., Hayes, R.B., Chanock, S.J., and Amundadottir, L. (2010a). A comprehensive resequence analysis of the KLK15-KLK3-KLK2 locus on chromosome 19q13.33. Hum. Genet. 127, 91–99.

Parikh, H., Wang, Z., Pettigrew, K.A., Jia, J., Daugherty, S., Yeager, M., Jacobs, K.B., Hutchinson, A., Burdett, L., Cullen, M., *et al.*, and Amundadottir, L. (2010b). Fine mapping the KLK3 locus on chromosome 19q13.33 associated with prostate cancer susceptibility and PSA levels. Hum. Genet. 129, 675–685.

Park, S.W., Kim, C.S., and Lee, G. (2010). Association of polymorphisms in the prostate-specific antigen (PSA) gene promoter with serum PSA level and PSA changes after dutasteride treatment in Korean men with benign prostatic hypertrophy. Korean J. Urol. 51, 824–830.

Pavlopoulou, A., Pampalakis, G., Michalopoulos, I., and Sotiropoulou, G. (2010). Evolutionary history of tissue kallikreins. Plos One 5, e13781.

Pflueger, D., Terry, S., Sboner, A., Habegger, L., Esgueva, R., Lin, P.C., Svensson, M.A., Kitabayashi, N., Moss, B.J., MacDonald, T.Y., Cao, X., Barrette, T., Tewari, A.K., Chee, M.S., Chinnaiyan, A.M., Rickman, D.S., Demichelis, F., Gerstein, M.B., and Rubin, M.A. (2011). Discovery of non-ETS gene fusions in human prostate cancer using next-generation RNA sequencing. Genome Res. 21, 56–67.

Puvabanditsin, S., Garrow, E., Brandsma, E., Savla, J., Kunjumon, B., and Gadi, I. (2009). Partial trisomy 19p13.3 and partial monosomy 1p36.3: Clinical report and a literature review. Am. J. Med. Genet. A. 149A, 1782–1785.

Roach, J.C., Glusman, G., Smit, A.F., Huff, C.D., Hubley, R., Shannon, P.T., Rowen, L., Pant, K.P., Goodman, N., Bamshad, M., Shendure, J., Drmanac, R., Jorde, L.B., Hood, L., Galas, D.J. (2010). Analysis of genetic inheritance in a family quartet by whole-genome sequencing. Science 328, 636–639.

Rowley, J.D. (1973). Letter: A new consistent chromosomal abnormality in chronic myelogenous leukaemia identified by quinacrine fluorescence and Giemsa staining. Nature 243, 290–293.

Salbert, B.A., Solomon, M., Spence, J.E., Jackson-Cook, C., Brown, J., and Bodurtha, J. (1992). Partial trisomy 19p: case report and natural history. Clin. Genet. 41, 143–146.

Schluth-Bolard, C., Till, M., Rafat, A., Labalme, A., Le Lorc'h, M., Banquart, E., Angei, C., Cordier, M.P., Romana, S.P., Edery, P., and Sanlaville, D. (2008). Monosomy 19pter and trisomy 19q13-qter in two siblings arising from a maternal pericentric inversion: clinical data and molecular characterization. Eur. J. Med. Genet. 51, 622–630.

Schrock, E., du Manoir, S., Veldman, T., Schoell, B., Wienberg, J., Ferguson-Smith, M.A., Ning, Y., Ledbetter, D.H., Bar-Am, I., Soenksen, D., Garini, Y., and Ried, T. (1996). Multicolor spectral karyotyping of human chromosomes. Science 273, 494–497.

Shan, S.J., Scorilas, A., Katsaros, D., and Diamandis, E.P. (2007). Transcriptional upregulation of human tissue kallikrein 6 in ovarian cancer: clinical and mechanistic aspects. Br. J. Cancer 96, 362–372.

Shaw, J.L., and Diamandis, E.P. (2007). Distribution of 15 human kallikreins in tissues and biological fluids. Clin. Chem. 53, 1423–1432.

Shinoda, Y., Kozaki, K., Imoto, I., Obara, W., Tsuda, H., Mizutani, Y., Shuin, T., Fujioka, T., Miki, T., and Inazawa, J. (2007). Association of KLK5 overexpression with invasiveness of urinary bladder carcinoma cells. Cancer Sci. 98, 1078–1086.

Storchova, Z., and Pellman, D. (2004). From polyploidy to aneuploidy, genome instability and cancer. Nat. Rev. Mol. Cell. Biol. 5, 45–54.

Tan, D.S., Lambros, M.B., Natrajan, R., and Reis-Filho, J.S. (2007). Getting it right: designing microarray (and not 'microawry') comparative genomic hybridization studies for cancer research. Lab. Invest. 87, 737–754.

Thompson, S.L., Bakhoum, S.F., and Compton, D.A. (2010). Mechanisms of chromosomal instability. Curr. Biol. 20, R285–295.

Tomlins, S.A., Rhodes, D.R., Perner, S., Dhanasekaran, S.M., Mehra, R., Sun, X.W., Varambally, S., Cao, X., Tchinda, J., Kuefer, R., Lee, C., Montie, J.E., Shah, R.B., Pienta, K.J., Rubin, M.A., and Chinnaiyan, A.M. (2005). Recurrent fusion of TMPRSS2 and ETS transcription factor genes in prostate cancer. Science 310, 644–648.

Tomlins, S.A., Mehra, R., Rhodes, D.R., Smith, L.R., Roulston, D., Helgeson, B.E., Cao, X., Wei, J.T., Rubin, M.A., Shah, R.B., and Chinnaiyan, A.M. (2006). TMPRSS2:ETV4 gene fusions define a third molecular subtype of prostate cancer. Cancer Res. 66, 3396–3400.

Vilar, E., and Gruber, S.B. (2010). Microsatellite instability in colorectal cancer-the stable evidence. Nat. Rev. Clin. Oncol. 7, 153–162.

Weaver, B.A., and Cleveland, D.W. (2007). Aneuploidy: instigator and inhibitor of tumorigenesis. Cancer Res. 67, 10103–10105.

Weaver, B.A., and Cleveland, D.W. (2008). The aneuploidy paradox in cell growth and tumorigenesis. Cancer Cell 14, 431–433.

Weaver, B.A., and Cleveland, D.W. (2009). The role of aneuploidy in promoting and suppressing tumors. J. Cell Biol. 185, 935–937.

Yousef, G.M., Luo, L.Y., and Diamandis, E.P. (1999). Identification of novel human kallikrein-like genes on chromosome 19q13.3-q13.4. Anticancer Res. 19, 2843–2852.

Yousef, G.M., Chang, A., Scorilas, A., and Diamandis, E.P. (2000). Genomic organization of the human kallikrein gene family on chromosome 19q13.3-q13.4. Biochem. Biophys. Res. Commun. 276, 125–133.

Yousef, G.M., and Diamandis, E.P. (2001). The new human tissue kallikrein gene family: structure, function, and association to disease. Endocr. Rev. 22, 184–204.

Yousef, G.M., Bharaj, B.S., Yu, H., Poulopoulos, J., and Diamandis, E.P. (2001). Sequence analysis of the human kallikrein gene locus identifies a unique polymorphic minisatellite element. Biochem. Biophys. Res. Commun. 285, 1321–1329.

Konstantinos Mavridis, Manfred Schmitt, and Andreas Scorilas

10 Kallikrein-related Peptidases as Biomarkers in Personalized Cancer Medicine

10.1 Introduction

Personalized medicine may sound as a contemporary scientific term, but the truth is that it was already introduced some 2,500 years ago by Hippocrates (460 BC-370 BC), a Greek physician, broadly celebrated as the "father" of western medicine. He was the first to point out the immense variations in the constitution of individuals (referred to as "great natural diversities"), the unique features of disease manifestations in each case, and the importance of providing different treatment to different patients. He therefore suggested a beyond any doubt revolutionary therapeutic approach, which states that each patient's personal characteristics, along with the particularities of his/her environment, should be taken into account, in order to provide a customized and effective treatment plan (Sykiotis *et al.*, 2005). Deriving from that, the U.S. National Cancer Institute provides the current definition of personalized medicine as follows: "a form of medicine that uses information about a person's genes, proteins, and environment to prevent, diagnose, and treat disease" (NCI, 2011). Taking into account the scientific limitations at the time, it is obvious that Hippocrates established the central principle of what is nowadays defined as "Personalized Medicine".

Twenty-five centuries after this perception, and following the completion of the "Human Genome Project", we are now at the beginning of elucidating the entire molecular profile of individuals. This exciting approach will certainly provide novel strategies for estimating disease predisposition, as well as for predicting disease progression and response to treatment (Hong and Oh, 2010; Offit, 2011; Sikora, 2007). During the last decade, significant advances in genomic science have introduced refined analyses of the DNA, RNA, protein, and metabolites of an individual, as an indispensable clinical tool for determining how to cure a patient and how to organize the associated clinical management. These novel approaches are expected to customize medical care to every single patient, in order to allow for individualized and tailored molecular medicine (Ginsburg and Willard, 2009; Hong and Oh, 2010).

Cancer, even though it is characterized by known, common molecular features, can also be described as a multifaceted accumulation of various disease pathophysiologies. Cancer continues to constitute an enormous burden for public health, throughout the globe (Jemal *et al.*, 2011). One of the most essential characteristics of this disease is that its pathobiology at each stage, from carcinogenesis to invasion and metastasis, is manifested *via* extremely dissimilar and often difficult-to-predict disease patterns. This unfortunate phenomenon is largely responsible for the failure of current broadband anticancer treatment strategies for conquering the disease (Hanahan and Weinberg, 2011). Otherwise, given the heterogeneity of human malignancies, cancer patients are thought to be the perfect candidates to benefit from a personalized health care system (Diamandis *et al.*, 2010; Overdevest *et al.*, 2009).

Recently developed research strategies, based on or stimulated by the concept of "personalized cancer medicine", have enriched routine clinical practice with some inspiring success stories regarding individualized cancer management (Duffy *et al.*, 2011). More precisely, personalized cancer medicine has already been applied to the confrontation with human malignancies, using four central strategies: a) prediagnostic genetic screening, b) early detection, c) tumor classification for prognosis and prediction of patient response to treatment, and d) targeted anticancer therapy (Overdevest *et al.*, 2009). To allow this, at first, the personal geno- and phenotype of individuals has to be exploited both at the cellular and the molecular level.

From a pre-diagnostic point of view, genetic linkage studies in families with hereditary breast, ovarian, or colon cancers have provided important results, that are now translated into preventive tests for patients with a strong family history background related to these types of cancer. For example, women bearing mutations in the breast cancer susceptibility gene 1 (*BRCA1*) or the breast cancer susceptibility gene 2 (*BRCA2*) have an elevated risk of developing breast and/or ovarian cancer. Likewise, individuals with mutations in the DNA mismatch repair *MLH1* and *MSH2* genes are likely to develop colorectal malignancies. These high-risk individuals are advised to follow the established medical guidelines for precautionary purposes and are encouraged to subject themselves to careful monitoring (Diamandis *et al.*, 2010; Duffy *et al.*, 2011). In this respect, it is worth mentioning that several *kallikrein-related peptidase* (*KLK*) single nucleotide polymorphisms (SNP) are associated as well with an elevated risk of breast, prostate, or ovarian cancer (see Chapter 2, Volume 1).

A widely-known paradigm with regard to applied personalized cancer medicine through early detection is the use of the cancer biomarker PSA (prostate specific antigen, KLK3) for screening of the male population, to uncover otherwise occult prostate cancer (Stephan *et al.*, 2007; Ulmert *et al.*, 2009). Comprehensive molecular profiling of tumors, focusing on DNA, mRNA, microRNA (miRNA), or protein (signatures) can provide far more prognostic information than the currently used traditional histological and/or morphological classification of tumors (tumor type, nuclear grade, stage), often leading to a misjudged course of the cancer and, thus, to erroneous decisions regarding the extent of surgery and systemic (neo)adjuvant treatment (Diamandis *et al.*, 2010; van't Veer and Bernards, 2008).

Currently, this aspect of personalized cancer medicine is most evident in breast cancer management. Laboratory tests for estrogen and progesterone receptor (ER, PR), human epidermal growth factor receptor 2 (*HER2*), and urokinase-type plasminogen activator (uPA)/plasminogen activator inhibitor-1 (PAI-1) as well as gene expression signature assays like Oncotype DX® (Genomic Health, Redwood City, CA, USA), MammaPrint® (Agendia, Amsterdam, The Netherlands), and Endopredict® (Sividon Diagnostics, Cologne, Germany) are currently used in clinical practice for prognostic information and treatment decision-making in treating breast cancer patients. Analogous tests (ColoPrint, Agendia; Oncotype DX colon cancer assay, Genomic Health) have been developed for risk assessment of colorectal cancer patients, and are expected to be provided for other malignancies (e.g. lung cancer, malignant melanoma, prostate cancer) as well, in the near future (Lu *et al.*, 2012; Markert *et al.*, 2011; Mehan *et al.*, 2012; Overdevest *et al.*, 2009; Schramm and Mann, 2011).

The molecular clarification of the biological dissimilarities of tumors has also provided necessary information in order to develop novel drugs that target particular elements of cancer development and progression (van't Veer and Bernards, 2008). A fine example of targeted anticancer therapy is the development, clinical testing, and subsequent fast-track approval by the U.S. Food and Drug Administration (FDA) of the cancer drug imatinib mesylate (Gleevec®, Novartis, Basel, Switzerland), which targets the ABL protein tyrosine kinase activity in chronic myeloid leukemia and gastrointestinal stromal malignancies (GIST). Another inspiring case, that shows the helpful application of personalized medicine in cancer drug development, is the humanized antibody trastuzumab (Herceptin®, Roche, Penzberg, Germany) used to treat breast cancer patients, in order to tackle the overexpression of the oncoprotein HER2. In both cases, only those patients that are characterized by these specific molecular abnormalities are recommended for treatment with these drugs. It should also be mentioned that non-small cell lung cancers (NSCLC) with mutations in the tyrosine kinase domain of the epidermal growth factor receptor (EGFR) respond more adequately to the small-size synthetic tyrosine kinase inhibitors gefitinib (Tarceva®, OSI Pharmaceuticals, Melville, NY, USA) or erlotinib (Iressa®, AstraZeneca, London, UK), whereas colorectal cancer patients bearing *KRAS* mutations are resistant to the EGFR-directed, antibody-based drugs cetuximab (Erbitux®, Bristol-Myers Squibb, New York, NY, USA) and panitumumab (Vectibix®, Thousand Oaks, CA, USA). Consequently, molecular determination of *EGFR* and *KRAS* statuses should be carefully taken into account before administrating any of these drugs to cancer patients (Diamandis *et al.*, 2010; Duffy *et al.*, 2011; Ginsburg and Willard, 2009; Sikora, 2007).

Despite the recent promising innovations in the design of targeted anticancer drugs, the majority of cancer patients is still treated with more conventional therapeutic schemes, according to (inter)national guidelines. Advances in the surgical management of cancer patients, systemic therapy, and radiation have undoubtedly aided achieving prolongation of life and/or palliation of symptoms suffered by cancer patients (Lee and McLeod, 2011; Peterson, 2011). Nonetheless, and despite all ongoing efforts, there currently is no effective methodological approach for selecting the most appropriate therapeutic scheme for an individual patient, based solely on the prediction of its effects (Lee and McLeod, 2011; Lippert *et al.*, 2011). As a result, unexpected disease recurrences and/or unfavorable chemotoxic responses represent a frequent phenomenon in cancer treatment.

One of the main objectives of personalized cancer medicine is to stratify those patients who were given the same diagnosis, but would respond differently to the same systemic adjuvant therapy. The information required for this classification could arise from gene polymorphisms of cancer patients and/or from cancer-related gene or protein expression profiles (Sikora, 2007; van't Veer and Bernards, 2008). For example, genetic variations in genes related to drug metabolism, such as those encoding for parts of the cytochrome P450 system can be used to calculate the optimal dosage of the drugs and predict more effective treatment results (Diamandis *et al.*, 2010).

Besides, it is widely known that aberrant gene expression is involved in cancer development and progression, and to a great extent determines the characteristics of cancers (Hanahan and Weinberg, 2011; Sikora, 2007; Ullah and Aatif, 2009). This consideration leads to the hypothesis that pharmacodynamic mechanisms of anticancer drugs could rely, at least in part, on their ability to affect cancer-related gene expression. This theory might easily explain the pleiotropic effects of anticancer drugs on different cell types or individuals. These effects could be predicted and/or monitored by assessment of cancer-related gene expression profiles, before systemic (neo)adjuvant treatment, as a therapeutic molecular tool.

10.2 The role of KLKs as cancer biomarkers for predicting and monitoring response to chemotherapy or endocrine therapy

Discovering novel cancer biomarkers for the prediction and monitoring of individual cancer patients' responses to anticancer drug treatment is still a major clinical concern. The significant role of KLKs as promising screening, diagnostic, or prognostic clinical tools in various human malignancies is presented in Chapters 1–9. In this chapter, we describe the potential role of KLKs as cancer biomarkers, mainly focusing on aspects concerning individual patients' cancer therapy response prediction assessments.

Tab. 10.1 Association of KLKs with the response of cancer patients to therapy.

	Malig-nancy	Treatment modalities	Clinical utility	References
KLK3	Pro-state cancer	Androgen deprivation therapy, chemotherapy	Monitoring/predicting patients' course of the disease.	Ulmert *et al.*, 2009
	Breast cancer	Tamoxifen	High protein levels of cytosolic KLK3 are associated with poor response.	Foekens *et al.*, 1999
KLK4	Ovarian cancer	Paclitaxel	High KLK4 protein expression in cancer tissues is associated with therapy resistance.	Xi *et al.*, 2004

Tab. 10.1 (continued)

	Malig-nancy	Treatment modalities	Clinical utility	References
KLK5	Ovarian cancer	Platinum-based chemotherapy	Drop in serum KLK5 after chemotherapy is associated with positive response. Serum-based multiparametric biomarker panel for response prediction.	Oikonomopoulou *et al.*, 2008 Oikonomopoulou *et al.*, 2008
KLK6	Ovarian Cancer	Platinum-based chemotherapy	High pre-surgical serum KLK6 levels are associated with poor response to therapy. High pre-therapy KLK6 serum levels predict unfavorable therapy response. Enhances the discriminatory potential (responders/non-responders) of a multiparametric biomarker panel assessed in cytosolic extracts. Serum-based multiparametric biomarker panel for response prediction.	Diamandis *et al.*, 2003 Oikonomopoulou *et al.*, 2008 Zheng *et al.*, 2007 Oikonomopoulou *et al.*, 2008
KLK7	Ovarian Cancer	Chemotherapy Platinum-based chemotherapy	High *KLK7* expression levels are associated with therapy resistance. Serum-based multiparametric biomarker panel for response prediction.	Dong *et al.*, 2010 Oikonomopoulou *et al.*, 2008
KLK8	Ovarian Cancer	Platinum-based chemotherapy	High pre-therapy KLK8 serum levels predict unfavorable therapy response. Enhances the discriminatory potential (responders/non-responders) of a multiparametric panel assessed in cytosolic extracts.	Oikonomopoulou *et al.*, 2008 Zheng *et al.*, 2007
KLK10	Breast cancer Ovarian Cancer	Tamoxifen Platinum-based chemotherapy	High KLK10 levels are predictive of treatment resistance. High pre-surgical serum KLK10 levels are associated with no response. High pre-therapy KLK10 serum levels predict unfavorable response.	Luo *et al.*, 2002 Luo *et al.*, 2003 Oikonomopoulou *et al.*, 2008
KLK11	Ovarian Cancer	Chemotherapy	KLK11 protein levels are associated with chemotherapy response.	Borgoño *et al.*, 2003
KLK13	Ovarian Cancer	Platinum-based chemotherapy	KLK13 protein levels are associated with enhanced response.	Zheng *et al.*, 2007
KLK14	Ovarian Cancer	Chemotherapy	*KLK14* mRNA levels are associated with good response.	Yousef *et al.*, 2003

Data presented in the scientific literature highlights the importance of KLKs for predicting the course of certain cancer diseases and/or the response to drug-based therapeutic schemes (summarized in **Tab. 10.1**). KLKs have been widely associated with both molecular and clinical manifestations of malignancy (Borgoño and Diamandis, 2004; Sotiropoulou *et al.*, 2009). Recently-published data reveals the association between cancer-related KLK expression and response to chemotherapy or endocrine therapy, in patients afflicted with cancer of the prostate, breast, or ovary (Avgeris *et al.*, 2010; Mavridis and Scorilas, 2010).

10.2.1 Prostate cancer

PSA (KLK3) is regarded as one of the most extensively used cancer biomarkers. It is mainly used as a tool for screening for and monitoring of prostate cancer (Stephan *et al.*, 2007; Ulmert *et al.*, 2009) (see Chapter 4). In fact, it is thought that PSA testing provides far more advantages as a treatment monitoring tool than as a screening one. Any abrupt increase in PSA levels after surgery or radiation therapy, often referred to as biochemical relapse, is a signal of possible disease recurrence. This information offers physicians the opportunity to take prompt and full measures to fight the disease, in order to evade otherwise unmanageable tumor progression, and thus to enhance the survival probability of prostate cancer patients.

These measures may include endocrine therapy (androgen deprivation) for hormone-sensitive and chemotherapy for aggressive hormone-refractory prostate cancer patients. Subsequent PSA measurements should be employed for monitoring purposes. A decline in PSA is considered to be a criterion for effective response to chemotherapy or endocrine therapy, and has been determined to be a sign of improved survival (Berry, 2005; Fleming *et al.*, 2006; van Poppel *et al.*, 2009). Similar to PSA, post-treatment augmented KLK2 in serum is also a sign of disease recurrence (Ulmert *et al.*, 2009). In addition, the mRNA expression levels of *KLK4*, *14*, and *15* can be used as molecular indicators of poor outcome, while those of *KLK5* can serve as a biomarker of favorable prognosis (Mavridis and Scorilas, 2010).

Recent studies have implicated the role of *KLK* SNPs in prostate cancer prognosis, showing that knowledge of certain SNPs of the *KLK2* gene can significantly improve the prediction of biochemical disease recurrence after initial cancer treatment, especially when used in combinatorial models that take into account both clinical and pathological data (Kohli *et al.*, 2010; Morote *et al.*, 2010). Additionally, two other SNPs, one in the *KLK3* gene and the other one in the *KLK2–KLK3* intergenic region, are strongly associated with prostate cancer-specific survival (Gallagher *et al.*, 2010). These studies institute the role of *KLK* SNPs in stratifying prostate cancer patients who may be eligible for systemic adjuvant therapy.

10.2.2 Breast cancer

Several members of the *KLK* family exhibit a significant prognostic potential in breast cancer (see Chapter 6). Of particular interest to this malignancy is that KLK3 (Foekens *et al.*, 1999) and KLK10 (Luo *et al.*, 2002) were reported to be useful biomarkers for predicting response to endocrine treatment with the antiestrogen tamoxifen. Tamoxifen is administered to tumor tissue estrogen receptor-positive (ER-positive) breast cancer patients after surgical removal of the tumor (Brauch *et al.*, 2009). Breast cancer patients suffering from recurrent breast cancer, who are characterized by high levels of cytosolic PSA, are more likely to respond inadequately to tamoxifen therapy (Foekens *et al.*, 1999). Similar to KLK3, high cytosolic KLK10 levels are also predictive of lack of response, or resistance to tamoxifen (Luo *et al.*, 2002).

As a result, KLK3 and KLK10 might represent useful protein biomarkers for the stratification of breast cancer patients who may benefit from tamoxifen therapy, given that only this selection of ER-positive patients would respond effectively (Foekens *et al.*, 1999; Luo *et al.*, 2002). Interestingly, it was shown previously that an SNP located on the proximal androgen response element (ARE) of the *KLK3* gene does affect overall survival of breast cancer patients. Nonetheless, since a multivariate analysis that included additional factors did not endorse such an association, further studies are needed to evaluate the clinical potential of these two biomarkers (Bharaj *et al.*, 2000).

10.2.3 Ovarian cancer

Recent studies have proven a strong connection between survival of ovarian cancer patients and presence of *KLK3* or *KLK15* SNPs (Batra *et al.*, 2011; O'Mara *et al.*, 2011). More important, early prediction of platinum- or taxane-based chemotherapy response/failure for early or advanced ovarian cancer patients is a pressing clinical priority since, currently, no effective biomarker is in use in routine clinical practice that predicts therapy response in these patients. Platinum-based cancer drugs, such as carboplatin or cisplatin, are broadly used as a therapeutic option, alone or in combination with other agents such as the mitotic inhibitor paclitaxel, a member of the taxane drug category, for first-line treatment of ovarian cancer (McGuire and Markman, 2003). Taxanes stabilize microtubules and, as a result, interfere with the normal breakdown of microtubules during cell division (Fu *et al.*, 2009). Platinum-based drugs exert their mechanism of action through cross-linking to DNA, disturbing DNA synthesis and transcription, and ultimately promoting programmed cell death (Bose, 2002). Nevertheless, a significant number of ovarian cancer patients develops resistance to these types of chemotherapy (Gillet *et al.*, 2012). Disease recurrences that develop within six months after completion of a platinum-containing regimen are generally considered platinum-refractory or platinum-resistant cases. For these patients, alternative options are to follow modified chemotherapeutic plans, such as the administration of paclitaxel, topotecan (a topoisomerase inhibitor), or other cyto-static drugs (Baird *et al.*, 2010).

Many KLKs are implicated in ovarian cancer initiation, progression, and metasta-sis (Dong *et al.*, 2010; Prezas *et al.*, 2006; White *et al.*, 2010b; Zhang *et al.*, 2012). The majority of KLKs evolved as potential diagnostic and prognostic molecular tools, in order to characterize and monitor human ovarian malignancies (see Chapter 7). Inter-estingly, various members of the KLK family, namely KLK4-8, 10, 11, 13, and 14 have been associated with chemotherapy response, mostly platinum-based treatment, in ovarian cancer patients (Avgeris *et al.*, 2010; Yousef and Diamandis, 2009). Increase in KLK4 protein expression, as assessed by immunohistochemistry of ovarian cancer tissues, is associated with resistance to paclitaxel (Xi *et al.*, 2004). KLK6, initially iden-tified as a CNS-expressed serine protease termed neuropsin/myelencephalon-specific protease (MSP) (Scarisbrick *et al.*, 1997), is now known as an aberrantly-expressed serine protease, produced by ovarian tumors. It is regarded as a promising ovarian cancer biomarker (Yousef and Diamandis, 2009). An increase in presurgical serum KLK6 is associated with poor response to platinum-based chemotherapy (Diamandis *et al.*, 2003). As a result, this presurgical KLK6 serum can be utilized to identify poor chemotherapy responders among ovarian cancer patients. Similar to KLK6, high pre-surgical serum KLK10 is also present in ovarian cancer patients who are refractory to platinum-based chemotherapy (Luo *et al.*, 2003).

Recent data shows that high *KLK7* mRNA expression levels are associated with resistance to chemotherapy. Interestingly, it has been proposed that KLK7 may have a positive regulatory effect on $\alpha_5\beta_1$ integrin pathways, which leads to the promotion of multicellular aggregation and paclitaxel chemoresistance (Dong *et al.*, 2010). In contrast, KLK11 protein expression in ovarian tumors, as determined by an immu-noassay, is related to good chemotherapy response (Borgoño *et al.*, 2003). Likewise, significantly elevated *KLK14* mRNA expression levels, determined in tumor tissues, have been observed in ovarian cancer patients that have responded to platinum-based cancer drugs, compared to those who did not respond to this type of treatment (Yousef *et al.*, 2003). A parallel analysis of several of the KLKs and of other cancer bio-markers, and incorporation of the derived information into multifactorial statistical models, has been applied to ovarian cancer response prediction in two independent studies (Oikonomopoulou *et al.*, 2008; Zheng *et al.*, 2007).

Zheng *et al.* (2007) found that elevated tumor tissue KLK13 is positively related to platinum-based chemotherapy response. This finding was confirmed using mul-tivariate logistic regression analysis, adjusted for clinicopathological variables, which revealed the independent predictive strength of KLK13. This predictive value was found to be greater than any other biochemical or clinical variable used in the model. Receiver operating characteristic (ROC) curve analysis showed that this KLK member has a significant discriminatory potential for categorizing ovarian cancer patients into good responders (presenting fractional or complete response) and non-responders (presenting disease progression or no change at all). Interestingly, when a panel of markers, including KLK6 and KLK8, were used in combinatorial models, the discriminatory capacity of this model was increased. The inclusion of standard clinical information (e.g. clinical stage and debulking outcome) made the predictive power of this new panel of markers even greater, compared to any other combination.

The second study (Oikonomopoulou *et al.*, 2008) was conducted using ovarian cancer patients' sera. It suggests that increased levels of KLK6, 8, and 10, plus several non-KLK molecular markers (VTCN1, B7-H4, CA125, and REG4) in serum samples obtained before systemic cancer therapy, can predict response/failure to platinum-based chemotherapy alone or in combination with paclitaxel. When proper combinations of KLKs (e.g. KLK5, KLK6, or KLK7) and B7-H4 were integrated into a multiparametric model, this statistical model allowed for prediction of a possible response of ovarian cancer patients to chemotherapy.

10.3 Modulation of expression levels of *KLK* genes upon chemotherapy administration *in vitro*

Recent data revealed the clinical utility and the remarkable potential of targeted anticancer therapy based on the inhibition of specific tumorigenic pathways (Ginsburg and Willard, 2009; Sikora, 2007). Nevertheless, for the majority of human malignancies, conventional systemic chemotherapy with the use of common cytostatic drugs still remains the main anticancer drug treatment option (Peterson, 2011). The cytostatic cancer drugs currently in use exert their mechanisms of action *via* dissimilar pathways. Some may act through topoisomerase inhibition (e.g. topotecan, etoposide), while others work *via* mitosis disruption (e.g. paclitaxel, docetaxel), DNA intercalation/crosslinking (e.g. platinum-based drugs, doxorubicin, mitoxantrone), or impeding of the cell's metabolism (e.g. methotrexate, 5-fluorouracil). The exact biochemical pathways that are induced or inhibited upon anticancer drug treatment remain to be elucidated. It is also certain that response of cancer cells to chemotherapy is a multi-factorial phenomenon that depends on both acquired and intrinsic tumor cell features.

Modulation of gene expression is a key event in tumor growth, aggressiveness, and progression (Sikora, 2007; van't Veer and Bernards, 2008). Recent data shows that important alterations in the mRNA levels of apoptosis-related genes occur *in vitro*, upon treatment of various cell lines, derived from leukemia (Thomadaki *et al.*, 2009a) or solid tumors, with cytostatic drugs (Thomadaki and Scorilas, 2008). Expression of *KLK* genes is also modulated *in vitro* in prostate, breast, and gastric cancer cells upon treatment with anticancer agents normally used in routine clinical practice. Modulation of *KLK* expression in the presence of chemotherapeutics may sustain cancer progression through the selection of molecularly-altered subgroups of resistant cells. This modulation implicates KLKs as part of a chemotherapy-induced defense mechanism of cancer cells, making the KLKs ideal targets for targeted drugs. Also, these modulations encourage development of KLK-directed tools to help predict and monitor a cancer patient's response to chemotherapy.

10.3.1 Prostate cancer cells

Endocrine therapy (androgen deprivation therapy) remains one of the first-line options for treating prostate cancer. Nonetheless, prostate cancer cells are known to evolve into an extremely aggressive, androgen-independent state. At this point, the only effective treatment strategy available is conventional chemotherapy (Berry, 2005). A recent study, conducted using reverse transcription PCR and the androgen-independent PC3 prostate cancer cell line as an *in vitro* model system, has shown that modulation of the *KLK5* gene occurs upon treatment of these cells with the chemotherapeutic drugs etoposide, doxorubicin, carboplatin, or mitoxantrone. More specifically, *KLK5* mRNA levels were upregulated upon cancer drug treatment, in parallel with cell cytotoxicity progression which derived mainly from a decrease in the cell proliferation/viability capacity (Thomadaki *et al.*, 2009b). These results were also confirmed with another widely-used prostate cancer cell line (DU145), which also mimics androgen independence. Using a highly sensitive, quantitative, real-time PCR (qPCR) method, it has been found that mitoxantrone and docetaxel can upregulate *KLK5* mRNA expression of DU145 cells (Mavridis *et al.*, 2010).

As for *KLK5*, *KLK15* gene transcription is affected by certain anticancer compounds, such as mitoxantrone and docetaxel, which will result in an increase of *KLK15* mRNA. Notable upregulations were also observed during epirubicin and methotrexate administration. The chemotherapeutic drugs paclitaxel, navelbine, or gemcitabine, however, did not affect *KLK15* transcription levels (Mavridis *et al.*, 2011). Interestingly, both *KLK5* and *KLK15* mRNA levels may serve as valuable cancer biomarkers in order to indicate favorable and unfavorable, respectively, course of the malignancy (Mavridis and Scorilas, 2010; Mavridis *et al.*, 2011).

10.3.2 Gastric cancer cells

Chemotherapy is one of the main treatment options for gastric cancer, a malignancy with high worldwide mortality rates. Currently, there is no effective biomarker that can be used adequately for predicting the clinical course of gastric cancer (Gallo and Cha, 2006). Previously, it was shown that *KLK13* mRNA expression is an indicator of favorable outcome of gastric cancer patients (Konstantoudakis *et al.*, 2010). Cytostatic drug treatment influences *KLK13* mRNA expression in the AGS gastric cancer cell line. The anticancer drugs epirubicin, oxaliplatin, or methotrexate, used in concentrations corresponding to their IC_{50} values, resulted in a decrease of cell viability. This phenomenon is attributed to the induction of programmed cell death, given that the extent of any necrotic occurrence is restricted to ~ 10%. Data from qPCR shows that *KLK13* expression profiles are characteristically modified upon exposure of gastric cancer cells to either epirubicin or methotrexate. *KLK13* mRNA levels are modified through a gradual, time-dependent increase, elicited by epirubicin or methotrexate treatment. No significant differences in *KLK13* mRNA levels were observed after applying the platinum-based drug oxaliplatin (Florou *et al.*, 2012).

10.3.3 Breast cancer cells

In breast cancer, the effect of using chemotherapeutic agents in systemic (neo)adjuvant regimens may vary, depending on many factors, such as the histomorphological characteristics of the tumor and the individual's own risk of disease recurrence. Given the vast heterogeneity of breast cancers, which also includes dissimilar responses to chemotherapy, novel, individualized, predictive strategies have to be developed and clinically tested (McArthur and Hudis, 2007). With respect to this, it has been reported that for instance *KLK4, 5,* and *14* are differentially expressed in breast carcinoma and are associated with poor prognosis of this disease (Mavridis and Scorilas, 2010).

KLK4, 5, and *14* mRNA levels in the BT-20 breast cancer cell line are also significantly affected by the cytostatic drugs docetaxel, epirubicin, or methotrexate. The extent and type of *KLK* modulation (up- or downregulation) takes place in a drug- and time-dependent manner. Interestingly, the observed changes in *KLK* expression are associated with changes in the expression levels of molecules that comprise part of the apoptotic machinery. Immunocytochemical analysis of these treated cells revealed the marked presence of the pro-apoptotic protein BAX but the absence of the anti-apoptotic protein BCL2 (Papachristopoulou *et al.*, 2011).

KLKs have been previously proposed to play a role in the complex procedures of programmed cell death. KLK1-3, and 11 can cleave insulin-like growth factor binding proteins (IGFBP), thereby freeing insulin-like growth factor 1 (IGF1), a mitogenic factor that prevents apoptosis but favors cell proliferation. On the contrary, KLK3 can induce apoptosis of osteoclast precursors in a proteolysis-dependent manner and can promote the generation of reactive oxygen species (ROS), which can stimulate apoptosis in prostate cancer cells *via* a proteolysis-independent mechanism (Borgoño and Diamandis, 2004; Emami and Diamandis, 2007; Sano *et al.*, 2007).

10.3.4 The role of microRNAs (miRNAs) that target *KLK* expression and the methylation status of *KLK* genes in the *in vitro* response of cancer cells to chemotherapy

Lately, microRNAs (miRNA) have drawn the attention of the cancer research community (see Chapter 8). They regulate oncogene (Slack, 2012) or tumor suppressor gene expression and they play key-roles in carcinogenesis, tumor progression, invasion, and metastasis (Jiang *et al.*, 2012; Keklikoglou *et al.*, 2011; Kent and Mendell, 2006; Liu *et al.*, 2012). Additionally, miRNAs have been identified as promising diagnostic and prognostic tools for a number of human malignancies (Osaki *et al.*, 2008; White *et al.*, 2011).

Recently, published *in vitro* studies have revealed significant modulations of various miRNAs upon treatment of tumor cells with chemotherapy. Of particular interest is microRNA 224 (miR-224), which is downregulated in human colon cancer cells treated with the cancer drug 5-fluorouracil (Rossi *et al.*, 2007). Additionally, miR-224 has been found to be downregulated in human colon cancer cells resistant to methotrexate, another antimetabolite that is widely used in cancer treatment. In addition, miR-224 is thought to be involved in the defense mechanisms that this drug induces in cancer cells (Mencia *et al.*, 2011). The target genes of miR-224 include, among others, *KLK1*, *KLK10* (validated targets), and *KLK15* (predicted target) (White *et al.*, 2010a and b). As mentioned previously, *KLK15* mRNA levels are also upregulated by methotrexate treatment of prostate cancer cells (Mavridis *et al.*, 2011). It has been reported that mRNA levels of predicted miR-224 targets are increased upon inhibition of miR-224 in colon cancer cells (Mencia *et al.*, 2011). Taken together, this data indicates a putative regulatory mechanism between miR-224 and *KLK15*, which becomes activated upon anticancer drug treatment. In summary, the data presented so far implicates a crucial role of *KLK*-targeting miRNAs in response of cancer cells to chemotherapy. Nonetheless, this assumption needs to be tested in studies centering on other *in vitro* systems.

DNA-methylation is an epigenetic mechanism that can affect the response of malignant cells to anticancer drug treatment. Aberrant DNA-methylation can be viewed as a valuable molecular marker for predicting chemotherapy response (Shen *et al.*, 2007). As far as *KLK* genes are concerned, it has been implied that *KLK10* hypermethylation does play a role in, for instance, chemotherapy resistance of hepatocellular carcinoma (HCC) cells. In particular, it has been shown that *KLK10* is downregulated *via* hypermethylation in HCC (both in clinical samples and cell lines) and that restoration of its expression sensitizes HCC cells to 5- fluorouracil and doxorubicin treatment *in vitro* (Lu *et al.*, 2009).

10.4 Conclusions and future directions

In order to treat a disease, conventional medicine would ask the question: "based on the symptoms that he/she presents, which disease does the patient have?" In the era of personalized medicine this question would be different, namely: "what are the markers that reveal the particular disease of each individual patient?". Definitely, given the complexity of human malignancies and the vast number of tumor subtypes that exist, modern views and principles of personalized medicine are more applicable to cancer than to any other kind of disease (Diamandis *et al.*, 2010; Overdevest *et al.*, 2009). Irrespective of the still ongoing debate about how much of an advantage the Human Genome Project has brought to the healthcare system (Marshall, 2011), several very spectacular examples of targeted therapy, fit for subgroups of cancer patients, have come to the attention of the health community (e.g. Tamoxifen®, Gleevec®, Tarceva®, and Avastin®) (Diamandis *et al.*, 2010; Sikora, 2007).

Within this framework, numerous independent studies have indicated that KLKs represent a rich source of serum and tissue biomarkers which allow molecular classification, early diagnosis, and prognosis of human malignancies. Furthermore, KLKs may be indicators of the efficacy of anticancer drugs (Avgeris *et al.*, 2010; Borgoño and Diamandis, 2004; Emami and Diamandis, 2007; Yousef and Diamandis, 2009). This novel, cancer biomarker-driven approach has an immense potential for predicting the clinical outcome of cancer patients, and for providing customized treatment options to those patients who are likely to respond, while directing those who would not benefit towards alternative therapeutic options. KLKs have also arisen as novel molecular targets for tailored cancer therapy. Thus, it is not a surprise that upcoming clinical trials aim at testing the efficacy of synthetic KLK inhibitors for the treatment of asymptomatic hormone refractory prostate cancer (Kündig *et al.*, 2011).

Recent data suggests a role for chemotherapy-induced changes to the expression levels of *KLKs* in the therapeutic response to such agents (Mavridis *et al.*, 2010; Thomadaki *et al.*, 2009b). However, in order to interpret this information, the exact mechanisms of action of KLKs in different tumor types and microenvironments will need to be further elucidated. By combining this data, we should be able to acquire a deeper knowledge regarding the exact KLK-related intracellular mechanisms that are activated upon anticancer therapy. Eventually, this information will allow us to exploit the KLKs as therapeutic targets, and/or as tools for predicting and monitoring the response to treatment of cancer cells.

One should bear in mind that current cancer diagnosis, prognosis, and treatment of cancer patients mostly rely on histomorphological tumor characteristics (e.g. TNM status, nuclear grade). Nonetheless, the information derived from this widely-established approach may not be sufficient to warrant a particular cancer therapy. More to the point, the determination of cancer biomarker expression levels has been shown to be valuable, additive tool to improve cancer screening, diagnosis, and prognosis, as well as therapy response of the patients (Diamandis *et al.*, 2010).

A good example is provided by the extensive use of PSA (KLK3) in routine clinical practice.

Unquestionably, KLKs have already proven their cancer biomarker capacity and thus represent valuable utensils in the tool box of personalized cancer medicine. However, validation of the clinical impact of the various KLKs has to be done in large-scale, multicenter studies, before any of these markers can be recommended for routine clinical practice.

Bibliography

Avgeris, M., Mavridis, K., and Scorilas, A. (2010). Kallikrein-related peptidase genes as promising biomarkers for prognosis and monitoring of human malignancies. Biol. Chem. 391, 505–511.

Baird, R.D., Tan, D.S., and Kaye, S.B. (2010). Weekly paclitaxel in the treatment of recurrent ovarian cancer. Nat. Rev. Clin. Oncol. 7, 575–582.

Batra, J., Nagle, C.M., O'Mara, T., Higgins, M., Dong, Y., Tan, O.L., Lose, F., Skeie, L.M., Srinivasan, S., Bolton, K.L., Song, H., Ramus, S.J., Gayther, S.A., Pharoah, P.D., Kedda, M.A., Spurdle, A.B., and Clements, J.A. (2011). A kallikrein 15 (KLK15) single nucleotide polymorphism located close to a novel exon shows evidence of association with poor ovarian cancer survival. BMC Cancer 11, 119.

Berry, W.R. (2005). The evolving role of chemotherapy in androgen-independent (hormone-refractory) prostate cancer. Urology 65, 2–7.

Bharaj, B., Scorilas, A., Diamandis, E.P., Giai, M., Levesque, M.A., Sutherland, D.J., and Hoffman, B.R. (2000). Breast cancer prognostic significance of a single nucleotide polymorphism in the proximal androgen response element of the prostate specific antigen gene promoter. Breast Cancer Res. Treat. 61, 111–119.

Borgoño, C.A., Fracchioli, S., Yousef, G.M., Rigault de la Longrais, I.A., Luo, L.Y., Soosaipillai, A., Puopolo, M., Grass, L., Scorilas, A., Diamandis, E.P., and Katsaros, D. (2003). Favorable prognostic value of tissue human kallikrein 11 (hK11) in patients with ovarian carcinoma. Int. J. Cancer 106, 605–610.

Borgoño, C.A., and Diamandis, E.P. (2004). The emerging roles of human tissue kallikreins in cancer. Nat. Rev. Cancer 4, 876–890.

Bose, R.N. (2002). Biomolecular targets for platinum antitumor drugs. Mini Rev. Med. Chem. 2, 103–111.

Brauch, H., Murdter, T.E., Eichelbaum, M., and Schwab, M. (2009). Pharmacogenomics of tamoxifen therapy. Clin. Chem. 55, 1770–1782.

Diamandis, E.P., Scorilas, A., Fracchioli, S., Van Gramberen, M., De Bruijn, H., Henrik, A., Soosaipillai, A., Grass, L., Yousef, G.M., Stenman, U.H., Massobrio, M., van der Zee, A.G., Vergote, I., and Katsaros, D. (2003). Human kallikrein 6 (hK6): a new potential serum biomarker for diagnosis and prognosis of ovarian carcinoma. J. Clin. Oncol. 21, 1035–1043.

Diamandis, M., White, N.M., and Yousef, G.M. (2010). Personalized medicine: marking a new epoch in cancer patient management. Mol. Cancer Res. 8, 1175–1187.

Dong, Y., Tan, O.L., Loessner, D., Stephens, C., Walpole, C., Boyle, G.M., Parsons, P.G., and Clements, J.A. (2010). Kallikrein-related peptidase 7 promotes multicellular aggregation via the alpha(5)beta(1) integrin pathway and paclitaxel chemoresistance in serous epithelial ovarian carcinoma. Cancer Res. 70, 2624–2633.

Duffy, M.J., O'Donovan, N., and Crown, J. (2011). Use of molecular markers for predicting therapy response in cancer patients. Cancer Treat. Rev. 37, 151–159.

Emami, N., and Diamandis, E.P. (2007). New insights into the functional mechanisms and clinical applications of the kallikrein-related peptidase family. Mol. Oncol. 1, 269–287.

Fleming, M.T., Morris, M.J., Heller, G., and Scher, H.I. (2006). Post-therapy changes in PSA as an outcome measure in prostate cancer clinical trials. Nat. Clin. Pract. Oncol. 3, 658–667.

Florou, D., Mavridis, K., and Scorilas, A. (2012). The kallikrein-related peptidase 13 (KLK13) gene is substantially up-regulated after exposure of gastric cancer cells to antineoplastic agents. Tumour Biol [Epub ahead of print].

Foekens, J.A., Diamandis, E.P., Yu, H., Look, M.P., Meijer-van Gelder, M.E., van Putten, W.L., and Klijn, J.G. (1999). Expression of prostate-specific antigen (PSA) correlates with poor response to tamoxifen therapy in recurrent breast cancer. Br. J. Cancer 79, 888–894.

Fu, Y., Li, S., Zu, Y., Yang, G., Yang, Z., Luo, M., Jiang, S., Wink, M., and Efferth, T. (2009). Medicinal chemistry of paclitaxel and its analogues. Curr. Med. Chem. 16, 3966–3985.

Gallagher, D.J., Vijai, J., Cronin, A.M., Bhatia, J., Vickers, A.J., Gaudet, M.M., Fine, S., Reuter, V., Scher, H.I., Hallden, C., Dutra-Clarke, A., Klein, R.J., Scardino, P.T., Eastham, J.A., Lilja, H., Kirchhoff, T., and Offit, K. (2010). Susceptibility loci associated with prostate cancer progression and mortality. Clin. Cancer Res. 16, 2819–2832.

Gallo, A., and Cha, C. (2006). Updates on esophageal and gastric cancers. World J. Gastroenterol. 12, 3237–3242.

Gillet, J.P., Calcagno, A.M., Varma, S., Davidson, B., Bunkholt Elstrand, M., Ganapathi, R., Kamat, A., Sood, A.K., Ambudkar, S.V., Seiden, M., Rueda, B.R., and Gottesman, M.M. (2012). Multidrug resistance-linked gene signature predicts overall survival of patients with primary ovarian serous carcinoma. Clin. Cancer Res., [Epub ahead of print].

Ginsburg, G.S., and Willard, H.F. (2009). Genomic and personalized medicine: foundations and applications. Transl. Res. 154, 277–287.

Hanahan, D., and Weinberg, R.A. (2011). Hallmarks of cancer: the next generation. Cell 144, 646–674.

Hong, K.W., and Oh, B. (2010). Overview of personalized medicine in the disease genomic era. BMB Rep. 43, 643–648.

Jemal, A., Bray, F., Center, M.M., Ferlay, J., Ward, E., and Forman, D. (2011). Global cancer statistics. CA Cancer J. Clin. 61, 69–90.

Jiang, Y.W., and Chen, L.A. (2012). microRNAs as tumor inhibitors, oncogenes, biomarkers for drug efficacy and outcome predictors in lung cancer. Mol. Med. Rep. 5, 890–894.

Keklikoglou, I., Koerner, C., Schmidt, C., Zhang, J.D., Heckmann, D., Shavinskaya, A., Allgayer H., Gückel, B., Fehm, T., Schneeweiss, A., Sahin, O., Wiemann, S., and Tschulena, U. (2011). MicroRNA-520/373 family functions as a tumor suppressor in estrogen receptor negative breast cancer by targeting NF-κB and TGF-β signaling pathways. Oncogene, [Epub ahead of print].

Kent, O.A., and Mendell, J.T. (2006). A small piece in the cancer puzzle: microRNAs as tumor suppressors and oncogenes. Oncogene 25, 6188–6196.

Kohli, M., Rothberg, P.G., Feng, C., Messing, E., Joseph, J., Rao, S.S., Hendershot, A., and Sahsrabudhe, D. (2010). Exploratory study of a KLK2 polymorphism as a prognostic marker in prostate cancer. Cancer Biomark. 7, 101–108.

Konstantoudakis, G., Florou, D., Mavridis, K., Papadopoulos, I.N., and Scorilas, A. (2010). Kallikrein-related peptidase 13 (KLK13) gene expressional status contributes significantly in the prognosis of primary gastric carcinomas. Clin. Biochem. 43, 1205–1211.

Kündig, C., Kishi, T., and Deperthes, D. (2011). Engineered serine protease inhibitor MDPK67B to treat asymptomatic hormone refractory prostate cancer patients with rising PSA. Proceedings of the 4th International Symposium on Kallikreins and Kallikrein-Related Peptidases (ISK 2011), OP26, 56.

Lee, S.Y., and McLeod, H.L. (2011). Pharmacogenetic tests in cancer chemotherapy: what physicians should know for clinical application. J. Pathol. 223, 15–27.

Lippert, T.H., Ruoff, H.J., and Volm, M. (2011). Current status of methods to assess cancer drug resistance. Int. J. Med. Sci. 8, 245–253.

Liu, Y., Yin, B., Zhang, C., Zhou, L., and Fan, J. (2012). Hsa-let-7a functions as a tumor suppressor in renal cell carcinoma cell lines by targeting c-myc. Biochem. Biophys. Res. Commun. 417, 371–375.

Lu, C.Y., Hsieh, S.Y., Lu, Y.J., Wu, C.S., Chen, L.C., Lo, S.J., Wu, C.T., Chou, M.Y., Huang, T.H., and Chang, Y.S. (2009). Aberrant DNA methylation profile and frequent methylation of KLK10 and OXGR1 genes in hepatocellular carcinoma. Genes Chromosomes Cancer 48, 1057–1068.

Lu, Y., Wang, L., Liu, P., Yang, P., and You, M. (2012). Gene-expression signature predicts postoperative recurrence in stage I non-small cell lung cancer patients. PLoS One. 7, e30880.

Luo, L.Y., Diamandis, E.P., Look, M.P., Soosaipillai, A.P., and Foekens, J.A. (2002). Higher expression of human kallikrein 10 in breast cancer tissue predicts tamoxifen resistance. Br. J. Cancer 86, 1790–1796.

Luo, L.Y., Katsaros, D., Scorilas, A., Fracchioli, S., Bellino, R., van Gramberen, M., de Bruijn, H., Henrik, A., Stenman, U.H., Massobrio, M., van der Zee, A.G., Vergote, I., and Diamandis, E.P. (2003). The serum concentration of human kallikrein 10 represents a novel biomarker for ovarian cancer diagnosis and prognosis. Cancer Res. 63, 807–811.

Markert, E.K., Mizuno, H., Vazquez, A., and Levine, A.J. (2011). Molecular classification of prostate cancer using curated expression signatures. Proc. Natl. Acad. Sci. USA 108, 21276–21281.

Marshall, E. (2011). Human genome 10th anniversary. Waiting for the revolution. Science 331, 526–529.

Mavridis, K., and Scorilas, A. (2010). Prognostic value and biological role of the kallikrein-related peptidases in human malignancies. Future Oncol. 6, 269–285.

Mavridis, K., Talieri, M., and Scorilas, A. (2010). KLK5 gene expression is severely upregulated in androgen-independent prostate cancer cells after treatment with the chemotherapeutic agents docetaxel and mitoxantrone. Biol. Chem. 391, 467–474.

Mavridis, K., Stravodimos, K., and Scorilas, A. (2011). Quantitative analysis and study of the microRNA 224 (mir-224) and its target, KLK15 gene, in prostate tumors: investigation of their clinical significance. Proceedings of the 4th International Symposium on Kallikreins and Kallikrein-Related Peptidases (ISK 2011), OP18, 50.

McArthur, H.L., and Hudis, C.A. (2007). Breast cancer chemotherapy. Cancer J. 13, 141–147.

McGuire, W.P., 3rd, and Markman, M. (2003). Primary ovarian cancer chemotherapy: current standards of care. Br. J. Cancer 89 (Suppl 3.), S3–8.

Mehan, M.R., Ayers, D., Thirstrup, D., Xiong, W., Ostroff, R.M., Brody, E.N., Walker, J.J., Gold, L., Jarvis, T.C., Janjic, N., Baird, G.S., and Wilcox, S.K. (2012). Protein signature of lung cancer tissues. PLoS One. 7, e35157.

Mencia, N., Selga, E., Noe, V., and Ciudad, C.J. (2011). Underexpression of miR-224 in methotrexate resistant human colon cancer cells. Biochem Pharmacol. 82, 1572–1582.

Morote, J., Del Amo, J., Borque, A., Ars, E., Hernandez, C., Herranz, F., Arruza, A., Llarena, R., Planas, J., Viso, M.J., Palou, J., Raventós, C.X., Tejedor, D., Artieda, M., Simón, L., Martínez, A., and Rioja, L.A. (2010). Improved prediction of biochemical recurrence after radical prostatectomy by genetic polymorphisms. J. Urol. 184, 506–511.

NCI (2011). Definition of pesonalized medicine by the U.S. National Cancer Institute http://www.cancer.gov/dictionary?CdrID=561717, Access date: September 2011.

O'Mara, T.A., Nagle, C.M., Batra, J., Kedda, M.A., Clements, J.A., and Spurdle, A.B. (2011). Kallikrein-related peptidase 3 (KLK3/PSA) single nucleotide polymorphisms and ovarian cancer survival. Twin Res. Hum. Genet. 14, 323–327.

Offit, K. (2011). Personalized medicine: new genomics, old lessons. Hum. Genet. 130, 3–14.

Oikonomopoulou, K., Li, L., Zheng, Y., Simon, I., Wolfert, R.L., Valik, D., Nekulova, M., Simickova, M., Frgala, T., and Diamandis, E.P. (2008). Prediction of ovarian cancer prognosis and response to chemotherapy by a serum-based multiparametric biomarker panel. Br. J. Cancer 99, 1103–1113.

Osaki, M., Takeshita, F., and Ochiya, T. (2008). MicroRNAs as biomarkers and therapeutic drugs in human cancer. Biomarkers 13, 658–670.

Overdevest, J.B., Theodorescu, D., and Lee, J.K. (2009). Utilizing the molecular gateway: the path to personalized cancer management. Clin. Chem. 55, 684–697.

Papachristopoulou, G., Stamatopoulou, S., Talieri, M., Ardavanis, A., and Scorilas, A. (2011). Study of KLK4, KLK5 and KLK14 mRNA levels in breast cancer cells after treatment with antineoplastic agents. Proceedings of the 4th International Symposium on Kallikreins and Kallikrein-Related Peptidases (ISK 2011), OP16, 48.

Peterson, C. (2011). Drug therapy of cancer. Eur. J. Clin. Pharmacol. 67, 437–447.

Prezas, P., Arlt, M.J., Viktorov, P., Soosaipillai, A., Holzscheiter, L., Schmitt, M., Talieri, M., Diamandis, E.P., Krüger, A., and Magdolen, V. (2006). Overexpression of the human tissue kallikrein genes KLK4, 5, 6, and 7 increases the malignant phenotype of ovarian cancer cells. Biol. Chem. 387, 807–811.

Rossi, L., Bonmassar, E., and Faraoni, I. (2007). Modification of miR gene expression pattern in human colon cancer cells following exposure to 5-fluorouracil in vitro. Pharmacol. Res. 56, 248–253.

Sano, A., Sangai, T., Maeda, H., Nakamura, M., Hasebe, T., and Ochiai, A. (2007). Kallikrein 11 expressed in human breast cancer cells releases insulin-like growth factor through degradation of IGFBP-3. Int. J. Oncol. 30, 1493–1498.

Scarisbrick, I.A., Towner, M.D., and Isackson, P.J. (1997). Nervous system-specific expression of a novel serine protease: regulation in the adult rat spinal cord by excitotoxic injury. J. Neurosci. 17, 8156–8168.

Schramm, S.J., and Mann, G.J. (2011). Melanoma prognosis: a REMARK-based systematic review and bioinformatic analysis of immunohistochemical and gene microarray studies. Mol. Cancer Ther. 10, 1520–1528.

Shen, L., Kondo, Y., Ahmed, S., Boumber, Y., Konishi, K., Guo, Y., Chen, X., Vilaythong, J.N., and Issa, J.P. (2007). Drug sensitivity prediction by CpG island methylation profile in the NCI-60 cancer cell line panel. Cancer Res. 67, 11335–11343.

Sikora, K. (2007). Personalized medicine for cancer: from molecular signature to therapeutic choice. Adv. Cancer Res. 96, 345–369.

Slack, F.J. (2012). MicroRNAs regulate expression of oncogenes. Clin. Chem., [Epub ahead of print].

Sotiropoulou, G., Pampalakis, G., and Diamandis, E.P. (2009). Functional roles of human kallikrein-related peptidases. J. Biol. Chem. 284, 32989–32994.

Stephan, C., Jung, K., Lein, M., and Diamandis, E.P. (2007). PSA and other tissue kallikreins for prostate cancer detection. Eur. J. Cancer 43, 1918–1926.

Sykiotis, G.P., Kalliolias, G.D., and Papavassiliou, A.G. (2005). Pharmacogenetic principles in the Hippocratic writings. J. Clin. Pharmacol. 45, 1218–1220.

Thomadaki, H., and Scorilas, A. (2008). Molecular profile of breast versus ovarian cancer cells in response to treatment with the anticancer drugs cisplatin, carboplatin, doxorubicin, etoposide and taxol. Biol. Chem. 389, 1427–1434.

Thomadaki, H., Floros, K.V., and Scorilas, A. (2009a). Molecular response of HL-60 cells to mitotic inhibitors vincristine and taxol visualized with apoptosis-related gene expressions, including the new member BCL2L12. Ann. NY Acad. Sci. 1171, 276–283.

Thomadaki, H., Mavridis, K., Talieri, M., and Scorilas, A. (2009b). Treatment of PC3 prostate cancer cells with mitoxantrone, etoposide, doxorubicin and carboplatin induces distinct alterations in the expression of kallikreins 5 and 11. Thromb. Haemost. 101, 373–380.

Ullah, M.F., and Aatif, M. (2009). The footprints of cancer development: Cancer biomarkers. Cancer Treat. Rev. 35, 193–200.

Ulmert, D., O'Brien, M.F., Bjartell, A.S., and Lilja, H. (2009). Prostate kallikrein markers in diagnosis, risk stratification and prognosis. Nat. Rev. Urol. 6, 384–391.

van't Veer, L.J., and Bernards, R. (2008). Enabling personalized cancer medicine through analysis of gene-expression patterns. Nature 452, 564–570.

van Poppel, H., Joniau, S., Van Cleynenbreugel, B., Mottaghy, F.M., and Oyen, R. (2009). Diagnostic evaluation of PSA recurrence and review of hormonal management after radical prostatectomy. Prostate Cancer Prostatic Dis. 12, 116–123.

White, N.M., Bui, A., Mejia-Guerrero, S., Chao, J., Soosaipillai, A., Youssef, Y., Mankaruos, M., Honey, R.J., Stewart, R., Pace, K.T., Sugar, L., Diamandis, E.P., Doré, J., and Yousef, G.M. (2010a). Dysregulation of kallikrein-related peptidases in renal cell carcinoma: potential targets of miRNAs. Biol. Chem. 391, 411–423.

White, N.M., Chow, T.F., Mejia-Guerrero, S., Diamandis, M., Rofael, Y., Faragalla, H., Mankaruous, M., Gabril, M., Girgis, A., and Yousef, G.M. (2010b). Three dysregulated miRNAs control kallikrein 10 expression and cell proliferation in ovarian cancer. Br. J. Cancer 102, 1244–1253.

Xi, Z., Kaern, J., Davidson, B., Klokk, T.I., Risberg, B., Trope, C., and Saatcioglu, F. (2004). Kallikrein 4 is associated with paclitaxel resistance in ovarian cancer. Gynecol. Oncol. 94, 80–85.

Yousef, G.M., Fracchioli, S., Scorilas, A., Borgoño, C.A., Iskander, L., Puopolo, M., Massobrio, M., Diamandis, E.P., and Katsaros, D. (2003). Steroid hormone regulation and prognostic value of the human kallikrein gene 14 in ovarian cancer. Am. J. Clin. Pathol. 119, 346–355.

Yousef, G.M., and Diamandis, E.P. (2009). The human kallikrein gene family: new biomarkers for ovarian cancer. Cancer Treat. Res. 149, 165–187.

Zhang, R., Shi, H., Chen, Z., Feng, W., Zhang, H., and Wu, K. (2012). Effects of kallikrein-related peptidase 14 gene inhibition by small interfering RNA in ovarian carcinoma cells. Mol. Med. Report 5, 256–259.

Zheng, Y., Katsaros, D., Shan, S.J., de la Longrais, I.R., Porpiglia, M., Scorilas, A., Kim, N.W., Wolfert, R.L., Simon, I., Li, L., Feng, Z., and Diamandis, E.P. (2007). A multiparametric panel for ovarian cancer diagnosis, prognosis, and response to chemotherapy. Clin. Cancer Res. 13, 6984–6992.

Index